T0179072

Artificial Intelligence

Artificial Intelligence

Technologies, Applications, and Challenges

Edited by

Lavanya Sharma
Amity University, India

Pradeep Kumar Garg
IIT Roorkee, India

CRC Press
Taylor & Francis Group
Boca Raton London New York

CRC Press is an imprint of the
Taylor & Francis Group, an **informa** business

A CHAPMAN & HALL BOOK

First edition published 2022

by CRC Press
6000 Broken Sound Parkway NW, Suite 300, Boca Raton, FL 33487-2742

and by CRC Press
2 Park Square, Milton Park, Abingdon, Oxon, OX14 4RN

CRC Press is an imprint of Taylor & Francis Group, LLC

Library of Congress Cataloging-in-Publication Data
Names: Sharma, Lavanya, editor.
Title: Artificial intelligence : technologies, applications, and challenges
/ Lavanya Sharma, Pradeep Kumar Garg.
Description: First edition. | Boca Raton : Chapman & Hall/CRC Press, 2022.
| Includes bibliographical references and index.
Identifiers: LCCN 2021019494 (print) | LCCN 2021019495 (ebook) | ISBN
9780367690809 (hardback) | ISBN 9780367690823 (paperback) | ISBN
9781003140351 (ebook)
Subjects: LCSH: Artificial intelligence.
Classification: LCC Q335 .A78784 2022 (print) | LCC Q335 (ebook) | DDC
006.3--dc23
LC record available at https://lccn.loc.gov/2021019494
LC ebook record available at https://lccn.loc.gov/2021019495

ISBN: 978-0-367-69080-9 (hbk)
ISBN: 978-0-367-69082-3 (pbk)
ISBN: 978-1-003-14035-1 (ebk)

DOI: 10.1201/9781003140351

Typeset in Times
by SPi Technologies India Pvt Ltd (Straive)

Dedicated to My Dada Ji (Late. Shri Ram Krishan Choudhary Ji)
Ek prerna mayeh Vyaktitavh

Dr. Lavanya Sharma

Dedicated to my Parents (late Shri Ramgopal
Garg and Late Smt. Urmila Garg)

Prof. Pradeep K. Garg

Contents

Section I   Introduction to Artificial Intelligence

Section II   Artificial Intelligence: Tools and Technologies

Section III   Artificial Intelligence–Based Real-Time Applications

Preface

This book provides an overview of the basic concept of artificial intelligence tools from historical background to real-time applications domains, related technologies, and their possible solutions to take up future challenges. It offers detailed descriptions with practical ideas of using AI to deal with the dynamics, the ecosystem, and challenges involved in surpassing diversified field, image processing, communications, integrity, and security aspects. The AI, in combination for outdoor and indoor scenarios, proved to be most advantageous for the companies and organizations to efficiently monitor and control their day-to-day processes such as design, production, transportation, maintenance, implementation, and distribution of their products.

This book consists of four important parts that provide an overview of artificial intelligence, critical applications domains, tools, and technologies. In addition, it provides insights to undertake the research work in future challenging areas. Overall, this publication would help the readers understand the needs of artificial intelligence for individuals as well as organizations.

Acknowledgments

I am especially grateful to *my dada ji*, *my parents*, *my husband*, and *my beautiful family* for their continuous support and blessings. I would like to thank my husband *Dr. Mukesh (general and laparoscopic surgeon)* for his continuous motivation and support throughout this project. Apart from his busy schedule, he always motivated and supported me.

I owe my special thanks to *Ms. Samta Choudhary ji* and *Late Shri Pradeep Choudhary ji* for their invaluable contributions, cooperation, and discussions. In the journey of my life, he was much more than my mentor. He was a kind-hearted person, someone I could trust, someone who was open-minded, non-judgmental, aware, and had a great sense of humor. Even though I lost him this year, I feel he always looks after me. Moreover, he has left me with the most valuable of all guide-blessings.

I am very much obliged to *Prof. Pradeep K Garg*, the second editor of this book, for his motivation and support. This book would not have been possible without the blessings and valuable guidance of Prof. Garg.

Above all, I express my heartiest thanks to God (The One to Whom We Owe Everything) *Sai Baba of Shirdi* for all blessings, guidance, and help by you and only you. I would like to thank God for believing in me and being my defender. Thank you, God Almighty.

Dr. Lavanya Sharma

I am extremely grateful to my family Mrs. Seema Garg, Dr. Anurag Garg, Dr. Garima Garg, Mr. Hansraj Aggrawal, Ms. Pooja Aggrawal, and Master Avyukt Garg, and all relatives and friends for their understanding, continuous encouragement, moral support, and well wishes. Above all, I express my gratitude to Almighty God for offering all blessings and giving me enough strength to work hard to complete the book on time, as planned.

I am also thankful to the entire team at CRC Press for the timely publication of this book.

Prof. Pradeep K. Garg

Editors

Dr. Lavanya Sharma completed her M.Tech (Computer Science and Engineering) in 2013 from Manav Rachna College of Engineering, affiliated with Maharshi Dayanand University, Haryana, India. She completed her Ph.D. from Uttarakhand Technical University, India, as a full-time Ph.D. scholar in the field of digital image processing and computer vision in April 2018, and received a TEQIP scholarship for the same. Her research work is on motion-based object detection using background subtraction technique for smart video surveillance. She received several prestigious awards during her academic career.

She is the recipient of several prestigious awards during her academic career and qualified certification courses from ISRO Dehradun unit, India. She has 40+ research papers to her credit, including Elsevier (SCI Indexed), Inderscience, IGI Global, IEEE Explore, and many more. She has published three books and two books with Taylor & Francis, CRC Press in 2019 and 2020. She also has two patents in her account on object detection in visual surveillance. She has also contributed as an Organizing Committee member of Springer's ICACDS conferences 2016, Springer's ICACDS 2018, Springer's ICACDS 2019, Springer's ICACDS 2020, ICRITO 2021, and Springer's ICACDS 2021. Presently, she is the editorial member/reviewer of various journals of repute and active program committee member of various IEEE and Springer conferences also. Her primary research interests are Digital Image Processing and Computer Vision, Artificial Intelligence, Machine learning, deep learning, and Internet of Things. Her vision is to promote teaching and research, providing a highly competitive and productive environment in academic and research areas with tremendous growing opportunities for the society and her country.

Professor Pradeep Kumar Garg has worked as the Vice-Chancellor, Uttarakhand Technical University, Dehradun (2015–2018). Presently, he is working as a professor in the Department of Civil Engineering, IIT Roorkee. He completed B.Tech (Civil Engineering) in 1980 and M.Tech (Civil Engineering) in 1982, both from the University of Roorkee (now IIT Roorkee). He is a recipient of the Gold Medal at IIT Roorkee for securing the highest marks during the M.Tech program, the Commonwealth Scholarship Award for doing Ph.D. from University of Bristol (UK), and the Commonwealth Fellowship Award to carry out post-doctoral research work at the University of Reading (UK). He joined the Department of Civil Engineering at IIT Roorkee in 1982, and gradually advancing his career, rose to the position of Head of the Department in 2015 at IIT Roorkee.

Professor Garg has published more than 310 technical papers in national and international conferences and journals. He has undertaken 27 research projects and provided technical services to 85 consultancy projects on various aspects of Civil Engineering, generating funds for the Institute. He has authored five textbooks on (Remote Sensing, Geomatics Engineering, Digital Soil Mapping, UAV, and Digital Surveying Methods), and edited two books on Environmental Monitoring and Video Surveillance. He has developed several new courses and practical exercises in geomatics engineering. Besides supervising a large number of undergraduate projects, he has guided about 72 M.Tech and 26 Ph.D. theses. He is instrumental in prestigious Ministry of Human Resource Development (MHRD)-funded projects on e-learning, Development of Virtual Labs, Pedagogy, and courses under the National Programme on Technology Enhanced Learning (NPTEL). He has served as an expert on various national committees, including Ministry of Environment and Forests, National Board of Accreditation (All India Council of Technical Education), and Project Evaluation Committee, Department of Science and Technology, New Delhi.

Professor Garg has reviewed a large number of papers for national and international journals. Considering the need to train the human resources in the country, he has successfully organized 40 programs in advanced areas of surveying, photogrammetry, remote sensing, geographic information system (GIS), and global positioning system (GPS). He has successfully organized ten conferences and workshops. He is a life member of 24 professional societies, out of which, he is a fellow member of eight societies. For academic work, Professor Garg has travelled widely, nationally and internationally.

Contributors

Varshini Balaji
SSN College of Engineering
Anna University
Chennai, India

Alok Bhardwaj
Civil Engineering Department
Indian Institute of Technology Roorkee
Roorkee, India

D. K. Bhattacharyya
Department of Space
North Eastern Space Application Centre
Umiam, India

Thierry Bouwmans
Laboratory MIA
University of La Rochelle
France

Mukesh Carpenter
Department of Surgery
Alshifa Hospital, Okhla
New Delhi, India

Avinash Chouhan
Department of Space
North Eastern Space Application Centre
Umiam, India

Dibyajyoti Chutia
Department of Space
North Eastern Space Application Centre
Umiam, India

Pradeep Kumar Garg
Civil Engineering Department
Indian Institute of Technology Roorkee
Roorkee, India

Rahul Dev Garg
Geomatics Engineering Group, CED
Indian Institute of Technology Roorkee
Roorkee, India

Jhony H. Giraldo
Laboratory MIA, University of La Rochelle
France

Deepa Gupta
Amity Institute of Information Technology
Amity University
Noida, India

B. Jokanović
SGL Carbon GmBH
Meitingen, Germany

V. Jokanović
ALBOS doo and
Institute of Nuclear Science "Vinča"
Belgrade, Serbia

Mudit Kapoor
Geomatics Engineering Group, CED
Indian Institute of Technology Roorkee
Roorkee, India

Vallidevi Krishnamurthy
Department of Computer Science and Engineering
SSN College of Engineering
Anna University
Chennai, India

Ashish Maharjan
Department of Computer Science and Engineering
Sikkim Manipal Institute of Technology
Sikkim Manipal University
Majitar, India

Sneha Mishra
Department of CSE
Galgotias University
Greater Noida, India

Shankar Narayanan
Department of Computer Science and Engineering
Sri Sivasubramaniya Nadar College of
    Engineering
Chennai, India

Nilay Nishant
Department of Space
North Eastern Space Application Centre
Umiam, India

Ashis Pradhan
Department of Computer Science and Engineering
Sikkim Manipal Institute of Technology
Sikkim Manipal University
Majitar, India

P. L. N. Raju
Department of Space
North Eastern Space Application Centre
Umiam, India

V. Sanjay Thiruvengadam
Department of Computer Science and
    Engineering
Sri Sivasubramaniya Nadar College of
    Engineering
Chennai, India

K. Sarvani
Department of Zoology
Narayana Junior College
Telangana, India

Sudhriti Sengupta
Amity Institute of Technology
Amity University
Noida, India

K. Spandana
Department of Computer Science
Sreenidhi Institute of Science and Technology
Telangana, India

K. Sreeram
Department of Computer Science and Engineering
Sri Sivasubramaniya Nadar College of
    Engineering
Chennai, India

Priyanka Srivastava
Perception and Cognition Group
Cognitive Science Lab
Kohli Research Centre for Intelligent Systems
IIIT Hyderabad, India

P. Subhash Singh
North Eastern Space Application Centre
Department of Space
Umiam, India

T. Venkat Narayana Rao
Department of Computer Science and Engineering
Sreenidhi Institute of Science and Technology
Hyderabad, India

R. C. Vignesh
Department of Computer Science and Engineering
Sri Sivasubramaniya Nadar College of
    Engineering
Kalavakkam, India

Dileep Yadav
Department of CSE
Galgotias University
Greater Noida, India

Section I

Introduction to Artificial Intelligence

1

Overview of Artificial Intelligence

Pradeep Kumar Garg
Indian Institute of Technology Roorkee, India

CONTENTS

1.1 Introduction

Since the invention of computers, humans have been developing various approaches to increase operational speed and decrease physical size in diverse types of hardware and applications. While expanding the uses of computer systems, humans were interested in exploring whether a machine can think, work and behave like a human (McCarthy, 2019). This curiosity gave rise to the growth of artificial intelligence (AI), creating computer-controlled machines (e.g., robot) almost as intelligent as human beings. AI can be defined as "a science and a set of computational techniques that are inspired by the way in which human beings use their nervous system and their body to feel, learn, reason, and act" (McCarthy, 2019, pp. 1, 2–10).

AI is composed of two words, "artificial" and "intelligence," where "artificial" stands for "human-created' and "intelligence" stands for "thinking power." In other words, AI is *"a man-made object with thinking power'*. The intelligence is intangible which may be described as "the ability of a system to calculate, reason, perceive relationships and analogies, learn from experience, store and retrieve information from memory, solve problems, comprehend complex ideas, use natural language fluently, classify, generalize, and adapt new situations" (Iyer, 2018).

AI allows machines or computers to perform in an intelligent manner. For AI to work, availability of "data" is the main key (Joshi, 2020). Humans need some device or software that can process and handle the large amounts of data with minimum effort and speed. This handling of data and processing is known as data science. Data science can be defined as the "scientific study of data, that stores, records and analyses data for the benefits of society" (Joshi, 2020, pp. 1–5). Humans can learn faster and process certain things faster even with a limited amount of data, but AI-based systems need massive amounts of data to generate any useful inferences. The answers are present in the data, which can be obtained by applying AI to get them out. The AI techniques speed up the implementation of the complex programs. AI is currently being applied in a variety of fields, ranging from playing chess and music to making complex decisions, creating models, predicting patterns, and even self-driving cars (Iyer, 2018).

1.2 Definitions of AI

According to the father of AI, John McCarthy, artificial intelligence is "the science and engineering of making intelligent machines, especially intelligent computer programs" (McCarthy 2019, pp. 1–2). In other words, AI can be defined as "a branch of computer science by which we create intelligent machines which can think like human, act like human, and able to make decisions like human" (McCarthy, 2019, pp. 2–3). AI in a sense is the simulation/replication of intelligence processes by computer systems that can think and act rationally in the way similar to humans. There are many definitions and explanation available in literature about AI, as summarized in Table 1.1.

1.3 History of AI

The concept of inanimate constructs that can operate independently of humans is not new; in fact, it has been known since ancient times. The Greek god Hephaestus has been depicted forging robot-like servants out of gold. The modern computers were developed in the late nineteenth and early twentieth centuries. With the advent of modern, high-speed computers, it became possible to develop and test the ideas of machine intelligence. The pioneer project was conceived back in the 1950s. Since then, every industry has been trying to develop and/or make use of AI. Table 1.2 summarizes the systematic development of AI tools and technology.

TABLE 1.1

Various Definitions of Artificial Intelligence

S. No.	Authors	Definitions/Explanation
1	Bellman (1978)	"The automation of activities that we associate with human thinking, activities such as decision making, problem solving, learning."
2	Charniak and McDermott (1985)	"The study of mental faculties through the use of computational models."
3	Haugeland (1985)	"The exciting new effort to make computers think machines with minds, in the full and literal sense."
4	Schalkoff (1990)	"A field of study that seeks to explain and emulate intelligent behavior in terms of computational processes."
5	Kurzweil (1990)	"The art of creating machines that perform functions that require intelligence when performed by people."
6	Rich and Knight (1991)	"The study of how to make computers do things at which, at the moment, people are better."
7	Winston (1992)	"The study of the computations that make it possible to perceive, reason, and act."
8	Luger and Stubblefield (1993)	"The branch of computer science that is concerned with the automation of intelligent behavior."
9	Dean et al. (1995)	"The design and study of computer programs that behave intelligently. These programs are constructed to perform as would a human or an animal whose behavior we consider intelligent."
10	Nilsson (1998)	"Many human mental activities, such as writing computer programs, doing mathematics, engaging in common sense reasoning, understanding language, and even driving an automobile, are said to demand intelligence. We might say that (these systems) exhibit artificial intelligence."

TABLE 1.2

Summary of Developments in AI

Activity	Year	Particulars
First computer-related developments	1836	Charles Babbage, mathematician at Cambridge University, and Augusta Ada Byron first developed a programmable machine.
	1923	Karel Čapek's play *Rossum's Universal Robots* opened in London, where the word "**robot**" was used first time.
	1940s	John Von Neumann, mathematician at Princeton University, conceived the architecture for a computer that included a program and its processed data that can be stored in the computer's memory.
Maturation of Artificial Intelligence	1943	Warren McCulloch and Walter Pits carried out the first work that is now known as AI. They suggested a model of **artificial neurons**. The foundation for neural networks was laid out.
	1945	Isaac Asimov, a Columbia University alumnus, coined a term "**robotics.**"
	1949	Donald Hebb developed a new rule, called **Hebbian learning**, for modifying the strength between neurons.
	1950	Alan Turing, a British mathematician, World War II code-breaker, and a pioneer in machine learning, published *Computing Machinery and Intelligence*. He introduced the **Turing Test** for evaluation of intelligent behavior of the machines equivalent to human intelligence. Claude Shannon published *Detailed Analysis of Chess Playing*.
The birth of artificial intelligence	1955	Allen Newell and Herbert A. Simon developed the **first artificial intelligence program**, naming it the "**Logic Theorist**." This program was capable of proving 38 out of 52 mathematics theorems, as well as develop new proofs for several problems.
	1956	American computer scientist John McCarthy at the Dartmouth College Conference first used the term "**artificial intelligence**." During that time, computer languages, such as FORTRAN, LISP, and COBOL, were invented. Demonstration of the first running AI program was done at Carnegie Mellon University. It attracted lot of government and industry support.
The golden years – early enthusiasm	1965	Robinson's complete algorithm for logical reasoning was introduced.
	1966	Algorithms for solving mathematical problems were developed. Same year, Joseph Weizenbaum created the first **chatbot**, named **ELIZA,** which laid the foundation for the chatbots used today.
	1969	**Shakey**, a robot having locomotion, perception, and problem-solving capabilities, was developed by Stanford Research Institute.
	1972	The first intelligent humanoid robot, named **WABOT-1**, was built in Japan.
	1973	Edinburgh University's robot, called **Freddy**, could use vision technology to locate and assemble models.
The first AI winter	1974	The beginning of a period, which would last until the end of the decade, during which computer developers experienced a severe shortage of government fund for research work, leading to a decrease in interest in AI.
	1979	**Stanford Cart**, the first computer-controlled autonomous vehicle, was built.
A boom of AI	1980	AI came back using new techniques of deep learning, including Edward Feigenbaum's **Expert Systems** that replicated the decision-making capability of human experts. That year, the American Association of Artificial Intelligence organized its first national conference at Stanford University.
	1985	**Aaron**, the drawing program, was created by Harold Cohen.
	1986	Popularity of neural networks.
The second AI winter	1987	Private investment and government funding for AI research dry out once again due to huge costs and not enough return on investment. However, the **XCON** Expert System proved very cost effective.
	1990	Many advances in AI took place, such as machine learning, Web crawler, scheduling, data mining, multi-agent planning, natural language understanding and translation, case-based reasoning, games, vision, and virtual reality.
	1991	AI logistics planning and scheduling program that involved up to 50,000 vehicles, cargo, and people was adopted by US forces during the First Gulf War.

(Continued)

TABLE 1.2 (*Continued*)

Summary of Developments in AI

Activity	Year	Particulars
The emergence of intelligent agents	1995	The emergence of intelligent agents.
	1997	**IBM Deep Blue** defeated world chess champion **Gary Kasparov**, the first computer to defeat a human world chess champion.
	2000	Interactive robot pets developed. **Kismet**, a robot with a face capable of expressing emotions, was developed by researchers at MIT. Another robot, called **Nomad** was used to explore remote areas of Antarctica and located meteorites.
	2002	For the first time, AI **Roomba**, a vacuum cleaner, found application in the home.
	2006	AI used by business firms such as Facebook, Twitter, and Netflix.
Deep learning, big data, and artificial general intelligence	2011	IBM's **Watson**, a program capable of understand natural language and solving complex questions quickly, ultimately won **Jeopardy**, a quiz show.
	2012	An Android app, called **Google now**, was launched, which could be used as a prediction tool.
	2014	The **Eugene Goostman**, a chatbot, won a competition in the infamous **Turing test**.
	2018	**Project Debater** developed by IBM could be used to debate complicated topics with two master debaters, and outperformed them. Google developed a virtual assistant, **Duplex**, which made a call to book hairdresser appointment, with a human receptionist on the other end of the line not realizing she was talking to a computer program.
	Present	Increased computational power and volume of available data has increased the use of AI in the late 1990s, and this trend is accelerating. AI has enhanced the use of natural language processing, computer vision, robotics, machine learning, deep learning, etc. AI is useful in controlling vehicles, diagnosing diseases, and predicting behaviors. Recently, the 18-times historic defeat of World Go champion Lee Sedol by Google DeepMind's **AlphaGo** has proved the capabilities of intelligent machines.

1.4 The Importance of AI

AI can automate repetitive learning through the datasets. But AI has some basic differences from hardware-driven automation, as it can perform continuous, large-volume tasks reliably (Iyer, 2018). For such automation, some human intervention is still required to initialize the system. Automation, communication platforms, and machines can be integrated together with massive data to apply to several new applications. Given that AI adds intelligence to existing processes, it cannot be viewed as an independent application. For example, in new-generation Apple products, the Siri is included as a useful feature.

AI uses progressive learning algorithms that allow the data to carry out the programming. It can find structure and irregularities in the data to be used in classification and/or a prediction. For example, the AI-based program can teach itself to playing chess, and it can also be used to recommend the next product for online buyers. In the same way the models continue to adapt with the input of new data. The back-propagation technique allows the algorithm to refine itself, with the help of training data and new data, if the predicted results are not accurate. AI can analyze large data with hidden layers of neural networks. It can obtain higher accuracy through deep neural networks (https://www.javatpoint.com). The DL models require Big Data to train, as they learn directly from a dataset. The more data is fed to models, the more accurately they predict the results. For example, Alexa, Google Search, and Google Photos are all using the DL approach; the more we utilize them, the more accurate they become. In the medical field, AI-based DL, image classification, and object recognition techniques can be employed to possibly detect the disease on MRIs with almost as much reliability as when it's done by trained radiologists.

AI is not going to replace humans, but it supplements human abilities so they can be performed better. As AI algorithms learn entirely differently from humans, they ought to perceive things differently, and can easily visualize the relationships and patterns that cannot be seen by humans (McCarthy, 2019; Joshi, 2020). Thus the human–AI partnership can offer many opportunities:

(i) It can provide further support to our existing abilities, and allow for better perception and understanding.
(ii) It can introduce analytics to industries in which AI is currently being used.
(iii) It can be used to improve the analytic technologies such as computer vision, time-series analysis, etc.
(iv) It can bridge the economic, language, and translation barriers.
(v) It provides know-how of ML to be used to build predictive models for AI.
(vi) It can learn how software is to be utilized to process, analyze, and derive meanings from natural language.
(vii) It can process images and videos for several real-time applications.
(viii) It can build intelligent systems to provide interactive communications between humans and AI systems.

1.5 Processes Involved with AI

The AI programs will have cognitive skills: reasoning, problem solving, learning, perception, and self-correction, as given below (McCarthy, 2019):

1. **Reasoning process:** The AI program here focuses on selecting the most appropriate algorithm to achieve the required results. It is the process that is used for making judgments, decisions, and predictions. Reasoning processes are mainly categorized as inductive reasoning and deductive reasoning.
2. **Learning process**: Its function is acquiring data and creating rules in order to devise actionable information from data. Learning improves understanding of the subjects under study. The rules, also called algorithms, help provide sequences of instructions to perform a task using computing devices. It involves acquiring knowledge by way of study, practice, and gaining experience. Humans, some animals, and AI-based systems have the ability to learn (Rouse, 2020).
3. **Problem-solving process**: It is used to get the required solution from the current situation by taking another approach. Problem solving may include decision-making, i.e., selecting the best out of several possible alternatives to get the objectives.
4. **Perception process**: It includes selecting, acquiring, interpreting, and ultimately analyzing the information. In case of humans, perception is supported by sensory organs. Perception mechanisms in AI place the sensors data together in a useful manner.
5. **Self-correction process**: It is designed to continually refine the algorithm so that it determines the most accurate results.

1.6 AI as an Interdisciplinary Tool

AI is a technology that encompasses many areas including computer science, biology, psychology, sociology, philosophy, mathematics, and neuron science. One or more areas may be required to create an AI system. From an interdisciplinary perspective, the AI domains include explicit knowledge, language aptitude, verbal and numerical reasoning, creative and critical thinking, as well as working memory, as shown in Figure 1.1.

AI today is one of the growing technologies in computer science or data science, which has created a revolution globally by developing intelligent machines and tools (Shankar, 2020). AI is developed in a way similar to the operation of a human brain, specifically the way a human learns, decides, and works while attempting to solve a problem, and then using this outcome to develop intelligent machines and software. AI includes the use of expert systems, machine learning (ML), deep learning (DL), natural language processing (NLP), neural network, and fuzzy logic, as shown in Figure 1.2.

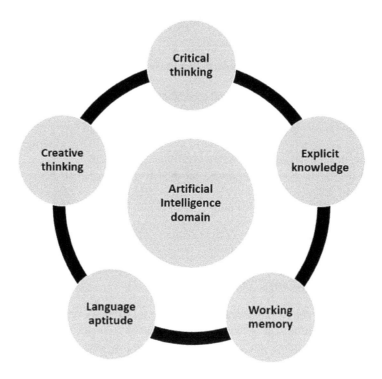

FIGURE 1.1   Various interdisciplinary domains of AI.

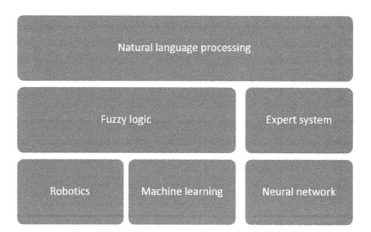

FIGURE 1.2   Various technologies used in AI.

The ML is about instructing a computer by providing it with data so it learns several things on its own, even when it has not been explicitly programmed. It is part of the expanding collection of AI tools that helps people make smarter, more logical decisions. The ultimate aim of ML is to allow independent decision-making by machines. The ML-based AI has several applications in education, medicine, search engine results, digital marketing, and more. Such AIs have a big demand in businesses, as they utilize ML to enhance users' experience, like for Amazon and Flipkart. The ML techniques have made significant progress in the past, and the commonly used are: (i) supervised learning, (ii) unsupervised learning, and (iii) reinforcement learning (Tutorials Point, 2021).

Robotics, a subset of AI, includes different fields of engineering and sciences, which involve design and manufacture of robots as well as their applications. They are often used to undertake difficult tasks that are not possible for humans, or to perform repetitive work. The AI-based robots work by studying the objects in their surroundings and by taking relevant actions. The automation tools with AI technologies can be used for repetitive work, as well as rule-based data processing tasks that are usually done by humans. For example, robots can be used in production of goods or for moving, spraying, painting, precision checking, drilling, cleaning, coating, carving, surgery, nursing, etc. (Shankar, 2020).

The ML-based AI applications can take large volumes of data and quickly transform them into actionable information. The robots combined with the ML can automate larger jobs and respond to process about the changes. The ML is also used to develop robots that are used to interact in social settings. Artificial neural networks (ANN) and DL technologies are also gaining popularity, as AI can process huge amounts of data more quickly and make more accurate predictions than humans can possibly do (Rouse, 2020). Some neural networks based applications include recognition of pattern, face, character, and handwriting. They can be used to manage the real-world problems and devise their solutions quickly.

An expert system can mimic the decision-making capability of humans. Expert systems integrate software, machine, reasoning, explanation, and actions to the users. Table 1.3 presents a scenario comparing programming without AI and programming with AI. The examples of expert systems include flight-tracking systems, predicting systems, clinical systems, etc.

Fuzzy logic approach can be used to compute based on "degrees of truth" rather than "true or false" (1 or 0) Boolean logic, on which the modern computers are based. The binary logic is not able to solve complex problems. Most of the processes are nonlinear in nature, and no specific model would be suitable to every situation. Fuzzy logic controllers are popular globally, especially with unstructured information (Shankar, 2020). The examples include consumer electronics and automobiles, among others.

The NLP requires AI methods that analyze the natural human languages to derive useful insights to solve problems. Existing approaches to the NLP are using ML. The NLP may include sentiment analysis, speech recognition, and text translation (Tucci, 2020). A well-known example of NLP is spam detection, which can interpret the subject title and body of an e-mail to determine the presence of "junk" content. Virtual assistants such as Alexa and Siri are good examples of computer applications helping people with daily tasks. These assistants can ask a few questions from the user to know what he/she wants, instead of analyzing huge amounts of data to understand a request, therefore drastically reducing the time to get the desired answer.

A correlation between AI, ML, ANN, and DL is shown in Figure 1.3. The broad differences are given in Table 1.4.

TABLE 1.3

Programming without AI and with AI

S. No.	Programming not using AI	Programming using AI
1	Without AI, any computer program may be able to answer only the *specific* questions.	With AI, any computer program may be able to answer the *generic* questions.
2	Modifications in the program would require changes in its basic structure.	AI programs can easily adapt new changes by having independent modules together, so any module can be modified without changing its basic structure.
3	Changes in the program are time-consuming, and may affect the program entirely.	Modification in the program is quick and easy.

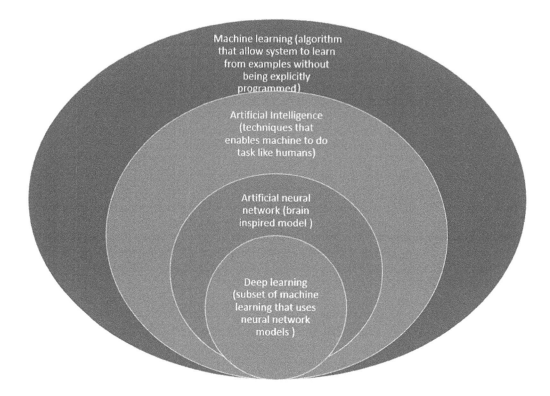

FIGURE 1.3   Relationship between AI, ML, ANN, and DL.

TABLE 1.4

Major Difference between AI, ML, ANN, and DL

AI	ML	ANN	DL
It originated around the 1950s	It originated around the 1960s	It originated around the 1950s	It originated around the 1970s
It is a subset of data science	It is a subset of data science and AI	It is a subset of data science, AI, and ML	It is a subset of data science, AI, and ML
It represents simulated intelligence in machines, and its aim is to build machines that can think like humans.	Computer can work/act without programming. Its aim is to make machines learn through data so that they can solve problems	These are the set of algorithms, modeled just like the human brain Their objective is to tackle complex problems	It is the process of automation of predictive analytics. It uses neural networks to automatically identify the patterns for feature extraction.
	Google search engine is used for speech recognition, image search, translation, etc. For example, Amazon and Flipkart are providing personalized services to individuals based on their likes and dislikes.		Some deep learning examples include self-driven vehicles, face recognition on phone, computer vision, and tagging on Facebook.

1.7 Types of AI

AI can be classified into seven types depending on the performance of machines (https://www.javatpoint.com): reactive machines, limited memory machines, theory of mind, self-aware, ANI, AGI, and ASI, as briefly explained below (Rouse, 2020).

1. **Reactive machines**: Reactive machines are conventional types of AIs that possess only limited capability to simulate the ability of human mind. Reactive machines work without memory-based functionality, and so are unable to correct their present actions based on their past experiences. Therefore, these machines are not capable of "learning." They study the surroundings and select the best solution among the possible ones. A well-known example is Deep Blue, the IBM chess program that defeated Garry Kasparov in the 1990s (Joshi, 2020). The Deep Blue can recognize pawn on the chessboard to make a move, but it cannot retain any memory as well as incorporate past experiences for making present decisions.

2. **Limited memory**: As is clear from the name, these AI systems have a small amount of memory, and thus very limited capacity to apply past experiences to new decisions. This group includes, among others, chatbots, virtual assistants, and self-driving vehicles. Many existing applications fall under this category of AI. These machines can retain data for a short time, limited by the capacity of their memory. In addition to having the capabilities of reactive machines, limited memory machines are capable of learning from the historical data to make certain decisions. The AI systems using DL require large volumes of data for training, which they can store in their memory for solving the current/future problems. For example, an image recognition AI can be trained on a large number of images and their features to identify the objects it has scanned. Any new image will make use of the training images and, based on its "learning experience," would label the new image with better results. A self-driven vehicle constantly detects the movements of all other vehicles around it and adds them to its memory. It can store the speed and pattern of changing lanes, etc., of vehicles around it, and can safely navigate on the basis of these data.

3. **Theory of mind**: This is a psychology term. Theory of mind is the future AI systems that are presently planned to be developed (Tucci, 2020). When applied to AI, these systems are expected to have the social intelligence to understand emotions. The two types of AI mentioned previously comprise the majority of modern systems, with this type and self-aware type of AIs being developed as a concept, and the work is still in progress. The main purpose of building such an AI is to simulate human emotions and beliefs through computers that can impact future decisions. For example, if two individuals plan to work together, they should interact to work effectively.

   Various models are used to understand human behavior, but one with a mind of its own is yet to be created. These systems can understand human requirements and predict behavior. Such systems can assist in the future based on human expectations. Such AI will have the ability to understand humans by interacting with them and identifying their needs, emotions, and requirements. For example, Bellhop Robot is being developed for hotels, with the ability to assess the demands of people wishing to come stay at the hotel.

4. **Self-aware**: These AI systems have a sense of self and possess human-like consciousness and reactions. Machines with self-awareness will be able to understand their own current state, and thus be conscious about themselves, and will use information to infer the emotions of others. This is the expected next stage of AI development. It is believed that this type of AI will achieve the ultimate goal of AI development. It will have emotions, needs, beliefs, and potentially desires of its own. Such AI will operate like a human and start predicting its own needs and demands.

   The self-aware AI is expected to enhance the output many times, but it can also lead to disaster. Such AI would have dangerous ideas, like self-preservation, which may not always coincide with the wishes, or even the actual physical well-being, of humans. Such machines although will have the capability to develop self-driven actions. This is the type of AI associated with every apocalyptic prediction of the end of the human civilization.

5. **Artificial Narrow Intelligence (ANI)**: The ANI is also known as Weak AI, that is, the one designed and trained to undertake only one particular type of work (Rouse, 2020). This definition includes all the existing AIs, including the most complicated ones. Any AI that utilizes ML and DL to teach itself may be called an ANI. Since the ANI performs only a specific task autonomously due to its programming limitations, it has a very limited or narrow set of competencies. These systems correspond to all the reactive and limited memory AIs. Examples include industrial robots and virtual personal assistants, which use weak AI. Speech recognition AI is another example of a weak AI, which identifies spoken words and converts them into a machine-readable format.

6. **Artificial General Intelligence (AGI)**: The AGI is also known as Strong AI. Its program can replicate the cognitive abilities of the human brain. It can perform a variety of tasks, as well as learn and improve itself. It is a self-teaching system that can outperform humans in a large number of disciplines. It provides the ability to perceive, understand, learn, and function, just as human beings do. The AGI systems employ fuzzy logic to apply domain knowledge and find a solution automatically to an unknown task. Such systems are able to reduce substantially the time required for training. Examples include the Pillo Robot that can answer questions related to health, or AlphaGo, a computer program to play the board game Go, which has defeated Lee Sedol, a South Korean professional gamer.

7. **Artificial Super Intelligence (ASI)**: The ASI will probably be the future AI research area, as it would be the most capable intelligence in the world. The ASI will not only replicate the intelligence of human beings but also have much higher storage (i.e., memory), faster data analysis, and better decision-making powers. The capabilities of ASI are expected to supersede that of humans. The AGI and ASI are expected to create a big revolution in the future, but they also may threaten our way of life. An example of ASI includes the Alpha 2, which is the first humanoid ASI Robot (Rouse, 2020).

1.8 Advantages and Disadvantages of AI

Every technology has some merits and demerits (https://www.javatpoint.com; Tucci, 2020). AI has many more advantages than disadvantages, as discussed in the following section.

Advantages

1. **Better accuracy**: The AI-based machines help analyze patterns and trends by accurately assessing the needs of the users. An AI-enabled machine is responsible for selecting the input data and values as per past experience or information, reducing human error and providing high accuracy. For instance, if a firm is more dependent on the data that is fed to a system manually, the chances of 100% correctness of data entered into the system are lower than if the input is automated. By contrast, a machine that can analyze its surroundings to capture the data automatically into the system is considered to be more accurate, eliminating the possibility of a manual error.

2. **Higher speed**: The AI systems are very fast and can make predictions with a higher degree of accuracy than is possible for humans.

3. **Better decision-making**: Human perception, understanding, and decision-making are often affected by personal bias and current emotional state. Since the machines are not affected by bias or emotions, AI-enabled systems could provide the most optimal decisions and solutions without any personal prejudices. One of the first examples of this is the loss of Garry Kasparov – a chess grand champion but still prone to human error – to IBM's Deep Blue back in the 1990s.

4. **High reliability**: AI-equipped machines are capable of performing repeating actions with an unchangingly high degree of accuracy.

5. **Day-night working**: The AI systems can work continuously for long periods of time, without the need for break for sleep, food, elimination, or recreation, all of which humans need.

6. **Dealing with complexities**: While many people tout their ability to "multitask" on their resumes, it is actually impossible for humans to handle several tasks at the same time with the same degree of focus given to all of them. Machines, on the other hand, can process large amounts of data required for several tasks to be performed simultaneously, without any confusion and consequent errors.

7. **Working in risky areas**: AI-equipped machines are very useful in actions that are hazardous to humans, such as defusing a bomb, exploring the nuclear sites, cleaning up a toxic spill, and the like.

8. **Optimization of resources**: The AI systems have the capabilities to assess and interpret multiple data streams at the same time, from handling databases of products and customers to analyzing the patterns of purchase. Humans are not physically able to accomplish these multiple tasks simultaneously. Thus, these machines would help in the resource optimization.

9. **Digital assistant**: For example, the AI technology is used by various e-commerce companies to display the products per customer's need.

10. **Working as a public utility**: AI is helpful in public utilities, self-driving cars, regulation of traffic, facial recognition, natural language processing, etc.

Disadvantages

1. **High cost**: An AI system consisiting of hardware and software is very costly, and it also requires recurring expenses for maintenance and upgrades to meet day-to-day needs. In addition, it may be costly to process the voluminous information required by AI programming.

2. **No original creativity**: Humans are always creative and full of new ideas, but AI machines are not creative and imaginative to beat the human intelligence.

3. **No out-of-box thinking**: Even smarter AI-based machines cannot think or work out of context, but will perform the task they have been trained on.

4. **No feelings and emotions**: Even the best-performing AI machines do not have feelings, so they fail to make any kind of emotional attachment with humans. These machines, in fact, may be harmful to users if they are not used properly.

5. **Dependency on machines**: With the advancements in technology, humans are becoming dependent on gadgets/devices/machines/software, and thus may not use much of their mental capabilities.

1.9 Some Examples of AI

Intelligent gadgets can make everyday tasks simple and fast. For example, Alexa is capable of keeping a record of our daily appointments, list of items to be purchased, play the desired music, read news, and play innovative games (Shankar, 2020). Some other examples include the following:

1. **Echo**: Echo, launched by Amazon, is a cloud-based voice assistant, Alexa. It is capable of hearing, comprehending, and responding to commands or questions of the users and offer possible solutions. For example, you can ask Alexa if you need an umbrella before going out, and it might suggest you take one, as it may to rain in the afternoon.

2. **Flipkart**: Flipkart, an e-commerce shopping platform, can be used to suggest items to its customers based on their past purchase or viewing history of items.

3. **Pandora**: The Pandora platform uses AI to determine the music the users require. It does not, however, provide any song choices.

4. **Netflix**: Netflix is the most popular Over The Top (OTT) platform today, and is also known as Other Than Television platform, among which are Amazon Prime, Hulu, and others. The OTT

platforms provide services that deliver content to its customers over the internet by paying a subscription fee. They also recommend additional content based on the user's previous choices.

5. **Siri**: Developed by Apple, Siri is a voice-activated interactive assistant. It uses ML technology to understand the ways the users are navigating through their phones, sending messages, and making phone calls. To use this feature, begin by saying, "Hello Siri," followed by an action request.

1.10 Applications of AI

AI has wide applications. More and more industries, such as education, health care, travel, entertainment, finance, and marketing, rely heavily on its ability to solve complex problems and perform complex functions efficiently (Sharma and Garg, 2020). It is also being used in military planning, intelligent vehicle movement, credit card transaction monitoring, robots, credit card fraud detection, automobiles, etc. (Tutorials Point, 2021; Tucci, 2020). The AI is trying to make users' daily lives much more easy and comfortable. The following are some areas having potential applications of AI:

1. **AI as a Service (AIaaS)**: The deployment of an AI platform may be expensive, as it involves the cost of hardware, software, and staff. Therefore, many firms are incorporating AI in their products to provide access to AIaaS platforms (Tucci, 2020). The AIaaS allows to experiment with various AI platforms for businesses and applications before investing heavily in an AI platform. Popular AI-based cloud offerings include IBM Watson Assistant, Amazon AI, Google AI, Microsoft Cognitive Services, etc.

2. **Automobiles**: Many automobile industries are providing AI-based virtual assistants to their users for better driving performance, such as TeslaBot by Tesla. The AI is now being applied toward development of driverless cars. These cars, with the help of AI systems would be able to apply brakes, change lanes, navigate, etc. Such cars will study the patterns of other surrounding cars moving on the road and implement the moves necessary for safe driving autonomously. Autonomous vehicles use computer vision, image recognition, and DL to navigate a vehicle in a given lane and at the same time avoid obstructions like dividers, pedestrians, light-poles, animals, etc.

3. **Agriculture**: The AI is emerging in the fields of agriculture, which requires various resources for obtaining the best yields. Agriculture robotics is being applied in agriculture for crop monitoring and predictive analysis to help farmers. The AI techniques for farming help increase productivity and yield.

4. **Banking**: Banks are using chatbots to provide services and offers to their customers, and to deal with the transactions without human involvement. The AI virtual assistants improve the services and cut down the costs of establishments. Financial organizations make use of AI to improve decision-making for loans, keep track of approved loans, set credit limits, as well as highlight the investment opportunities to their customers (Tucci, 2020).

5. **Business**: Business can use AI-based solutions to assess the weaknesses and strengths in order to improve its financial and customer relationship management (CRM), among other things. AI can help in automating the works, saving considerable time and manpower requirements. The ML algorithms that can better serve customers are integrated into analytics and CRM platforms to. Manufacturing units can improve the quantity and quality of its production by using AI required to assess the demand and supply, assembling the parts, etc.

   AI is being used in the e-commerce business in a big way to provide competition to e-commerce industry. It is helping its customers to find out the related products with suggested size, color, or brand. Chatbots are being used in websites of companies to provide almost instant customer service. For example, McDonald's has been using AI to analyze customers' ordering trends. Further, customers can place orders directly by using kiosks or interactive terminals instead of dealing with a live cashier; this has reduced order errors and increased sales.

6. **Data security**: In digital worlds, cyberattacks are growing very fast, and the security of data has become crucial for all organizations. AI is being used to make this data safer and more secure. AI and ML in cybersecurity products are providing added value to identify malware attacks. The AI is capable of assessing new malware attacks much faster than the human operators. The AI-based security technology gives organizations advanced information to take precautions against threats before real damage occurs. The technology, such as AEG bot or AI2 platform, is playing an important role in helping organizations fight with cyberattacks; they can also be used to determine software bugs that allow cyberattacks to happen (McCarthy, 2019).

7. **Education**: AI can adapt the learning as required by each student, and deliver a good learning experience. In addition, it provides universal access to all students, as well as helps them work at their own pace. The system also automates examination grading systems by reducing the involvement of educators, providing them more time to teach. An AI chatbot, as a teaching assistant, can communicate effectively with students. An AI tutor can teach the subject as required by the students. The AI can work as a personal virtual tutor for students in future, which will be easily accessible to students at anywhere any time.

8. **Entertainment**: AI-based applications such as Netflix or Amazon are providing entertainment services all over the world. With the help of ML-based AI algorithms, these services also recommend specific programs or shows for its users.

9. **Finance**: Finance applications require collection of personal data of individuals and provide help, advice, and suggestions related with finances, and can even help doing securities trading. Today, trading on Wall Street is done through AI software (McCarthy, 2019). The finance industries are employing ML in the automation, chatbot, adaptive intelligence, algorithm trading, etc. into financial processes. The AI systems, such as Intuit Mint or TurboTax, are being used by financial institutions for personal financial applications, while other programs, such as IBM Watson, are being used to buy homes.

10. **Gaming**: AI can be used for gaming purpose to generate alternative solutions in a game based on decisions taken by the users in the game, such as player movements, pathfinding, etc. AI-based programming is used by many video games today, such as *Minecraft* and *Tom Clancy's Splinter Cell* (Tutorials Point, 2021). The AI machines can play crucial roles in games, such as poker, chess, etc.

11. **Government**: Governments are using AI to draw suitable policies and services, analyze road accidents, and find solutions for many other problems. The AI-based applications are reducing costs, minimizing errors, taking heavy workloads, and helping bust the backlogs.

12. **Health care**: AI is assisting doctors in many ways and providing faster recovery to the patients (Iyer, 2018). AI can help doctors and patients with diagnoses and inform the latest conditions to the patients, and, if the condition is serious, ensuring medical help reach patients faster. AI has several advantages and is expected to have a positive impact on the health care industry. The AI robots are being developed that will be able to care for the elderly and remind them to take their medicine and even locate the misplaced items like eyeglasses. Various AI applications may include use of online virtual health assistants and chatbots by the patients, collection of medical history, fixing of appointments, and helping with administrative tasks.

The AI technologies are also helpful to understand pandemics, such as COVID-19. For example, BlueDot, a Canadian company, used AI technology to detect COVID-19 outbreak in Wuhan, China, soon after the first few cases were detected. The IBM Watson can understand the natural language and provide responses to the queries. The system can mine the data of patients to develop a framework for presenting the results with a relative score. But while the predictive algorithms could be helpful in controlling pandemics or other global threats, the ultimate impact of AI is impossible to predict (Tucci, 2020).

It is known that robots are increasingly assisting the surgeons in an operating room. Specialized robots are being manufactured to carry out experimentation and provide life-like experiences without carrying out any hands-on experimentation on patients. For example, Gaumard, a health care education company, is now producing robots that can be used to perform various experiments by medical students and medical professionals to do practical learning. These life-like robots can interact with care providers and simulate facial expressions and other physical responses to the questions and actions of doctors or medical students when prompted, spoken to, or touched. Not only can medical professionals interact with the robots, but the robots also can be operated on to teach the procedure and also to take corrective steps if any errors are made during an operation. Using such AI-based system, medical students can easily make incisions, conduct surgeries, draw blood, monitor breathing, etc. (McCarthy, 2019).

13. **Law**: The use of AI is proving to be time-saving to automate the labor-intensive processes of the legal industry, and thus help improve the services of clients. Law firms and professionals make use of ML-based AI to analyze the data and predict the outcomes. In addition, computer vision is used to extract information and the NLP is used to interpret requests for information.

14. **Natural language processing**: The NLP utilizes the capabilities of machines to understand natural languages. Two of the most commonly used examples of NLPs, available in many smartphones and computer software, are spell check and autocorrect. In 2019, two AIs created by Alibaba and Microsoft defeated a team of persons in a Stanford reading-comprehension test (McCarthy, 2019). The algorithms could "read" a series of Wikipedia entries on the topic, and successfully answered a number of questions about the topic more precisely than the human participants could do.

15. **Personal assistant**: An AI-based personal assistant can perform several tasks based on verbal or written commands, such as navigating the records or assessing if some person suffered a heart attack during an emergency call services. This is a good example of weak AI, as the algorithm has been created to perform a specific task. The best-known examples of AI assistants are Google, Alexa, and Siri (Kowalewskisays, 2019). One of the most advantageous points about an AI assistant is that it serves as a great help in various applications of AI. As more and more consumers are using Virtual Personal Assistants, speech recognition has become essential in our lives. Phones, computers, and home appliances are increasing our dependence on AI and ML through voice. According to recent statistics, the AI assistant market is going to expand further and will become worth USD 25 billion by 2025 (Businesswire, 2019).

16. **Robotics**: AI has a remarkable role in robotics. Manufacturing industries are adapting to incorporate the use of robots into their workflows. Earlier, the industrial robots were separated from human workers and programmed to perform single tasks. Today industrial robots function as cobots, which are smaller and multitasking robots. Such cobots can be used to take up the jobs in warehouses, industries, and other workspaces. Normally, robots are programmed to perform tasks that are repetitive in nature, but AI-based robots are used to perform several tasks with their own previous experience, and even without preprogramming (Tutorials Point, 2021). Humanoid robots are best examples of AI-based intelligent robots; like Erica and Sophia can talk and behave like human-beings. Their sensors can detect physical data from the real world, such as light, sound, temperature, movement, and pressure, and these systems can learn from their past and apply that knowledge to the new environment (Tschopp, 2018).

    Industrial robots are used in the manufacturing fields as an alternative to humans. For example, such robots have been in use in the automobile manufacturing sector for quite some time, as some processes in car making may not be safe for humans. In 1961, Unimate, the first-ever industrial robot, was used by General Motors on an assembly line. Currently, the robots are used in warehouses for many other duties also (McCarthy, 2019). In 2014, Amazon has deployed Kiva robots in their centers' warehouses, which are helping employees to fill orders very quickly (15 minutes) that humans alone can manage (90 minutes). These robots can pick up the items and transport the inventory directly to human workers. Programmed with object detection technology, these robots can move freely throughout the warehouse, avoiding potential collisions with other Kiva robots or human workers.

17. **Social media**: AI can be used to organize and manage large volumes of data efficiently. Social media sites like Facebook, Twitter, and Snapchat may contain profiles of large number of users, which are required to be stored and managed efficiently. The AI can analyze this huge block of data to identify the latest trends, hashtags, and requirements, among other things, of different users.

18. **Supermarkets (retail):** Some large industries in the retail sector have started using AI-based robots to handle the tasks previously carried out by human customer associates (Iyer, 2018). Stock inventories generally are time-consuming and require multiple employees to track items that need to be restocked so they can be reordered. Several supermarkets and other retail markets are now using robots to take stock inventory. For example, Walmart, a retail industry giant, and Bossa Nova, a robotics company, have teamed up to create a supermarket application. The Bossa Nova robot would be used to scan the shelves in real time to collect product data, doing so much faster than a human employee could. Such a robot aims to improve product availability, enhance customer experience, and reduce the workload of customer associates.

19. **Transportation and travel:** Demand for AI is also growing in travel industries. In addition to AI being used in autonomous vehicles, it is used to manage traffic, estimate flight delays, and many other tasks (Tucci, 2020). It is also used in the travel insurance sector to file claims faster and more efficiently after the accidents. The AI can be used for making travel arrangements and suggesting accommodations, flights and best routes to its customers. Travel companies are employing AI-powered chatbots for faster response and better service for their customers.

20. **Vision systems:** Vision based algorithms are being developed to predict future actions of individuals (Tutorials Point, 2021). Machine vision can capture and analyze visual information using a camera and video and digital signal processing. These systems can understand, interpret, analyze, and display visuals. For instance, doctors can utilize expert system to operate on patients. Police can use them to recognize the faces of criminals based on drawings done by a forensic artist.

21. **Speech recognition:** Some AI-based systems can be used for hearing and comprehending the sentences and their meanings while a person is talking. These systems are capable of handling a variety of accents, slang words, background noise, change in a person's voice due to an illness, and many more aspects.

22. **Handwriting recognition:** The algorithm is able to read the text written on paper using a pen or on screen using a stylus. In addition, it can also recognize letter shapes and convert them into editable text (Tutorials Point, 2021).

1.11 Summary

The AI is one of the important areas of computer/data science allowing a machine to perform tasks in a way similar to a human performing them. Its main goal is giving machines the ability to process information and make decisions based on that information, the same ways humans do. However, the science and the industry of AI are far from being fully explored and developed. In particular, AI/ML/DL possess significant potential to make human living safer and easier. It is implied that in the future, AI will help humans with many more tasks currently in infancy, such as, for example, space travel.

Robots nowadays are performing many tasks that in the past were done by humans. But robots cannot function without human cotrol, programming, debugging, and analysis. The AI-based robots would make human lives more comfortable, and soon they are going to be an essential part of our daily lives in the same way computers have been since the 1980s. Still, even though AI is becoming increasingly prevalent in many applications, it is not going to completely replace human operators. In the long run, AI is expected to enhance human abilities and be the dominant technology of the future.

REFERENCES

Bellman, R. (1978), *An introduction to artificial intelligence: Can computers think?* San Francisco: Boyd and Fraser Publishing Company.
Businesswire (2019), Global Intelligent Virtual Assistant (IVA) Market 2019–2025: Industry Size, Share & Trends-ResearchAndMarkets.com, https://www.businesswire.com/news/home/20190822005478/en/Global-Intelligent-Virtual-Assistant-IVA-Market-2019-2025
Charniak, E. and McDermott, D. (1985), *Introduction to artificial intelligence.* Reading: Addison-Wesley.

Dean, T., Allen, J. and Aloimonos, Y. (1995), *Artificial intelligence: Theory and practice*. New York: Benjamin Cummings.

Haugeland, J., (Ed.). (1985), *Artificial intelligence: The very idea*. Cambridge, MA: MIT Press.

History of Artificial Intelligence. Available at: https://www.javatpoint.com/history-of-artificial-intelligence [accessed on 23 April 2020a].

History of Artificial Intelligence. Available at: https://blog.solvatio.com/en/from-deep-blue-to-alexa-the-history-of-artificial-intelligence [accessed on 23 April 2020b].

History of Artificial Intelligence. Available at: https://blog.solvatio.com/en/from-deep-blue-to-alexa-the-history-of-artificial-intelligence [accessed on 23 April 2020c].

History of Artificial Intelligence. Available at: https://www.javatpoint.com/history-of-artificial-intelligence [accessed on 23 April 2020d].

Iyer, Ananth (2018), Artificial Intelligence, April 22, https://witanworld.com/article/2018/04/22/witan-sapience-artificial-intelligence/

Joshi, Navee (2020), Types of Artificial Intelligence, COGNITIVE WORLD, https://www.forbes.com/sites/cognitiveworld/2019/06/19/7-types-of-artificialintelligence/#3e68129b233e

Kowalewskisays, Michał (2019), 4 Amazing Ways AI Personal Assistants Can Impact Your Business, March, Artificial Intelligence, https://www.iteratorshq.com/

Kurzweil, R. (1990), *The age of intelligent machines*. Cambridge: MIT Press.

Luger, G. and Stubblefield, W. (1993), *Artificial intelligence: Structures and strategies for complex problem solving*. Redwood City: Benjamin/Cummings.

McCarthy, John (2019), Artificial Intelligence Tutorial – It's your time to innovate the future, Dataflair Team, November 27, https://data-flair.training/blogs/artificial-intelligence-ai-tutorial/

Nilsson, N.J. (1998), *Principles of artificial intelligence*. Palo Alto, CA: Tioga Publishing Company.

Rich, E. and Knight, T. (1991), *Artificial intelligence*. New York: McGraw-Hill.

Rouse, Margaret (2020), Artificial intelligence, https://searchenterpriseai.techtarget.com/definition/AI-Artificial-Intelligence

Schalkoff, R. (1990), *Artificial intelligence: An engineering approach*. New York: McGraw-Hill.

Shankar, Ramya (2020), Future of Artificial Intelligence, 9 April, https://hackr.io/blog/future-of-artificial-intelligence

Sharma, L. and Garg, P. (Ed.). (2020), *From visual surveillance to Internet of Things*. New York: Chapman and Hall/CRC, https://doi.org/10.1201/9780429297922

Tschopp, Marisa (2018), On Trust in AI – a systemic approach, August 28, https://medium.com/womeninai/on-trust-in-ai-a-systemic-approach-d1e1bd112532

Tucci, Linda (2020), Ultimate guide to artificial intelligence to enterprise, https://searchenterpriseai.techtarget.com/definition/AI-Artificial-Intelligence

Tutorials Point (2021), Artificial Intelligence: Intelligent Systems, https://www.tutorialspoint.com/artificial_intelligence/artificial_intelligence_overview.htm [accessed on April 20 2020]

Winston, P. (1992), *Artificial intelligence*. Reading: Addison-Wesley. [accessed on April 23 2020].

2

Knowledge Representation in Artificial Intelligence: An Overview

Lavanya Sharma
Amity University, Noida, India

Pradeep Kumar Garg
Indian Institute of Technology, Roorkee, India

CONTENTS

2.1 Introduction

We humans are good at perception, logical thinking, and interpreting the knowledge. We know things – that is, we possess knowledge – and as per our thinking we act and perform different kinds of actions in real-time scenarios. In terms of machines, all the things done by human are termed knowledge representation and reasoning (KRR). KRR is a segment of AI that deals with thinking of AI agents and how it contributes to agents' intelligent behavior [1–3]. They are also responsible for representing the real-world facts and information so that a machine can understand and solve specific tasks such as theorem proving, gaming, medical imaging, natural language processing (NLP), and many more. Basically it is a method of describing the representation of knowledge in AI and also enabling intelligent machines to learn from this and act accordingly like a human [4–10].

In literature, various kinds of knowledge are presented that need to be represented in AI systems [9–14]. Some of them are:

- **Object:** It is about the fact in the real world, an entity, a thing: guitar, class, human, chair, etc. For example, a guitar has strings, or a classroom has a blackboard and chairs [15, 16].
- **Event:** An action that happens in the real world.
- **Meta-knowledge:** A portion of data that represents knowledge in the real world.
- **Facts:** Denotes a statement that is true in real world.
- **Knowledge base (KB):** Also known as a core component of the knowledge-based agents. It is a database used for sharing and management of knowledge.

DOI: 10.1201/9781003140351-2

Various KB are structured in such a way that they not only store data but also find solutions for additional problems using data from past experience stored in them. KB has a very important role in AI, which can be expressed by considering the methodology followed by AI systems [5, 7, 16, 17]. The procedure is as follows:

- **Perception block**: It contains set of senses for machines, or it is a core component using which interaction between a system and its environment takes places. It can be data, text, video, audial, temperature, or many other things.
- **Learning block**: In this block models are trained that are required for machine working. In this block the learning algorithms such as machine learning (ML) and deep learning (DL) are coded. This block is directly connected to the perception block in order to retrieve the information required for training of models.
- **Reasoning–Knowledge representation block**: It is the most critical block that effectively takes inputs from the perception block and extracts the important information from it. This block makes sure the knowledge is easily available and can be provided to the model or learning agent if required.
- **Planning and execution block**: It takes input from the above reasoning–knowledge block and provides a functional road map to the machine. It also specifies the action to be taken and expected outcome also.

2.2 Types of Knowledge

In AI, knowledge can be represented in various manner that depends on its structure or perspective of the designer or type of internal structure used [18–25]. An effective knowledge representation (KR) should include the knowledge that is required to solve specific real-time tasks as shown in Figure 2.1. It should

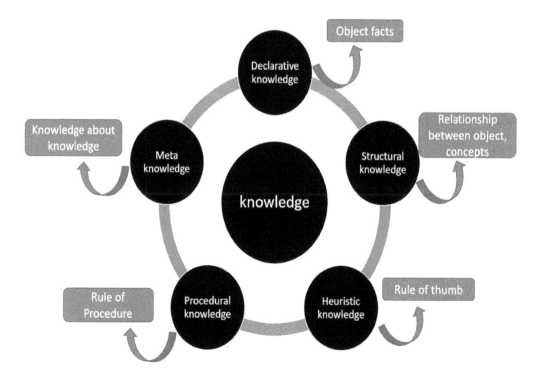

FIGURE 2.1 Different types of knowledge.

be natural, easy to maintain, and compact in nature [14–17, 26–28]. In literature, several kinds of knowledge are present, as follows:

- **Declarative knowledge**: It is to identify something and mainly consists of facts, objects, and concepts.
- **Procedural knowledge or imperative knowledge**: This knowledge is responsible for knowing how to do something. It consists of a set of rules, strategies, planning, agenda, and many more.
- **Heuristic knowledge**: This knowledge represents various experts in a particular domain. It is based on rules-of-thumb generated from past experiences, general awareness of techniques that are fruitful to work with but not representing complete certainties.
- **Structural knowledge**: This is a kind of basic knowledge to problem-solving and deals with the relationship between objects. It illustrates connections between different kinds of concepts such as parts, instances of, and groupings of something.

2.3 The Relation between Intelligence and Knowledge

In the realistic application domain, knowledge occupies a pivotal position for exhibiting intelligent behavior in AI agents. In AI, an agent can act accurately if it has past experience about the input or some knowledge of input. For example, there is one decision-maker that uses knowledge and acts accordingly by sensing its environment. If the prior knowledge is not present, then it cannot act accurately [11, 12, 29, 30]. (See Figure 2.2).

In the upcoming years, it is likely we will have a very large amount of KB that consist of information, data, and knowledge that will benefit various industries. The significance of data and text analytics is increasing day by day. With a lot of online KB and upcoming KB software, AI and knowledge management are gaining a huge thrust. But there is one major challenge, namely the presence of unstructured data. In the near future, organizations will have a proliferation of both structured and unstructured data. This forces humanity to discover paths that uncover knowledge from various resources. In this kind of scenarios, domains such as big data come into the consideration [15–17]. Cognitive computing (CC) is an important tool for drawing out information from Big Data.

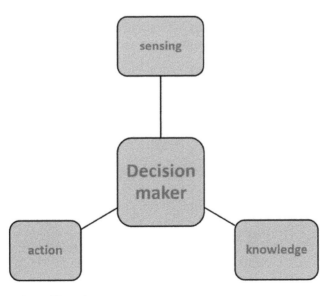

FIGURE 2.2 Decision-maker and its environment.

2.4 Life Cycle of Knowledge in AI

In an AI system, there are five major components involved in displaying intelligent behavior: perception, learning, KRR, planning, and execution (Figure 2.3). The given figure represents KRR as a main component. The interaction of real world and its components help the machine display intelligence. Using perception it retrieves information from its surroundings, which can be visual, audio, or any other form of sensory input. Second, the learning component (LC) is responsible for learning from data captured through perception. In this cycle, KRR is the major component [16, 26, 28]. These components are also involved in representing the human-like intelligence in machines and, while independent of each other, are also integrated simultaneously. In this, both design and implementation of AI depend on analysis of KRR.

2.5 Different Approaches to Knowledge Representation

In literature, various approaches to KR are present, such as simple relational knowledge, inheritable knowledge, procedural knowledge, and inferential knowledge [1, 3, 10]. Relational knowledge is considered the simplest way to store facts, and each fact about a set of objects is stored in a column in a systematic order. It uses the relational approach as shown in Figure 2.4.

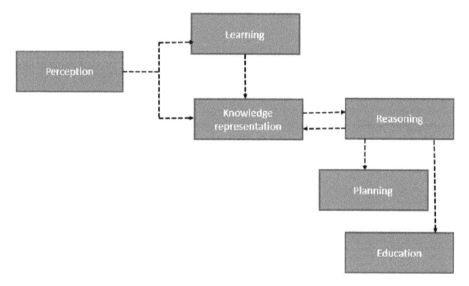

FIGURE 2.3 AI system and its components.

Player's	Weight	Age
Player1	60	25
Player2	29	18
Player3	70	26
Player4	78	29

FIGURE 2.4 Simple relational knowledge approach.

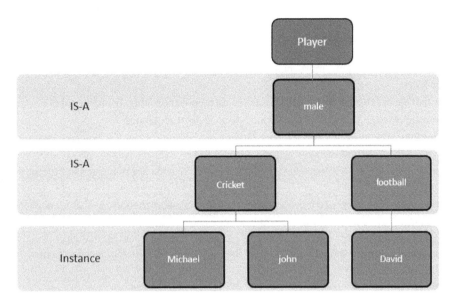

FIGURE 2.5 Inheritable knowledge approach.

In the inheritable knowledge method, data is stored in a hierarchical manner, and arrows are used to represent objects and their associated values. In this approach, a generalized or hierarchical approach is applied, and an inheritance concept is used, which means elements can inherit values from other class members [10]. This approach consists of inheritable knowledge that displays the connection between class and instance, known as instance relation. Each frame represents a group of attributes and their value, as shown Figure 2.5. Inferential knowledge presents knowledge as well-formed formulas (wff) and logics. This method guarantees correctness as compared to the other methods and also is used to derive more facts such as represented in Equation 2.1.

Let us consider the following statements:

- "Daphne is a Lady"
- "All ladies are mortal"

    From the above statements, a new statement can be generated:

$$
\begin{aligned}
&\text{Lady}(\text{Daph}) \\
&\forall x = \text{lady}(x) \rightarrow \text{mortal}(x)\text{s}
\end{aligned}
\tag{2.1}
$$

In case of procedural knowledge, small programs or segments of code are used that describe how to proceed with doing specific things. In this approach, the If–Then rule is used, and coding can be done using LISP and prolog programming language. A heuristic or domain-specific knowledge can be easily represented using this approach.

2.6 Basic Requirements for Knowledge Representation (KR) System

This section provides some of the general requirements for a KR system, and also explores the requirements in detail. The KR system means the complete package including KR language that contains semantics and syntax. Second, an inference engine is used to perform reasoning with a specific representation encoded in LISP or prolog to respond [3, 10]. There is one KB development environment that contains some editing tools possibly including knowledge editing and debugging tools. A good KR system must have the following characteristics:

- **Accuracy**: The system must possess the capacity for representing all kinds of required knowledge.
- **Acquisition efficiency**: It is the ability to obtain a new knowledge from the old one.
- **Semantic clarity**: The system should have well-defined semantics.
- **Scalability**: The system's performance should not degrade if the KB becomes large and should take less response time.
- **Expressiveness**: The KB language should be expressive and easy to understand for the knowledge engineer.
- **Naturalness**: The language representation should be system friendly, so that a knowledge person can express their requirement in a natural way.
- **Support for knowledge entry**: The KB should react to inconsistencies in data entered by an engineer, and a constraint should be in place at first to prevent the KB engineer from entering the data. It also helps the knowledge engineer perform debugging or correct the rules if an inconsistency occurred.
- **Foreign system interface**: The system should be able to link with other system or World-Wide Web (www).
- **Graphics**: It should have a user-friendly graphical editor to enter data, view data, and to apply knowledge.
- **Robustness**: The system should be free of bugs or errors.
- **Portability**: The system should be easily portable among various platforms.
- **Documentation**: The system should be well documented.
- **Cost**: The system should be less expensive

2.7 Techniques of Knowledge Representation

In today's world, humans are using computers in their day-to-day lives to solve increasingly more complex problem. The interaction between the human and the computer is increasing very rapidly, which allows an exchange of information. Both trends require the computer to be able to use a huge volume of knowledge. Researchers in the domain of AI are investigating how knowledge can be easily expressed in a computer system [1, 2, 10, 31–33]. There are four main techniques of KR, such as logical representation (LR), semantic network (Snet), frame, and production rules.

- **Logical representation**: It is a language having certain specific rules that gives with propositions. It means drawing a conclusion based on several conditions and contains precisely defined syntax and semantics that support the inference. This representation is split into two subcategories: propositional logic and predicate logic. Each sentence can be translated into logics using syntax and semantics [1, 2]. Syntax is the rules that help in constructing the legal sentences in the logic and also determines which symbol is to be used in knowledge representation. Semantics are the rules by interpretation of sentence in the logic can be done. They assign a meaning to each sentence.
- **Semantic networks**: are also important knowledge representation technique and are alternative to predicate logic. In this technique, knowledge can be represented in the form of graphical networks. This technique consists of nodes to represents objects and arcs to describe the relationship between objects [3, 10, 32, 34]. This network is easy to understand and extendable. This representation consists of mainly two types of relations: (a) IS-A relation (inheritance) and (b) kind-of-relation as shown in Figure 2.6 where knowledge is represented as nodes and arcs. Each object is connected to another object by some relation.

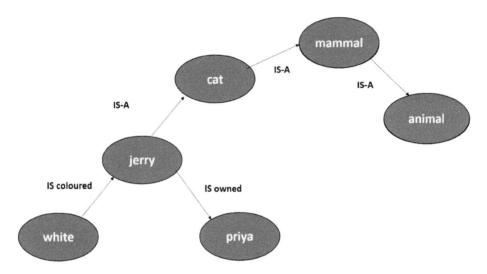

FIGURE 2.6 Semantic network technique.

- **Frames**: It is a database that consists of attributes along with their corresponding values to describe a real-world entity. It contains collection of slots and their associated values, which are known as facets. This technique consists of various number of slots, and a slot may include any number of facets [1–3, 31]. It may have any number of values as shown in Figure 2.7. Frames are also known as "slot-filter knowledge representation" and are derived from semantic net. In this, knowledge about any object or any event is stored in the knowledge base (KB).
- **Production rules system**: This system consists of conditions and actions pair that means "If condition then action." This system consists of three important parts: (1) a set of production rules, (2) working memory, and (3) a recognize act cycle. In this system, an AI agent checks the condition, and if it becomes true, then only rule fires and its corresponding action is carried out. Then a condition part of the rule checks which rule has to be applied to a particular problem, and on the other hand action part consists of problem-solving steps; this process is known as recognize act cycle [1–3]. The working memory consists of the existing state of problem solving and rules.

Slots	Filters
Title	From Visual Surveillance to Internet of Things
Genre	Computer Science
Editors	Lavanya sharma and Pradeep K Garg
Edition	FirstEdition
Year	2019
Page	257

FIGURE 2.7 Frames consisting of slots and values.

This knowledge match and can fire other rules. In case a new state is generated, more than one rule will be fired at a time, which is known as a "conflict set." Now, the agent can select a rule from the existing state, which is known as "conflict resolution" [33–35].

2.8 Real-Time Challenge

- **Key attributes:** Some basic attributes occurred in almost every problem domain [1, 2].
- **Relationship among attributes:** The relationship between attributes and objects should be very clear and representable [1–3].
- **Selecting granularity:** The amount of detailed knowledge required to represent is also very important, so the level of granularity plays a very important role and is one of the challenging issues in KR [1, 2].
- **Set of objects:** The correct representation of a set of objects is also very important. If a set of objects is not clearly defined, then knowledge represented will be partially and it over all effect on problem solving [1, 3, 10].
- **Selecting the proper structure:** All the information stored amounts to a huge volume of data, so accessing the relevant information is difficult [1, 3, 10, 17].

2.9 Conclusion

In terms of machines, all the things done by humans are termed knowledge representation and reasoning (KRR). It is a segment of AI that deals with thinking of AI agents and how it contributes to an agent's intelligent behavior. This chapter provides a basic overview of knowledge representation (KR) and how it is helpful for demonstrating the intelligent behavior in AI agent's various kinds of knowledge, as well as a relation between intelligence and knowledge. This chapter presents a detailed overview of a life cycle of knowledge in AI, basic requirements for a knowledge representation (KR) system, and different techniques and challenging issues of KR.

REFERENCES

1. Artificial Intelligence: Available at: http://www.hbcse.tifr.res.in/jrmcont/notespart1/node38.html [accessed September 12, 2020].
2. Types of Techniques in Knowledge Representation: Available at: https://link.springer.com/chapter/10.1007/978-3-642-73402-1_13 html [accessed September 12, 2020].
3. Lavanya Sharma, P. Garg (Eds.). (2020). *From Visual Surveillance to Internet of Things*. New York: Chapman and Hall/CRC, https://doi.org/10.1201/9780429297922
4. Lavanya Sharma, *"Introduction: From Visual Surveillance to Internet of Things," From Visual Surveillance to Internet of Things*, Taylor & Francis Group, CRC Press, Vol. *1*, pp. 14.
5. Lavanya Sharma, P. K. Garg, *"Block based Adaptive Learning Rate for Moving Person Detection in Video Surveillance," From Visual Surveillance to Internet of Things*, Taylor & Francis Group, CRC Press, Vol. *1*, pp. 201.
6. L. Fogel, A. Owens, M. Walsh, *"Artificial Intelligence Through a Simulation of Evolution,"* in *Biophysics and Cybernetic Systems Maxfield*, Washington, DC: Spartan Books, 1965.
7. Suraj Makkar, Lavanya Sharma, *"A Face Detection Using Support Vector Machine: Challenging Issues, Recent Trend, Solutions and Proposed Framework,"* in *Third International Conference on Advances in Computing and Data Sciences (ICACDS 2019, Springer)*, Inderprastha Engineering College, Ghaziabad, April12–13, 2019.
8. Lavanya Sharma, P. K. Garg, *"IoT and Its Applications," From Visual Surveillance to Internet of Things*, Taylor & Francis Group, CRC Press, Vol. *1*, pp. 29.

9. Lavanya Sharma, Annapurna Singh, Dileep Kumar Yadav, "Fisher's Linear Discriminant Ratio based Threshold for Moving Human Detection in Thermal Video," *Infrared Physics and Technology*, Elsevier, March 2016.

10. L. Sharma (Ed.). (2021). *Towards Smart World*. New York: Chapman and Hall/CRC. https://doi.org/10.1201/9781003056751

11. Lavanya Sharma, *"Human Detection and Tracking Using Background Subtraction in Visual Surveillance," Towards Smart World*. New York: Chapman and Hall/CRC, https://doi.org/10.1201/9781003056751, pp. 317–328, December 2020.

12. N. Abramson, D. Braverman, G. Sebestyen, "Pattern recognition," *IEEE Transactions on Information Theory*, vol. *IT-9*, pp. 257–261, 1963.

13. A. Newell, H. Simon, "The logic theory machine—a complex information processing system," *IRE Transactions on Information Theory*, vol. *IT-2*, pp. 61–79, 1956.

14. G. Pask, "A Discussion of Artificial Intelligence and Self Organization," in *Advances in Computers*, New York: Academic Press, vol. *5*, pp. 110–218, 1964.

15. Lavanya Sharma, Birendra Kumar, "An Improved Technique for Enhancement of Satellite Images," *Journal of Physics: Conference Series*, *1714*, 012051, 2021.

16. Supreet Singh, Lavanya Sharma, Birendra Kumar, "A Machine Learning Based Predictive Model for Coronavirus Pandemic Scenario," *Journal of Physics: Conference Series*, *1714*, 012023, 2021.

17. B. Garcia-Garcia, T. Bouwmans, A. J. R. Silva, "Background Subtraction in Real Applications: Challenges, Current Models and Future Directions," *Computer Science Review*, 35, 100204, 2020.

18. Knowledge Representation: Available at: https://www.javatpoint.com/propositional-logic-in-artificial-intelligence html [accessed September 12, 2020].

19. Akshit Anand, Vikrant Jha, Lavanya Sharma, "An Improved Local Binary Patterns Histograms Techniques for Face Recognition for Real Time Application," *International Journal of Recent Technology and Engineering*, vol. *8*, no. 2S7, pp. 524–529, 2019.

20. Lavanya Sharma, Dileep Kumar Yadav, "Histogram based Adaptive Learning Rate for Background Modelling and Moving Object Detection in Video Surveillance," *International Journal of Telemedicine and Clinical Practices, Inderscience*, June 2016.

21. S. Shubham, V. Shubhankar, K. Mohit, Lavanya Sharma, *"Use of Motion Capture in 3D Animation: Motion Capture Systems, Challenges, and Recent Trends,"* in *1st IEEE International Conference on Machine Learning, Big Data, Cloud and Parallel Computing (Com-IT-Con)*, India, pp. 309–313, 14–16 February.

22. R. Faulk, "An Inductive Approach to Mechanical Translation," *Communications of the ACM*, vol. *1*, pp. 647–655, November 1964.

23. T. Evans, *A Heuristic Program of Solving Geometric Analogy Problems*, 1963.

24. Lavanya Sharma, Nirvikar Lohan, *"Internet of things with object detection"*, in *Handbook of Research on Big Data and the IoT, IGI Global*, pp. 89–100, March, 2019. (ISBN: 9781522574323, DOI: 10.4018/978-1-5225-7432-3.ch006).

25. T. Bouwmans, A. Sobral, S. Javed, S. K. Jung, E. H. Zahzah, "Decomposition into Low-Rank Plus Additive Matrices for Background/Foreground Separation: A Review for a Comparative Evaluation with a Large-Scale Dataset," *Computer Science Review*, 23, 1–71, 2017.

26. Lavanya Sharma, Dileep Kumar Yadav, Sunil Kumar Bharti, *"An Improved Method for Visual Surveillance using Background Subtraction Technique," IEEE, 2nd International Conference on Signal Processing and Integrated Networks (SPIN-2015)*, Amity University, Noida, India, February 19–20, 2015.

27. Dileep Kumar Yadav, Lavanya Sharma, Sunil Kumar Bharti, *"Moving Object Detection in Real-Time Visual Surveillance Using Background Subtraction Technique," IEEE, 14th International Conference in Hybrid Intelligent Computing (HIS-2014)*, Gulf University for Science and Technology, Kuwait, December 14–16, 2014.

28. Lavanya Sharma, D. K. Yadav, Manoj Kumar, *"A Morphological Approach for Human Skin Detection in Color Images," 2nd National Conference on "Emerging Trends in Intelligent Computing & Communication"*, GCET, Gr. Noida, India, April 26–27 2013.

29. Lavanya Sharma, Nirvikar Lohan, "Performance Analysis of Moving Object Detection using BGS Techniques," *International Journal of Spatio-Temporal Data Science, Inderscience*, February 2019.

30. S. Amarel, "An Approach to Automatic Theory Formation," in *Principles of Self-Organization*, New York: Pergamon Press, 1962.

31. Knowledge Representation: Available at: https://www.cs.utexas.edu/users/pclark/working_notes/010.pdf html [accessed September 12, 2020].

32. Lavanya Sharma, Nirvikar Lohan, "Performance Analysis of Moving Object Detection Using BGS Techniques in Visual Surveillance," *International Journal of Spatiotemporal Data Science, Inderscience,* vol. *1,* pp. 22–53, 2019.

33. T. Bouwmans, "Traditional and Recent Approaches in Background Modeling for Foreground Detection: An Overview," *Computer Science Review, 11,* 31–66, 2014.

34. J. H. Giraldo, T. Bouwmans, GraphBGS: Background Subtraction via Recovery of Graph Signals. arXiv preprint arXiv:2001.06404, 2020.

35. T. Bouwmans, S. Javed, M. Sultana, S. K. Jung, "Deep Neural Network Concepts for Background Subtraction: A Systematic Review and Comparative Evaluation," *Neural Networks, 117,* 8–66, 2019.

3

Programming Languages Used in AI

Sudhriti Sengupta
Amity University, Noida, India

CONTENTS

3.1 Introduction

AI is the field of study in which a machine can be given the ability to perform tasks by using human-like intelligence or reasoning. AI nowadays is capable of playing games, helping marketers advertise to interested customers, playing music, painting, just to name a few tasks. AI achieves some knowledge-intensive tasks by applying human logic and reasoning powers combined with the strong diligence and veracity of a machine. AI offers less chance of an error, better performance standards, and more effective usage of resources. However, some drawbacks of AI are its present high cost, dependency on automated processes, inability to use emotions in its performance, etc. If AI is used in a malicious way, it has the potential to create havoc in the world. The promises of benefits of using AI application and devices overlay the concept of misuse of AI. In this chapter, an overview of AI is presented along with its role in different areas. One example of popular AI application is Smart Assistant, such as Alexa and Siri. Also the different programming languages used in AI are discussed, along with their advantages and disadvantages [1, 2, 10–14].

3.2 An Overview of AI

Artificial intelligence is a multidisciplinary approach of incorporating natural intelligence into machines so that they can perform tasks by applying logic and reasoning. AI aims to duplicate human intelligence in computing devices. Russel and Norvig define AI as "the study of agents that receive percept from the

DOI: 10.1201/9781003140351-3

environment and perform actions." Accordingly, there are four different ways for defining AI, namely thinking humanly, thinking rationally, acting humanly, and acting rationally. The first two terms deal with reasoning and processing whereas the other two deal with behavioral aspects.

The AI techniques are run by rules, machine learning, deep learning, etc. In 1943, McCullogh and Pitts proposed the model for building a neural network. The year 1950 was a very important one for the development of AI. In this year Turing proposed the Turing test to determine whether the machine is intelligent or not. SNARC (Stochastic Neural Analog Reinforcement Calculator), the first neural network computer, was developed and laws of robotics were published in this year. In 1952, Samuel developed a self-learning program for playing chess. In 1972, Prolog, the programming language for AI, was developed. Many other techniques were developed in the coming years. In 1955, McCarthy coined the term "AI." By 1974, computers became faster and more affordable, leading to the development of natural language processing (NLP). In 1966, Eliza, the first chatbot, was introduced. The first Expert was developed in 1980. In 1997, IBM Deep Blue beat a world champion at chess. In 2002, a vacuum cleaner called Roomta was developed. By 2006, organizations like Facebook, Netflix, and others began using AI in their operations. By using AI, software-enabled devices can solve real-world problems efficiently and accurately [15, 16]. Personal virtual assistants like Siri, Google Assistant, Cortana, and the like were created with the help of AI. AI-enabled devices also can be used in work deemed dangerous for human beings, such as mining. AI offers several advantages such as high accuracy, faster speed, optimized reliability, etc. However, some issues with AI are also present such as high cost, lack of emotion, creating dependencies on machine, etc. [3, 4].

AI is a multidisciplinary field of study covering not only computer science but also mathematics, biology, sociology, statistics, etc. (Figure 3.1).

AI is very versatile and does not limit itself to computer industry. It plays an important role in health care, banking, security surveillance, automobile industry, to name just a few. AI is likely to replace humans in a number of jobs. This are some positive aspects of this development, specifically the elimination of risk to life and health associated with some particularly hazardous professions. At the same time, however, this development is likely to increase unemployment and socioeconomic stresses. Which is why the integration of AI in those industries and positions must be accompanied by retraining of human personnel to initiate seamless transitions in the employment market. Perhaps those same workers can even be trained to handle and operate AI-enabled technology. AI can also be used to output and analyze tremendous volumes of data, leading to better expert systems and more informed policy making. Potential for human life improvement through AI also extend to such things like disaster management, health care, farming, environmental initiatives, crime prevention, managing of resources, etc. AI added to personal devices, like phones, promise to create a smoother life style through the use of visual assistant, chatbots, etc.

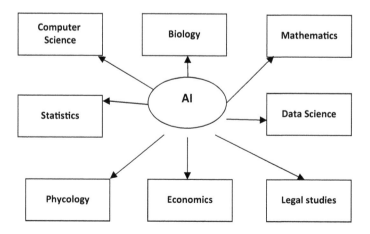

FIGURE 3.1 Multidisciplinary studies converge AI.

Hence, it can be understood that AI plays a significant role in all aspects of human life. Various applications are created in different domain to help uplift the human life and benefit society [16–18]. The next section discusses the role of AI in various fields such as, among others, agriculture, education, and health care.

3.3 The Role of AI

In today's society AI is used in many different types of applications used in homes, industry, educational institution, transport, games, etc. AI capabilities are being utilized by many different domains such as legal institutions, patent offices, banking, retail, etc. [7]. AI propels us toward the smart world that improves the quality of life. AI developed to perform some specific tasks is called Weak AI or Narrow AI. It is used for a wide range of things like remote sensing, automatized control, medical diagnosis, etc. In this section some of these applications are discussed.

3.3.1 AI in Agriculture

AI can be used in agriculture for monitoring of crops, predictive analysis of the weather, detection of diseases in crops, pest control, etc. AI helps in real-time analysis of various factors such as climate, water, and soil conditions, among others, to help farmers make decisions on farming technique, choice of feed and/or fertilizer, etc. This leads to enhanced harvesting. The use of AI to improve crop yield is called precision agriculture. Another role of AI is in the use of bots to perform such hazardous duties as weed elimination.

3.3.2 AI in Security

With the availability of Internet and devices using Internet, there has been a rise in cybercrime. AI has been successful in making data and resources safe and secure. Facial data can be captured for biometric recognition and analyses for authentication. AI is used in video surveillance to detect humans, cars, and other objects.

3.3.3 AI in Education

Educational institutions can use AI in many ways, such as automatic grading and doing routine admin job. This will allow teachers to spend more quality time on teaching and research. AI chatbots can also respond to various queries from students. AI can enable each student to work individually, at their own pace, and according to their own interest.

3.3.4 AI in Health Care

AI has a major impact on health care. It enables more efficient and quicker diagnosis of diseases. AI can be used to develop automatized medical image analysis [9]. It can analyze large volumes of patient data to identify patients at risk. Chatbots can be developed to provide primary care to patients in rural and far-away areas [19–21].

3.3.5 AI in Industry

In manufacturing, AI can be utilized to perform some repetitive tasks, saving human time and effort. It can be used to analyze customer preferences to predict trends in the retail sector. It can be used to develop chatbots to help customers have better and more personalized shopping experience. It can be used to predict risk and fraud in financial institutions and insurance companies.

The role of AI has been introduced in various fields like health care, education, autonomous cars, business, security, entertainment, gaming, and simulation. It has ushered us into the world of smart cities

[8, 17, 18]. To unleash the potential and promises of AI in their true sense, implementations for the betterment of society and lifestyle applications should be developed. Developing AI-enabled applications requires appropriate programming languages. In the next section, languages used in AI are discussed to provide a reference that will help users understand and select which language is to be used in what kind of application.

3.4 Languages Used in AI

The role of AI in bringing efficiency and increasing benefits for any organization or social group is immense. With the advent of AI, the world has reached a smarter place, various aspects of which are explored, tested, and implemented to satisfy human desire for improving quality of life. To usher the growth of AI into these domains, strong, robust, and powerful programming languages are required. These will help create and deploy AI-enabled applications.

There are many specialized languages developed for programming AI applications. The challenge is to find a language capable of providing the required specifications for AI to be developed. In this chapter, the popular languages are discussed, along with their advantages and disadvantaged, to let the researchers in this field have a ready reference. It has been identified that Python, Prolog, Lisp, Java, and C++ are among the major programming languages that can be used to fulfill the requirement of designing and developing different AI applications and software [14, 18, 19]. In the following subsections, a discussion of these programming languages is presented so that the programmers and developers can identify which language will be suitable for their software development process.

3.4.1 Java

Java is a popular programming language used for developing a variety of applications. Java is used for developing desktop computing, games, websites, server-side programs, mobile computing, etc. Java covers a wide area of application-building capabilities ranging from simple console for game development to super-computer programming. In creating programs for AI, Java also plays an eminent role. Java provides features such as easy debugging, maintainability, portability, security, and robustness, among others. Java is one of the common programming languages chosen for AI development. It supports machine learning, genetic programming, neural network, and other popular AI libraries. Some of the libraries for implementing AI are Tensor flow, Deep Java Library, Kubleflow, Open NLP, and Java Machine Learning Library. Java also supports an automatic memory manager that simplifies program development and deployment. Because of the use of Java Virtual Machine, it can be used on different varieties of application. Java is simple to use and to debug. Hence Java is one of the most widely used programming languages in AI development. However, Java is slower in execution and requires more response time as compared to programming languages like C++.

3.4.2 C++

C++ is widely preferred by AI developers because it provides the fastest execution. For building search engine, games, life-critical systems, etc., response time should be as short as practically possible. In this type of applications, C++ is used. C++ also provides extensive algorithms and is very efficient in statistical techniques implementation. In addition to this, C++ has object-oriented properties such as inheritance and data hiding. These properties ensure time saving, reusability of code, and security. Machine learning and neural network are also supported by C++. Complex AI programs can be solved by using rich library functions and tools present in C++. The main disadvantage of using C++ in AI development is that C++ lacks the ability to multitask efficiently. This makes C++ suitable for creating the core of a system only. Highly complex system building also is not preferable in C++, as it follows a bottom-up approach. C++ also does not support garbage collection.

3.4.3 Python

Python was introduced in 1991, and today it is one of the most popular programming languages in AI development. It is preferred by the beginners, as it is easy to learn and often provides a stepping-stone for data scientists and AI programmers. Besides providing huge library support, Python also ensures very good community support. Python provides extensive framework for deep learning and machine learning. Some of the popular Python libraries are:

- PyBrain – used for machine learning algorithm
- Theano – used for complex mathematical solution
- MXnet – used for deep learning applications
- PyTorch – used for computer vision and natural language processing

Other popular libraries are TensorFlow, Scikit-Learn. Keras, and SparkMLlib.

Python is portable across different platform such as Mac, Windows, and Linux. This provides integrating capabilities with other programing languages such as C, C++, Cobra, Java, etc. It supports features such as object-oriented paradigm, dynamic type checking, interfacing with databases, etc.

One of the issues in using Python in AI development is that it is slower in compilation and execution compared to C++ and Java. This happens due to the fact that an interpreter is used in Python. Python is not suitable for mobile computing. Moreover, developers using Python may find it difficult using other programming languages.

3.4.4 LISP (LISt Processing)

This is the oldest programming language used in AI, which was developed by John McCarthy. McCarthy coined the term "Artificial Intelligence" in 1956 at a Darmouth conference. LISP is a flexible program efficient in solving a variety of problem. LISP is a strong and dynamic programming language. It provides a macro framework as it facilitates implementation of different categories of programs like inductive logic project and machine learning. LISP is fast in execution and coding, as it uses compilers. It provides automatic garbage collection for memory management. Popular programing languages like R, Julia, and others are motivated by LISP. Rapid prototype creation and developing dynamic objects are also faculties in LISP used in AI development.

The main drawback with LISP is that it lacks a well-established library like the one present in other languages such as Java and Python. Its syntax is also not aesthetic to the programmers. Finally it requires rigorous configuration for usage in latest operating systems and devices.

3.4.5 Prolog

Prolog stands for **Pro**gramming **Log**ic. It was developed in 1972. The first chatbot, "Eliza," was developed using Prolog. It is based on mechanisms such as pattern matching and automatic back tracking. Database building is easy using Prolog. This language is suitable for creating fast prototype and representation tree–based data structure. In prolog programming, there are two ways for implementing AI:

1. Symbolic approach
2. Statistical approach

Symbolic approach covers rule-based expert system, constraint-based approach, etc. Data mining, machine learning, and neural network are supported by the statistical approaches of Prolog.

Prolog coding is different from, and hence more difficult to learn than, other conventional programming languages such as C++. Also, though it was developed earlier, many features and characteristics are not fully standardized. It makes the use of this languages by various developers across different platforms uncomfortable and cumbersome.

3.4.6 R

R was developed in 1995 by Ross Ihaka and Robert Gentleman for facilitation statistics and data analysis. R is preferred for AI programming, as it is very efficient in dealing with large numerical values. Also R supports various properties of programming like vector computation and object-oriented programming. R provides a large variety of packages and functionalities to support AI programming. It provides good facility to collaborate with other major programming languages such as C, C++, and Fortan. It has the capability of producing good-quality graphs. Popular package in R are:

- Gmodel – for facilitating model fitting
- Tm – for text mining
- ROBDC – database connectivity interface for R
- One R – for implementing machine learning application.

Also, R has a very active and strong community support. All these make R suitable for developing AI application. The main disadvantage of R is that it uses more memory as compared to other programming languages like Python. It lacks basic security, rendering it unsuitable for Web-based applications. People with no knowledge of programming language find difficult to learn R. It is difficult to implement algorithms by beginners. It is not suitable for graphics and its much slower in execution.

3.5 Conclusion

The emergence of AI promises changes that will penetrate the basis of human lifestyle by ushering changes in different fields such as media, health care, manufacturing, retail, utilities, automotive, etc. Hence, it is very understandable to know about the different programming language used in AI. There are several programming languages used in AI, which provides the researchers and developers with the ability to choose the most suitable one as per their requirements. All these programming languages provide some support, like intensive libraries for supporting AI. The different programming languages have their advantages and disadvantages. Java and Python are very popular but are slow in execution and may not be suitable in creating application interacting in critical real-time scenario. C++ is very fast, but the issue is that it is not apt in multitasking. So, C++ is preferred to write the core of the application. R and Prolog are very promising languages, but the programmers, especially the beginners, find it difficult to develop in it because of its different syntax. A language may be suitable for one particular type of project but not for others.

REFERENCES

1. Artificial Intelligence. Available at: https://builtin.com/artificial-intelligence [accessed March 20, 2020].
2. Historical Background Artificial Intelligence. Available at: https://data-flair.training/blogs/history-of-artificial-intelligence/ [accessed March 20, 2020].
3. Historical Background Artificial Intelligence. Available at: https://www.javatpoint.com/history-of-artificial-intelligence [accessed March 20, 2020].
4. Programming language overview of Artificial Intelligence. Available at: https://www.zfort.com/blog/best-programming-language-for-ai#: [accessed March 20, 2020].
5. Realtime applications of AI. Available at: https://www.newgenapps.com/blog/the-role-of-ai-in-our-daily-lives/ [accessed March 20, 2020].
6. https://elearningindustry.com/ai-is-changing-the-education-industry-5-ways
7. Tyagi, S., Sengupta, S., *Role of AI in Gaming and Simulation*, Lecture Notes on Data Engineering and Communications Technologies, 2020, *49*, pp. 259–266.
8. Kohli, D., Gupta, S.S., *Recent Trends of IoT in Smart City Development*. Lecture Notes on Data Engineering and Communications Technologies, 2020, *49*, pp. 275–280.

9. Singh, K., Sengupta, S., "Diagnostic of the Malarial Parasite in RBC Images for Automated Diseases Prediction," *Towards Smart World*. New York: Chapman and Hall/CRC, pp. 210–228, https://doi.org/10.1201/9781003056751.

10. Sharma, L., Garg, P. (Eds.), (2020). *From Visual Surveillance to Internet of Things*. New York: Chapman and Hall/CRC, https://doi.org/10.1201/9780429297922.

11. Sharma, L. (Ed.), (2021). *Towards Smart World*. New York: Chapman and Hall/CRC, https://doi.org/10.1201/9781003056751.

12. Sharma, L., "Human Detection and Tracking Using Background Subtraction in Visual Surveillance," *Towards Smart World*. New York: Chapman and Hall/CRC, pp. 317–328, https://doi.org/10.1201/9781003056751.

13. Sharma, L., Singh, A., Yadav, D.K., "Fisher's Linear Discriminant Ratio based Threshold for Moving Human Detection in Thermal Video," *Infrared Physics and Technology*, Philadelphia: Elsevier.

14. Sharma, L., Yadav, D.K., Bharti, S.K., "*An Improved Method for Visual Surveillance using Background Subtraction Technique*", *IEEE, 2nd International Conference on Signal Processing and Integrated Networks (SPIN-2015)*, Amity University, Noida, India, February 19–20, 2015.

15. Yadav, D.K., Sharma, L., Bharti, S.K., "*Moving Object Detection in Real-Time Visual Surveillance using Background Subtraction Technique*", *IEEE, 14th International Conference in Hybrid Intelligent Computing (HIS-2014)*, Gulf University for Science and Technology, Kuwait, December 14–16, 2014.

16. Abramson, N., Braverman, D., Sebestyen, G., "Pattern recognition," *IEEE Transactions on Information Theory*, vol. *IT-9*, pp. 257–261, 1963.

17. Newell, A., Simon, H., "The logic theory machine—A complex information processing system," *IRE Transactions on Information Theory*, vol. *IT-2*, pp. 61–79, 1956.

18. Faulk, R., "An inductive approach to mechanical translation," *Communications ACM*, vol. *1*, pp. 647–655, 1964.

19. Sharma, L., Garg, P.K., "IoT and its applications," *From Visual Surveillance to Internet of Things*, Taylor & Francis Group, CRC Press, Vol. *1*, pp. 29.

20. Sharma, L., Garg, P.K. "Smart E-healthcare with Internet of Things: Current Trends Challenges, Solutions and Technologies," *From Visual Surveillance to Internet of Things*, CRC Press, Taylor & Francis Group, pp. 215–234, October 2019.

21. Sharma, L., Garg, P.K. "A foresight on e-healthcare trailblazers", *From Visual Surveillance to Internet of Things*, CRC Press, Taylor & Francis Group, pp. 235–244, October 2019.

Section II

Artificial Intelligence

Tools and Technologies

4

Image Processing Using Artificial Intelligence: Case Study on Classification of High-Dimensional Remotely Sensed Images

Dibyajyoti Chutia
North Eastern Space Application Centre, Umiam, India

Avinash Chouhan
North Eastern Space Applications Centre (NESAC), Government of India, Umiam, India

Nilay Nishant
North Eastern Space Application Centre, Umiam, India

P. Subhash Singh
North Eastern Space Applications Centre (NESAC), Government of India, Umiam, India

D. K. Bhattacharyya
Tezpur University, Napaam, India

P. L. N. Raju
North Eastern Space Application Centre, Umiam, India

CONTENTS

4.1 Introduction

Traditionally, image processing is the process of either enhancing an image or extracting meaningful information. It concentrates more on the development of feature extraction techniques applied toward the statistical classification of visual imagery [1]. The need for image-processing technology is rising gradually with the fast growth of information technology, which enables a base for handling a variety of image-processing applications [2]. The majority of the image-processing techniques are based on mathematical and statistical theory. Image processing can be thought of as a computational process characterized by a set of algorithms used to process raw images received from the space or airborne sensors or even pictures taken by means of smart phones [3]. Image processing has a broad range of applications in the areas of electronics and telecommunications, medical science, remote sensing (RS), biotechnology, robotics, etc. Researchers have been trying to develop image analysis techniques as effective as the human vision system is for many decades. Large numbers of algorithms and techniques have been conceptualized and developed so far that cover various stages of image-processing and recognition [4] techniques. Specifically, RS technology has turned out to be a very important and widespread instrument in government agencies and industries [5–7] for the natural resources management and monitoring activities. This area has drawn a lot of research interest in the advancement of image-processing techniques. The advancement in sensor technology and constellation of earth-observing satellites around the globe have given enormous opportunities in modeling the earth surfaces for effective change detection and monitoring activities. The availability of high-resolution RS images has brought new challenges and opportunities in the development of advanced image-processing techniques for monitoring and change detection [8]. It allows the authorities to monitor and detect the changes in the landscape dynamics, including those areas that generally are not accessible or hazardous [9, 10].

In the terminology of ML [11], classification can be defined as the problem of identifying a new observation to a set of predefined observations on the basis of a set of training datasets. Current literature has revealed the remarkable potentials of AI-based techniques in the area of RS image processing. Previously widely recognized conventional techniques and models have been preponderated by the superior preprocessing and classification methods [12]. The rising profile of ML and deep learning (DL) techniques has drawn much attention for solving a broad range of tasks arising in image processing and classification, object detection, and image segmentation [2, 13]. ML-based classification techniques, mainly random forest (RF), artificial neural network (ANN), support vector machine (SVM), and fuzzy-based adaptive learning, have been dominating the other conventional classifiers in the remote sensing classification. They have tremendous potentials in classifying the VHR data, as they consider both spectral and spatial morphological characteristics of the image for prediction [12, 14]. Currently ensemble classification approaches have been found very successful in pattern and object detection applications, as the ensemble method employs a set of base classifiers to attain higher predictive ability [14, 15] than a conventional classifier.

This chapter highlights the issues and challenges of image-processing techniques with regard to the classification of VHR RS datasets. A case study on Airborne ROSIS-3 hyperspectral sensor image classification using an ML-based approach is demonstrated. Concluding remarks are given in the direction of classification of VHR RS data using DL techniques.

4.2 Issues and Challenges

Digital image processing is the exploitation of the numerical information on images to make it more appropriate for visual interpretation and digital processing. Image-processing techniques and approaches are well established because of the easy accessibility and availability of powerful computer systems, storage devices, software, etc. [16]. However, in spite of numerous advances in imaging technology, optical images captured by highly sensitive sensors often require advanced processing techniques to investigate [17]. Digital image processing for satellite images is one of the core activities in RS application areas such as agriculture, forestry, ecology, disaster management, and so forth. Nevertheless, image processing becomes complicated due to the huge dimensions of the RS images [18]. In spite of the application complexity, some basic techniques are common in RS-sensing applications [19]. The classification of RS data has become an essential and universal tool for monitoring environmental and natural resources in the last few decades. A typical classification system is developed based on the spectral-spatial resolution of RS data, selection of image processing and classification algorithms, user's criteria for classification, and time restriction. The sufficient number of ground-truth samples against each of the classes of interest is critical for RS classifications [20–23]. Training samples are mostly generated on a per-pixel basis to cut down the redundancy and autocorrelation among the targeted classes [24]. But it is difficult to collect training samples when the scene is very complex and heterogeneous, hence the choice of ground-truth samples must consider the spatial and spectral resolution of the RS data being utilized, ease of use of ground reference data, and the complexity of the terrain and landscapes in the study area [25]. The analysis of RS data is performed using a variety of image-processing techniques. Most of these processes are comprised of identification and restoration of faulty lines, geometric correction or geo-referencing, correction of radiometry of the image, and correction of errors induced by the atmospheric and topographic parameters. Atmospheric correction is not required while we are using single-date RS data [26], but it is mandatory if the study demands multi-temporal or multi-sensor data. A huge number of image preprocessing methods have been reported [27–33] for correction and normalization of radiometric and atmospheric errors. Topographic rectification is another important issue if the study area falls within a hilly region [34].

Conventional popular "hard" classifiers are often linked with the problem caused due to mixed pixel [35]. They have substantial complexity in handling the high-dimensional images with higher variances in the contents; they turn out with an inconsistent performance in term of predictive ability, and fail to identify the object of interest. Major weakness of the traditional classifiers has been reported [36]. Nonparametric ML-based classifiers are treated as most suitable approaches in RS for the classification of multisource data [37]. There are many issues and challenges that lie in the processing of hyperspectral RS image as compared to the multispectral RS image, as hyperspectral images are coming with a huge number of spectral bands containing enormous amounts of information. It asks for more sophisticated techniques to avoid the Hughes phenomenon and also to minimize the computational complexity induced by the higher spectral dimension of hyperspectral images. Identification of redundant and noisy bands is very crucial, and the choice of relevant and non-correlated spectral bands is one of the major tasks during the preprocessing phase [12]. The choice of the relevant set of features or attributes has been significant for receiving higher classification accuracy. As reported extensively in literature, elimination of unnecessary or noisy bands is key to getting an enhanced classification approach; however, this process may degrade the classification result. Many sources have been suggesting the advanced techniques in order to handle Hughes phenomena, but it is still an important concern that cannot be overlooked during classification [38]. The choice of the suitable bands for few applications desires specific consideration; however, this is an essential issue in hyperspectral image processing [39]. Usually, the region of interest contains one or more spectrally similar features that are difficult to define by a single Gaussian density function, hence the majority of the successful multispectral classifiers are unsuccessful to achieve good accuracy [40]. Processing of a huge number of hyperspectral bands also affects the computational performance of a classification system, which can lead to inconsistency in the results many applications.

4.3 Case Study on the Classification of Airborne ROSIS-3 Data Using ML Approach

This section demonstrates a case study on the classification of ROSIS-3 data using an ML-based approach. The architecture of the ML-based approach is comprised of the following components:

a) Removal of redundant spectral bands

b) Selection of optimal features

c) Calibration of model parameters for optimization of RF classifier

d) Evaluation of performance

4.3.1 ROSIS-3 Hyperspectral Dataset

A total of 103 atmospherically corrected spectral channels of ROSIS-3 with 0.43–0.86 μm spectral range with 1.3 m spatial resolution have been utilized in the investigation. The information on the training-test samples of classes for the University of Pavia, northern Italy dataset is given in Table 4.1. The Pavia dataset has been kindly provided by Prof. Paolo Gamba. The false color composite (FCC) image of the study area and distribution of training samples are depicted in the in Figure 4.1.

4.3.2 Removal of Redundant Spectral Bands of ROSIS-3 Dataset

Removal of redundant bands or the dimension reduction of spectral bands is a most important shortcoming of hyperspectral remote sensing when conventional image-processing techniques are utilized [41]. Minimum noise fraction (MNF) [42] and Independent Component Analysis (ICA) [43] fall under

TABLE 4.1

Details of Classes and Training-Test Samples for the Pavia Dataset

No	Class name	Train	Test
1	Asphalt	548	6,549
2	Bare Soil	532	5,139
3	Bitumen	375	1,326
4	Bricks	514	3,689
5	Gravel	392	2,102
6	Meadows	540	17,898
7	Metal sheets	265	1,334
8	Shadows	231	987
9	Trees	524	3,127
Total		3,921	42,151

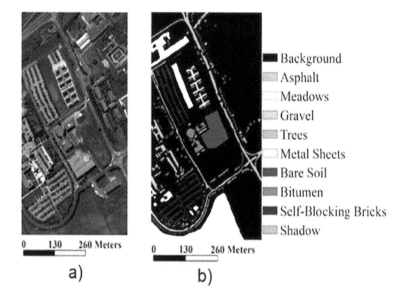

FIGURE 4.1 (a) FCC image of the study area and (b) ground truth image with nine land-cover classes.

the same family of Component Analysis (PCA). Both MNF and ICA are dominant dimension reduction techniques used in hyperspectral RS application. The MNF has been found very effective in reducing hyperspectral spectral bands [39, 44]. It needs past assessments of signal and noise covariance matrices and demands more computing resources than does PCA [14, 45–47]. However, it has advantages over the PCA because it takes the noise information in the spatial domain into consideration. On the other hand, ICA is computationally expensive, which restricts its applicability. In the current investigation, MNF was used for dimensionality reduction of the ROSIS-3 dataset. The MNF transformation was done through two successive PCA transformations. The first PCA uses the PCs of the noise-covariance matrix to decorrelate and rescale the pixels (noisy) in the image. The intrinsic dimensionality of the image is chosen based on the final eigenvalues and the related images obtained using the PCs. The bands with higher eigenvalues are chosen, because the images with eigenvalues nearer to 1 are mainly noise. The spectral dimension of the original image with 103 bands remains with only 27 bands after the dimension reduction process. The FCC of the dimension-reduced image is depicted in Figure 4.1(a).

4.3.3 Selection of Optimal Features

The ground-truth land-cover samples are characterized by a set of attributes or features through which the pattern of land-cover classes can be identified and separated. The selection of features for uniquely identification of land-cover classes in the feature space is not only crucial for the improvement of classification accuracy; selection of relevant feature can minimize the computational complexity. Features in RS data are predominantly associated with spatio-spectral properties. The wrappers [48] and filters [49] are extensively used as feature selection techniques in most of the ML applications. More often, the filter-based feature selection techniques are used, as they are computationally efficient and can handle the large feature datasets. We have adopted a correlation-based feature selection (CFS) technique [50] for the investigation as CFS, which has been effectively utilized in RS applications for the enhancement of performance of the ML techniques [12, 14, 47]. The CFS generates feature subsets, independent of the classification model, and is featured with efficient computational complexity [51]. The principle behind the CFS is filter where relevance of optimal set of features is identified by observing appropriate correlation measure. Let F be the feature set of original training data before applying CFS. A feature $f_r \in F$ is treated as relevant *iff* there exists some feature $f_i \in F$ for predicating a class of interest c for which $p(f_r = f_i) > 0$ such that

$$p\left(C = c \mid f_r = v_i\right) \neq p\left(C = c\right) \tag{4.1}$$

where c is the Pearson's correlation coefficient for each feature–feature inter-correlation and feature–class (C) correlation [14] computation. The feature subset that contains features with uncorrelated feature to the feature characteristics but is highly correlated to the feature to class predictive ability [14], is called an optimal subset of feature. If the mean correlation between the features and the class $(\overline{r_{cf}})$ is defined and the average inter-correlation between each pair of features $(\overline{r_{ff}})$ is given, then the "*Merit$_s$*" of a subset $S \in F$ with i features can be computed as:

$$Merit_s = \frac{i\overline{r_{cf}}}{\sqrt{i + i\left(i - 1\right)\overline{r_{ff}}}} \tag{4.2}$$

The relevance of a feature or feature subset will be based on the degree to which it identifies classes in feature instance space not already identified by the other features; that is represented by the *Merit$_s$*. The CFS algorithm computes the $\overline{r_{cf}}$ and $\overline{r_{ff}}$ for each $S \in F$ and searches for the next subset of feature in forward direction using the best first search algorithm by making local changes to the current feature. Forward searching stops after five successive fully expanded nodes show no increase in *Merit$_s$* [14]. The subset of feature with the highest *Merit$_s$* is treated as the optimal subset of feature and is used for defining each of the land-cover classes. Then the training dataset defined by the optimal set of features is evaluated using the C4.5 [52] classification algorithm before training the proposed ML-based classifier. The training dataset with the original feature set and training dataset with optimal subset of features are represented as T_F and T_S, respectively. For the ROSIS-3 dataset, total features $F = 147$ have been evaluated to get the optimal subset of a feature set defined by $S = 9$. The detailed experimental observation on the feature selection is illustrated in the "Results and Discussion" section. RF classifier will be trained with both T_F and T_S datasets in order to assess whether RF classifier trained with optimal subset of features has enhanced the overall performance of the RF classifier.

4.3.4 Calibration and Optimization of RF Classifier Model Parameters

In this experiment, the RF [53] classifier is optimized with two parameters: (i) number of random trees to build the RF and (ii) number of optimal sets of feature feeds to RF. The RF is characterized by the principles of bagging and random selection of best feature. It is one of the most dominant ensemble classifiers for classification of RS images [12, 14, 47, 54]. The RF is built with a set of tree-structured classifiers $\{h(x, \Theta_r), r = 1 \ldots R\}$ where the Θ_r are identically distributed independent random tree classifiers and each tree classifier casts a unit vote for the final classification of input x. Here, the kernel size of the RF that represents the number of Θ_r is represented by the R. The RF classification algorithm can be illustrated as follows:

ALGORITHM FOR RF

1. For $r = 1$ to R;
   a) Draw a training sample through bootstrap Z^* of size N from T_S
   b) Grow $h(x, \Theta_r)$ to the bootstrapped samples by repeating the subsequent steps for each node of the Θ_r until the minimum node size is reached
      i. Choose m features randomly from the S features ($m \ll S$)
      ii. Select the best feature among the m
      iii. Split the node
2. Output the ensemble of trees $\{\Theta_r\}_1^R$
3. To make a classification prediction at new point x; if $\hat{C}_r(x)$ is the class prediction of the r_{th} random tree, then $\hat{C}_{RF}^R(x) = \text{majority vote}\{\hat{C}_r(x)\}_1^R$

Once the ensemble of $h(x, \Theta_r)$ of size R is constructed, the majority voting for class prediction $\hat{C}_r(x)$ of the r_{th}RF can be done by aggregating the final class of each tree and corresponding weighted votes. Each tree is grown to the largest extent possible without any pruning of it. In each bootstrap training set, about one-third of the training instances are left for estimating the error of the RF. This is called as *out-of-bag* data and not used in the construction of r_{th} random tree. It is also called *bag-error* on the training dataset, which predicts test error during the construction of RF without using any cross-validation mechanism. It is also used to validate the relevance of feature in the randomly selected feature. Then RF is calibrated with the appropriate ensemble size (R) and the number of features (m) in optimizing the model to acquire higher performance in terms of classification accuracy and computational complexity. The calibration of the RF model parameters is carried out as illustrated below:

   i. The RF has been initialized with $R = 5$ with an increment of 5, maximum up to 200 or until it reaches the highest performance.
   ii. The value of R was defined as the one received with the highest accuracy of the earlier rating.
   iii. Similar like earlier test, RF is initialized with $m = 1$ and executed with an increment of 1 up to $m = 147$ for T_F and $m = 9$ for T_S.
      a. In each execution, the RF model is allowed to pick the best predicting features through random selection of features during each split of the node
      b. The m is set to those values from which the accuracy received were highest.
   iv. Finally R and m are calibrated with the optimized values received from the previous experiments for which it received the highest classification accuracy.

4.3.5 Evaluation of RF Classifier

The research was conducted on an Intel (R) Xeon 2X10 Cores E5-2480v2@3.20 GHz/48GB RAM workstation with NVIDIA P400 8GB GPU card. The training model time (t_{trg}), precision (P), Kappa statistics (K), overall accuracy (OA), and receiver operating characteristic (ROC) were used for the evaluation of classification performance. The same sets of parameters were also used for optimization of RF model parameters.

TABLE 4.2

Calibration of Model Parameters

	Model parameters		Performance parameters				
	R	m	t_{trg} *(sec)*	P	$\hat{K}$	ROC	OA *(%)*
T_F	12	11	13.01	0.86	0.87	0.96	89.09
T_S	10	8	6.09	0.96	0.94	0.98	96.45

4.3.6 Results and Discussion

The optimized RF classifier was analyzed with the both original training-test dataset (T_F) and dimensionality reduced training-test dataset (T_S). The calibration results with all the parameters are reflected in Table 4.2. In each execution, the RF model is permitted to select the best predicting features through random selection of features during each split of the node. The RF model was calibrated with an optimal subset of features of size m for which it is highest against the lowest value of *bag-error* with minimal R in a particular test. It is observed that the performance of optimized RF classifier has significantly enhanced with the T_S dataset as compared to the original dataset T_F (Table 4.2), even though the model was calibrated with both datasets. It achieved overall accuracy $OA = 9.45\%$ with $P = 0.96$, $\hat{K} = 0.94$ and $ROC = 0.98$ when the model was trained with T_S dataset.

The computational efficiency of the RF classifier was also enhanced with the optimal subset of features $m = 8$ than the original the original number of features $m = 147$. It took only 6.09 sec for building the training model. For the T_F dataset, the model was calibrated with $m = 11$, which also improves the performance of RF classifier with original data with full feature set (Table 4.2). The performance of optimized RF classifier is compared with three other powerful classifiers; multilayer perceptron (MLP), support vector machine (SVM), and rotation forest. The comparative performance of all the classifiers is given in Table 4.3. It is observed that the overall performance of all the classifiers is enhanced while they were trained with T_S dataset. The optimized RF classifier outperformed all other classifiers for both T_F and T_S datasets. Comparatively, both rotation forest and SVM classifiers have also shown satisfactory result than MLP (Table 4.3). The classification results of University of Pavia area (with training T_S dataset of ROSIS-3 sensor) using optimized RF classifier is depicted in Figure 4.2.

The strong point of the optimized RF classifier is that it cuts down the computational expenses, as it seeks the best feature randomly from the already selected optimal subset of features, not from the entire feature set. This also enables attaining higher performance for the instances where the datasets are associated with noise or extra features. The choice of the appropriate size of the RF classifier is important, as it decides the performance of both accuracy and computational expenses. The case study demonstrated on hyperspectral ROSIS-3 dataset using optimized RF classifier has been found characteristically better and effective in terms of predictive accuracy and execution time.

TABLE 4.3

Comparative Assessment of Classifiers

	T_F		T_S	
Classifiers	OA	$\hat{K}$	OA	$\hat{K}$
MLP	82.70	0.81	87.03	0.84
SVM	83.37	0.82	90.69	0.87
Rotation forest	82.06	0.80	91.08	0.88
enRF	89.09	0.87	**96.45**	**0.94**

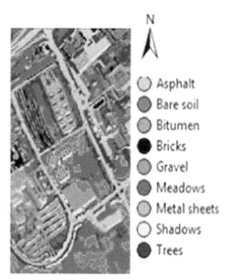

FIGURE 4.2   Classification university of Pavia area (with training T_S dataset of ROSIS-3 sensor) by optimized RF classifier classification method.

4.4 Conclusion and Future Directive

The technological advancement in the areas of image processing has increased the demands in many application domains. Significant improvements have taken place in the development of new algorithms and image-processing platforms. The recent development in the sensor technology has revolutionized the application domain, which indirectly demands the development of more sophisticated image-processing techniques. In addition, the availability of satellite sensors and the constellation of earth-observing satellites around the globe have given enormous opportunities in modeling the earth's surfaces for effective natural resources management and monitoring activities. The RS technology has become a vital and universal tool in the government setup including industries. This area has drawn a lot of research attention in the innovation of image processing techniques in the last two decades.

The increasing accessibility of time series unmanned aerial vehicle (UAV) imagery has augmented the frequency of monitoring and management of resource activity to a great extent. UAVs are very effectively utilized for real-time site monitoring, frequent surveys; they can capture the images quickly with enhanced quality [55]. During the last decade, the drone market has been flowering with an exponential escalation in all applications extent, particularly in the construction industry. Venturing new data-processing techniques or up-gradation of conventional remote sensing techniques are becoming the most important aspect to deal with multi-model analysis, object tracking in clutters, etc. [56]. Many application areas insist on automation and minimal human interaction; the AI/ML-based techniques are becoming more relevant in the current scenario. The AI and ML are key components and major drivers of hyper-automation along with other technologies like robotic process automation tools. The use of AI/ML is gradually more entwined with internet of thing (IoT) for a variety of applications. The AI, ML, and DL are already being employed to build IoT devices and services in a smarter and more secure fashion. In recent time, the DL-based approaches are becoming more appropriate for analyzing VHR images, as the efficiency of computational resources has enhanced to a great extent with the addition of a graphical processing unit (GPU). A different variant of DL models has been introduced for dealing with data with big data nature in the areas of RS.

High-dimensional imaging capability has brought new possibilities and challenges in the advancement of RS classification, specifically when dealing with the spectrally very similar land cover classes characterized by redundant and non-relevant features. The case study demonstrated in the chapter has emphasized the classification of VHR hyperspectral imagery using an ML-based ensemble approach. It involves a series of image-processing steps where dimension reduction in spectral and feature space has played a vital role in improving the classification accuracy. It reports on an optimized RF classifier based on the random selection of training samples and feature set. The process enhanced the diversity among the base classifiers, which results in significant improvement in the overall classification of the ROSIS-3 hyperspectral dataset as compared to the other advanced classifiers investigated here.

REFERENCES

1. Gilmore, J. F. 1985. *Artificial intelligence in image processing. Proceedings of the SPIE 0528, Digital Image Processing*, https://doi.org/10.1117/12.946419.
2. Xin, Z., Wang, D., 2019. Application of artificial intelligence algorithms in image processing. *Journal of Visual Communication and Image Representation*, *61*, 42–49, ISSN 1047-3203, https://doi.org/10.1016/j.jvcir.2019.03.004.
3. Pereira, C. S., Morais, R., Reis, M. J. C. S. 2017. *Recent advances in image processing techniques for automated harvesting purposes: A review. Intelligent Systems Conference (IntelliSys)*, London, pp. 566–575, doi:10.1109/IntelliSys.2017.8324352.
4. Frejlichowski, D. 2020. Special Issue on "Advances in Image Processing, Analysis and Recognition Technology". *Applied Sciences*, *10*, 7582.
5. Gross, J. E., Nemani, R. R., Turner, W., Melton, F. 2006. Remote sensing for the national parks. *Park Science*, *24*, 30–36.
6. Philipson, P., Lindell, T. 2003. Can coral reefs be monitored from space? *Ambio*, *32*, 586–593.
7. Phinn, S. R., Stow, D. A., Franklin, J., Mertes, L. A. K., Michaelsen, J. 2003. Remotely sensed data for ecosystem analyses: Combining hierarchy theory and scene models. *Environmental Management*, *31*, 429–441.
8. Anghel, A., Vasile, G., Boudon, R., d'Urso, G., Girard, A., Boldo, D., Bost, V. 2016. Combining space-borne SAR images with 3D point clouds for infrastructure monitoring applications. *ISPRS Journal of Photogrammetry and Remote Sensing*, *111*, 45–61. ISSN 0924-2716.
9. Li, Y., Liao, Q. F., Li, X., Liao, S. D., Chi, G. B., Peng, S. L. 2003. Towards an operational system for regional-scale rice yield estimation using a time-series of Radarsat ScanSAR images. *International Journal of Remote Sensing*, *24*, 4207–4220.
10. Schuck, A., Paivinen, R., Hame, T., Van Brusselen, J., Kennedy, P., Folving, S. 2003. Compilation of a European forest map from Portugal to the Ural Mountains based on earth observation data and forest statistics. *Forest Policy and Economics*, *5*, 187–202.
11. Ethem, A. 2010. *Introduction to Machine Learning*. MIT Press. p. 9. ISBN 978-0-262-01243-0.
12. Chutia, D., Bhattacharyya, D.K., Sarma, J., Raju, P.N.L. 2017. An effective ensemble classification framework using random forests and a correlation based feature selection technique. *Transactions in GIS*, *21*, 1165–1178. https://doi.org/10.1111/tgis.12268.
13. Kussul, N., Lavreniuk, M., Skakun, S., Shelestov, A. 2017. Deep learning classification of land cover and crop types using remote sensing data. *IEEE Geoscience and Remote Sensing Letters*, *14*(5), 778–782. doi:10.1109/LGRS.2017.2681128.
14. Chutia, D., Borah, N., Baruah, D. et al. 2020. An effective approach for improving the accuracy of a random forest classifier in the classification of Hyperion data. *Applied Geomatics*, *12*, 95–105, https://doi.org/10.1007/s12518-019-00281-8.
15. Rokach, L. 2010. Ensemble-based classifiers. *Artificial Intelligence Review*, *33*(1–2), 1–39.
16. Kannadhasan, S., Bhapith, V. B. 2014. Research issues on digital image processing for various applications in this world. *Global Journal of Advanced Research*, *1*(*1*), 46–55, ISSN: 2394-5788, http://gjar.org/publishpaper/vol1issue1/d5.pdf.
17. Cohen, E. A. K., Abraham, A. V., Ramakrishnan, S. et al. 2019. Resolution limit of image analysis algorithms. *Nature Communications*, *10*, 79, https://doi.org/10.1038/s41467-019-08689-x.
18. Asokan, A., Anitha, J., Ciobanu, M., Gabor, A., Naaji, A., Hemanth, D. J. 2020. Image processing techniques for analysis of satellite images for historical maps classification—An overview. *Applied Sciences*, *10*, 4207.

19. Namikawa, L., Castejon, E., Fonseca, L. 2009. *Digital Image Processing in Remote Sensing. Tutorials of the XXII Brazilian Symposium on Computer Graphics and Image Processing*, Rio de Janeiro, Brazil, pp. 59–71. doi:10.1109/SIBGRAPI-Tutorials.2009.1.3

20. Landgrebe, D. A. 2003. *Signal Theory Methods in Multispectral Remote Sensing*. John Wiley & Sons, Hoboken.

21. Hubert-Moy, L., Cotonnec, A., Le Du, L., Chardin, A., Perez, P. A. 2001. Comparison of parametric classification procedures of remotely sensed data applied on different landscape units. *Remote Sensing of Environment*, *75*, 174–187.

22. Chen, D., Stow, D. A. 2002. The effect of training strategies on supervised classification at different spatial resolution. *Photogrammetric Engineering and Remote Sensing*, *68*, 1155–1162.

23. Mather, P. M. 2004. *Computer Processing of Remotely-Sensed Images: An Introduction*. 3rd Ed., John Wiley & Sons, Chichester.

24. Gong, P., Howarth, P. J. 1990. An assessment of some factors influencing multispectral land-cover classification. *Photogrammetric Engineering and Remote Sensing*, *56*, 597–603.

25. Lu, D., Weng, Q. 2007. A survey of image classification methods and techniques for improving classification performance. *International Journal of Remote Sensing*, *28*(5), 823–870.

26. Song, C., Woodcock, C. E., Seto, K. C., Lenney, M. P., Macomber, S. A. 2001. Classification and change detection using Landsat TM data: when and how to correct atmospheric effect. *Remote Sensing of Environment*, *75*, 230–244.

27. Gilabert, M. A., Conese, C., Maselli, F. 1994. An atmospheric correction method for the automatic retrieval of surface reflectance from TM images. *International Journal of Remote Sensing*, *15*, 2065–2086.

28. Chavez, P. S. Jr. 1996. Image-based atmospheric corrections—Revisited and improved. *Photogrammetric Engineering and Remote Sensing*, *62*, 1025–1036.

29. Stefan, S., Itten, K. I. 1997. A physically-based model to correct atmospheric and illumination effects in optical satellite data of rugged terrain. *IEEE Transactions on Geoscience and Remote Sensing*, *35*, 708–717.

30. Tokola, T., Löfman, S., Erkkilä, A. 1999. Relative calibration of multitemporal Landsat data for forest cover change detection. *Remote Sensing of Environment*, *68*, 1–11.

31. Heo, J., Fitzhugh, T. W. 2000. A standardized radiometric normalization method for change detection using remotely sensed imagery. *Photogrammetric Engineering and Remote Sensing*, *66*, 173–182.

32. Du, Q., Chang, C. 2001. A linear constrained distance-based discriminant analysis for hyperspectral image classification. *Pattern Recognition*, *34*, 361–373.

33. Canty, M. J., Nielsen, A. A., Schmidt, M. 2004. Automatic radiometric normalization of multitemporal satellite imagery. *Remote Sensing of Environment*, *91*, 441–451.

34. Hale, S. R., Rock, B. N. 2003. Impacts of topographic normalization on land cover classification accuracy. *Photogrammetric Engineering and Remote Sensing*, *69*, 785–792.

35. Jensen, J. R. 1996. *Introduction to Digital Image Processing: A Remote Sensing Perspective*. 2nd Ed., Piscataway, NJ: Prentice Hall.

36. Foody, G. M., Mathur, A. 2004. A relative evaluation of multiclass image classification by support vector machines. *IEEE Transactions on Geoscience and Remote Sensing*, *42*, 1335–1343.

37. Chutia, D., Bhattacharyya, D. K., Sudhakar, S. 2012. Effective feature extraction approach for fused images of Cartosat-I and Landsat ETM+ satellite sensors. *Applied Geomatics*, *4*, 217–224. https://doi.org/10.1007/s12518-012-0088-y.

38. Hsu, P. H. 2007. Feature extraction of hyperspectral images using Wavelet and matching pursuit. *ISPRS Journal of Photogrammetry and Remote Sensing*, *62*(2), 78–92.

39. Harris, J. R., Ponomarev, P., Shang, J., Rogge, D. 2006. Noise reduction and best band selection techniques for improving classification results using hyperspectral data: application to lithological mapping in Canada's Arctic. *Canadian Journal of Remote Sensing*, *32*(5), 341–354.

40. Murat Dundar, M., Landgrebe, D. 2002. A model-based mixture-supervised classification approach in hyperspectral data analysis. *IEEE Transactions on Geoscience and Remote Sensing*, *40*(12), 2692–2699.

41. Mader, S., Vohland, M., Jarmer, T., Casper, M. 2006. *Crop classification with hyperspectral data of the HyMap sensor using different feature extraction techniques*. In *2nd Workshop of the EARSel SIG on Remote Sensing of Land Use & Land Cover*, edited by M Braun, Bonn, Germany, pp. 96–101.

42. Green, A. A., Berman, M., Switzer, P., Craig, M. D. 1988. A transformation for ordering multispectral data in terms of image quality with implications for noise removal. *IEEE Transactions on Geoscience and Remote Sensing, 26*(1), 65–74.

43. Hyvärinen, A., Oja, E. 2000. Independent component analysis: algorithms and applications. *Neural Networks, 13*(4), 411–430.

44. Yang, C., Everitt, J. H., Johnson, H. B. 2009. Applying image transformation and classification techniques to airborne hyperspectral imagery for mapping Ashe juniper infestations. *International Journal of Remote Sensing, 30*(11), 2741–2758.

45. Chutia, D., Bhattacharyya, D. K., Kalita, R., Sudhakar, S. 2014. OBCsvmFS: Object-based classification supported by support vector machine feature selection approach for hyperspectral data. *Journal of Geomatics, 8*(1), 12–19.

46. Anilkumar, R., Chutia, D., Goswami, J., Sharma, V., Raju, P. L. N. 2018. Evaluation of the performance of the fused product of Hyperion and RapidEye red edge bands in the context of classification accuracy. *Journal of Geomatics, 12*(1), 35–46.

47. Borah, N., Chutia, D., Baruah, D., Raju, P. L. N. 2017. *Dimension reduction of hyperion data for improving classification performance – An assessment. IEEE International Conference On Recent Trends In Electronics Information Communication Technology*, May 19–20, Bengaluru, India.

48. Kohavi, R. 1995. Wrappers for performance enhancement and oblivious decision graphs. PhD thesis, Stanford University.

49. Kohavi, R., John, G. 1996. Wrappers for feature subset selection. *Artificial Intelligence, 97*(1–2), 273–324. Special issue on relevance.

50. Hall, M. A. 1999. Correlation-based Feature Subset Selection for Machine Learning. Hamilton, New Zealand. PhD thesis.

51. Pushpalatha, K. R., Karegowda, A. G. 2017. *CFS based feature subset selection for enhancing classification of similar looking food grains – A filter approach. 2nd International Conference On Emerging Computation and Information Technologies (ICECIT)*, Tumakuru, pp. 1–6, doi:10.1109/ICECIT.2017.8453403.

52. Ross, Q. 1993. *C4.5: Programs for machine learning*, vol. *16*. Morgan Kaufmann Publishers, San Mateo, pp. 235–240.

53. Breiman, L. 2001. Random forests. *Machine Learning, 45*(1), 5–32.

54. Sonobe, R., Hiroshi, T., Xiufeng, W., Nobuyuki, K., Hideki, S. 2014. Random forest classification of crop type using multi-temporal TerraSAR-X dual-polarimetric data. *Remote Sensing Letters, 5*(2), 157–164.

55. Ham, Y., Han, K. K., Lin, J. J. et al. 2016. Visual monitoring of civil infrastructure systems via camera-equipped Unmanned Aerial Vehicles (UAVs): A review of related works. *Visualization in Engineering, 4*, 1. https://doi.org/10.1186/s40327-015-0029-z.

56. Yao, Y., Zhang, J., Hong, Y., Liang, H., He, J. 2018. Mapping fine-scale urban housing prices by fusing remotely sensed imagery and social media data. *Transactions in GIS, 22*, 561–581. https://doi.org/10.1111/tgis.12330.

5

Artificial Intelligence and Image Processing

Sudhriti Sengupta
Amity University, Noida, India

CONTENTS

5.1 Introduction

During the past years, artificial intelligence (AI) has led to many innovations and technological breakthrough in various areas of automation. The AI aims to automatize the systems for optimum performance and more efficient results. It is being used in many diverse areas like diagnosis of ailments in plants or animals, virtual chatbots, livestock managements, autonomous cars, medical image analysis, warehouse supply chain, analysis of sport, security or surveillance activities, etc. [1–3]. The capacity of computers, robots, or any machine to do work intelligently is called artificial intelligence. Here, machines or computers are equipped with characteristics of humans, e.g., reasoning and logic, generalizing, learning from experience, etc. The goal of AI can be termed as employing the full capabilities of memory and processing of a computer to match the human thinking power over a wide range of domain. Over the years, AI techniques have been developed to facilitate medical diagnosis, search engine, video surveillance, security, object recognition, etc. [4]. Various such techniques have been introduced in the field of image processing also for promoting, facilitating, and optimizing the different applications in this field. The ways to manipulate an image for enhancing its content or to retrieve information from it is called image processing. It is used in various applications, such as biometric, gaming, enforcement of law and order, visualization of medical condition, etc.

When applying AI to image processing, many types of development and activities can be done, like detection of objects, face recognition, prediction of diseases from images, finding patterns in images or video, etc. It also paves way for video surveillance [3, 5–10]. This chapter aims to study the concept of AI in the various fields of image processing. The second part of this chapter introduces the concept of

DOI: 10.1201/9781003140351-5

image processing; artificial intelligence is discussed in third part; and after this the role of AI in image processing is discussed. Finally, this chapter mentions the results, discussion, and conclusion in the subsequent sections.

5.2 Image Processing

Image processing is the collection of methods used to perform some techniques on image to derive useful information from the image or to enhance the visual cues in the image in terms of color, contrasts, etc. Image processing is a category of signal processing where images are the input and images or some of their related attributes are the output [11–14]. It also forms an important research area in computer science and multidisciplinary fields [15]. Image-processing techniques are broadly classified into two forms, i.e., the analog image processing and the digital image processing. In analog image processing, hard copy of the images like photographs, printouts, etc. is used as the processing medium. The output of analog image processing is always an image. Digital image processing refers to manipulating the images with the help of a computer. The output in this case may be images or attributes of images in form of masks, feature list, etc [16]. With the development and evolution of digital computer and devices associated with it, digital image processing has subjugated analog image processing in many areas of its application. Various overlaying fields associated with images are also introduced, such as computer graphics, computer vision, etc. The study of forming of images from different models in various devices is called computer graphics. Computer vision is associated with developing system in which the mean full information is retrieved from the input images. Scanning a face for identification is an example of computer vision application, whereas object drawing on a projector is an example of computer graphics. To discuss the concept of image processing and the usage of AI in this field, a brief introduction of image is done in the following sections.

5.2.1 Images

Image is two-dimensional, arranged in rows and columns. Mathematically, image can be defined as function $f(x, y)$ where x ix the horizontal coordinate and y is the vertical coordinate. The amplitude of f at any point (x, y) is the intensity of the point providing representation in the image.

The following function can be used to represent image:

$$f(x,y) = \begin{matrix} f(0,0) & f(0,1)..... & f(0,N-1) \\ f(1,0) & f(1,1)..... & f(1,N-1) \\ . & . & . \\ . & . & . \\ f(M-1,0) & f(M-1,1) & f(M-1,N-1) \end{matrix} \tag{5.1}$$

Here, $f(x, y)$ represents the image and the values $(f(0,0), f(0, 1), ..., f(M-1, 0)...f(M-1, N-1)$ represent each pixels in the image. In digital image the values of f(x,y) are finite, where each element has a particular position and value. These elements are called picture elements or pixels. A digital image can be of different types such as binary image, black-and-white image, 8-bit color format, 16-bit color format, 24-bit color format, etc. There are different phases of digital image processing. Image acquisition means capturing of image by a camera or other sensor and transforming it into a digital form. After acquiring, extraction of hidden details in the image and improvement of image appearance are done by image enhancement, image restoration, etc. Next, if necessary, the color of the image is used to extract the region of interest. Wavelet and multi-resolution processing, image compression, morphological transformation, segmentation process, and object recognition are the important steps that can be applied as per the requirement of the application.

Basically, image processing consists of three steps:

- Input image
- Analysis and manipulation of image
- Output in the form of image or report that is based on the analysis of image

Some of the important purpose of image processing are:

1. Visualization refers to the methods used to represent data in a form that can be understandable.
2. Restoring and sharpening process that improves the quality of the image.
3. Image retrieval helps in retrieval of image-based search.
4. Measurement involves measuring objects in background or foreground.
5. Pattern recognition distinguishes and classifies objects in an image [17].

The typical components of an image processing system are:

I. Image sensors that are used to acquire digital images.
II. Special hardware for image processing consisting of digitizer and an arithmetic and logic unit (ALU). Digitizers convert the output from sensors to digital formats. ALU carries arithmetic and logic application in the entire image.
III. The computer system, which can be a personal computer or a supercomputer.
IV. Image-processing software consisting of specific methods to perform tasks specific to images. Major programming languages, like MatLab, C++, Python, etc., contain modules related to image processing.
V. Storage area is very important in case of image, as the requirement of memory in this case is large. Short-term storage is used during processing, online storage for fast processing, and mass storage for permanent storage and infrequent access.
VI. Image displays are done by devices by color monitors and facilitated by graphical display unit, etc.
VII. Hardcopy devices include printers, plotters, etc. to obtain a hard copy of the image.
VIII. Network devices are essential due to the usage of computer communication; network devices are an inevitable part of an image processing system. One of the requirements of an image-processing application is high bandwidth for transmission.

5.2.2 Real-Time Usage of Image Processing

Image processing has introduced some major innovations in various fields such as agriculture, security, remote sensing, computer visions, diagnosis of diseases, etc. Precision agriculture is a farming technique that utilizes information technology to ensure maximum productivity of crops and optimum usage of resources [14, 18]. A computerized imaging system to determine the quantity of weeds present in crop cultivation has been developed [19]. Image processing plays an important role in security by providing and enhancing use of techniques such as digital watermarking, digital signature, biometric security, etc. [20]. Remote sensing refers to the area where information is collected with physical proximity. It acquires, detects, and processes the data to monitor information about a particular subject without physical contact. Some of the major uses of remote sensing are detection of road condition, forest area coverage, collection of Earth images to monitor atmospheric conditions, in agriculture, etc. Image processing has enhanced and introduced remote sensing to a vast extent by providing images under observation [21]. Some other applications of image processing are automatized photography, space imaging, medical imaging, automatic product optical sensing, robotic vision, unlock technology of devices, etc.

5.3 Artificial Intelligence

The AI is one of the emerging techniques allowing a machine to apply human-like logic and reasoning. In other words, AI gives a machine the ability to mimic the human brain [3, 6–10]. In 1956, the term "Artificial Intelligence was coined at Darmouth Conference, New Hampire. In 1966, Weizenbaum invented a computer program that could communicate with human operators. It was called Eliza. Subsequently, an expert system MYCIN was introduced for disease treatment in 1972. NETtalk was able to read, comprehend, and apply words for communication. A major breakthrough was the introduction of IBM Deep Blue for automated chess playing. It defeated word champion Garry Kasparov. In 2011, a question system called Watson was developed by IBM. Watson defeated champions Rutters and Jenning in a Quiz Match. The above example shows the development of artificial intelligence in successfully mimicking the human brain to do tasks typically requiring human thinking prowess. Today many types of problems can be solved by harnessing the advanced computer system with the development of AI. Several branches of artificial intelligence have evolved, such as deep learning, machine learning, etc. [22].

AI research is a multidisciplinary field of studies encompassing various area such as biological science, mathematics, sociology, computer science, etc. It aims to make decisions based on human logic with errors that may occur due to stress level, emotions, or any other factors affecting humans. AI will also speed up certain operations, allow a more precise analysis data, and maintain higher degree of security. It is categorized into three types based on their capabilities, namely narrow or weak AI, general AI, and super AI. Narrow AI performs some specific predefined and predetermined tasks such as playing chess. General AI aims to perform any task with an effectiveness of a human mind. At the present time, researchers are working on developing a general AI system. Super AI is a hypothetical idea where computers or machines could exceed human intelligence. Furthermore, on basis of functions, AI can be divided into reactive machine, limited memory of mind, and self awareness. Reactive memories are used for small tasks. They cannot learn from memory (e.g., IBM's Deep Blue). Limited machines use memory for decision-making, e.g., an autonomous automobile. Theory of mind and self-awareness type of AI does not exist at current stage. They deal with physiological effects such as feelings, emotions, etc.

AI relates to imparting to machines the ability to think and decide the way humans do. There are three main types of AI algorithms: classification algorithm, regression algorithm, and clustering algorithm. Classification techniques are used to categorize the different subjects into various class. An example is to find if a lesion in a medical image is fatal or nonfatal. Naïve Bayes, Decision Tree, Random Forest, Support Vector Machine, and K Nearest Neighbors are algorithms in this category. Regression algorithms are used to determine the resultant values depending on some input values. Forecasting the price of a stock is a popular example of regression techniques. Linear regression, Lasso regression, logistics regression, and multivariate regression are types of this algorithm. Clustering is the process of grouping data into sections, with each member in the section having some property. For example, claims with high risk in insurance industry can be grouped in one section. K means, fuzzy C means, and hierarchical clustering are examples of this type of algorithm. Classification and regression are types of supervised learning. Clustering is a type of unsupervised learning. In supervised learning, the model needs to use the technique with the help of an input variable and an output variable. In unsupervised learning, the output is determined by the input only without the effect of output. Suppose we have a group of vegetables and fruit. In case of a supervised technique, there will be labels such as "Fruit," "Vegetables," etc., and all the entities will fall into these categories [23, 24]. This is a classic case of supervised learning. However, in the case of unsupervised learning, there is no predefined label existing and depending on the characteristic each entity is grouped.

Researchers are using the technique of AI to solve issues in various domains or to optimize the effectiveness and efficiency in many fields. In finance and banking sectors, AI-based machines are developed to detect fraud related to credit card, create chatbots, etc. In field of farming and agriculture, new techniques such as digital agriculture and precision agriculture, among others, are developed. Detection of weeds from an area growing crops, prediction of time for yielding of crop, soil and water monitoring, etc. are some of the examples of usage of AI in agriculture. AI can be used for developing protocols for identification of potential threats, intruders, invalid access, etc. Marketing is one of the areas where AI plays a significant role. AI along with Big Data has created comprehensive and useful tools for market analysis. AI has proved its worth in the field of health care also. AI provides capabilities to the doctors and specialists to support research, analysis, better decision-making, etc. In the area of education, AI made some changes like auto-corrections. Better-quality teaching and learning materials are also created with the help of AI. The field of usages of AI is large and growing quickly. Some of the fields in which AI are used are medical image analysis, self-driving vehicles, warehouse supply chain management, livestock management, analysis of sport activities, surveillance, retail and marketing, etc.

5.4 Artificial Intelligence in Image Processing

For communication and interpreting of data, images are a primary tool [13, 14, 18]. As the number of image processing techniques has increased, the requirement of technologies to be incorporated in this area are also being developed. AI has also provided to be an inherent technology in the field of image processing [25]. For example, AI algorithm can be used for border detection of objects, to interpret images, recognize objects, etc. In 2006, Stoitsis et al. have introduced the usage of AI in developing computer-aided diagnostics (CAD) for CT images by focusing on liver tissues to classify into different categories. AI helps improve quality and interpretation of medical images and early diagnosis [26]. AI integrated with optical coherence tomography (OCT) was used to detect glaucoma. Glaucoma is a disease of eyes and can lead to vision loss if not treated correctly and timely. AI-based systems will focus on time and better chance of correct diagnosis [27]. AI-based classification techniques for diagnosis of skin disease also represent an area of research [28]. In finance and commerce applications, AI is also popular and integrated with image-processing techniques such as sophisticated chatbots, robots, prediction tools, etc. In 2011, Zhong et al. have developed an image classification system using AI for remote images [29]. AI integrated with computer vision, image processing, sensor technology, etc. has been used in efficient fruit and vegetable drying facility. The model and technique of drying are to be automatically controlled to solve issues, such as uneven drying, nutrient loss, damage of crops, etc. [30]. Mubarak et al. created a detection system for forest fire by applying temporal variations with rule-based image processing [31]. AI integrated with image processing is a promising area to develop various applications for enhancing the quality of life and sustainable development. In classification, feature extraction, edge detection, preprocessing, and other areas, AI has proven to be a very prominent optimizer of results. This led to development of various open-source libraries for processing of images and computer vision. Some of them are OpenCV (Open Source Computer Vision library), which can be integrated with major programming such as C++, Java, Phython, MatLab image-processing library, etc. Also free image databases, like ImageNet and Pascal VOC, are available for researchers to carry out research in the field of image processing. In the next section, an ant colony optimization (ACO)-based optimization technique for edge detection is discussed. A quantitative analysis of the ACO-based edge detection technique is also provided to show the improved result of an AI-integrated method.

5.5 Proposed Methodology

AI has proven its forte in various fields of image processing. It has successfully led to the creation of classifiers for different categories of images, enhanced quality of images, etc. One of the prominent usages is to improve the result of edge detection or segmentation in image. In this paper we have tried to prove the optimization of edge detection technique and threshold technique can be done by using the popular ACO technique. The entropy of the image obtained from the ACO-induced Sobel operator and threshold technique is higher than the entropy of the image obtained by Conventional Sobel operator and threshold method. This shows that an AI-induced image-processing technique produces optimized results as compared to a conventional technique.

5.5.1 Sobel Edge Detection

Sobel edge detection operator focuses on area of high frequency related to borders by doing a two-dimensional measurement. Sobel operator provides a smoothing effect to the natural borders in the input images forming the output images. Generally, this operator is used highlight area of high spatial frequency. Sobel operator is a popular edge detection techniques used to find the edges of an object within an image. This operator is less sensitive to noise but provides smoothing of the edge. Sobel operator is used in many applications as a technique for edge detection.

5.5.2 Threshold

Thresholding is the process of creating a binary image depending on some values, called threshold. The basic principal of this technique is that if the pixel value is larger than the threshold value, then it is set to white, otherwise if the pixel value is lesser than the threshold value, it is set to black. The goal of thresholding is to split the image into smaller segments or areas of interest. It reduces image complexity and helps in image classification.

5.6 Ant Colony Optimization

The ACO is a popular AI technique inspired by swarm intelligence. Animals like ants, and their foraging nature, are the inspiration behind this technique. The primary steps in this process are the construction of an initial stage where ants can be placed, followed by analysis of various paths taken by ants, and pheromone updating. Pheromones are elements that help create an optimal path that leads to an optimal solution. According to the proposed technique, ACO is introduced to optimize the result of the segmented image obtained by the threshold method and Sobel edge detection method. The borders or features of the test image are calculated by using the conventional techniques. Then the ants are introduced in the end points of the segmented image. The variation in each pixel leads to the movement of ants.

5.7 Entropy

Entropy is used to quantitatively measure the efficiency of the proposed ACO-based optimization techniques. It signifies the amount of data in the image. The quality of the image is high if its entropy is high. In this chapter, we have used entropy to measure the performance efficiency of the ACO-based technique as compared to a conventional method.

5.8 Result Analysis

In this chapter, a two-way approach is used to show the impact of AI techniques in image processing. The popular ACO optimization techniques are applied to the image edge detection technique and segmentation technique to observe if they provide better results. The experiment is done on Sobel edge detection technique and Prewitt edge detection technique to see the effect of introduction of AI techniques in them. For the first case we have considered test image (a), as shown in Figure 5.1, and applied a Sobel edge detection operator to it. The edge map obtained by the Sobel edge operator is enhanced using ACO-based optimization technique (Figure 5.2). Next we have calculated the entropy of the images obtained by using a conventional Sobel edge detector and ACO-induced Sobel edge operator.

In the case of threshold optimization, we have taken test image (a) and applied the threshold technique to obtain the segmented resultant image. To further optimize this, we have applied ACO to get an enhanced image. The validity of optimization by ACO technique is calculated by using Entropy of the image. Given below are the test images:

FIGURE 5.1   Test Image a.

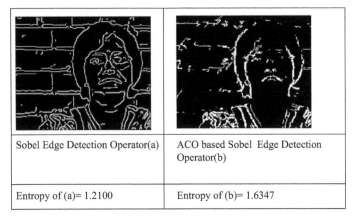

Sobel Edge Detection Operator(a)	ACO based Sobel Edge Detection Operator(b)
Entropy of (a)= 1.2100	Entropy of (b)= 1.6347

FIGURE 5.2   (a) Edge map obtained by a Sobel edge detection operator; (b) edge map obtained by an ACO-based Sobel edge detection operator.

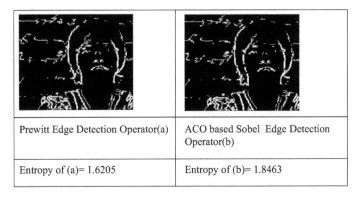

Prewitt Edge Detection Operator(a)	ACO based Sobel Edge Detection Operator(b)
Entropy of (a)= 1.6205	Entropy of (b)= 1.8463

FIGURE 5.3   AI segmentation techniques.

Figure 5.3 shows (a) a segmented image by threshold technique and (b) a segmented image by ACO-based threshold technique.

The result to observe the effect of use of ACO in the methods used in edge detection and segmentation can be perceived from Figure 5.4, which is drawn depending on the values of the entropy of the output images. These output images are obtained by applying the conventional techniques and ACO-based techniques.

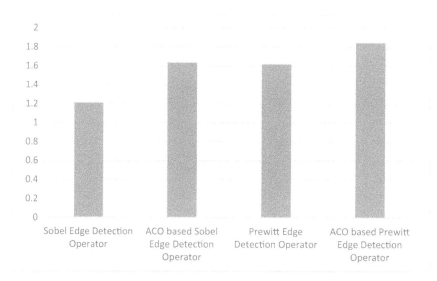

FIGURE 5.4   Performance evaluation of conventional technique vs. ACO-induced technique.

From Figure 5.4 it can be seen that the entropy of the output images obtained by conventional techniques is lesser than the entropy of the images obtained by applying AI-based ACO technique. The entropy of the images obtained by applying Sobel edge detection and Prewitt edge detection is 1.21 and 1.6205, respectively. The entropy of the images obtained by ACO-induced Sobel edge detection and ACO-induced Prewitt edge detection operator is 1.6347 and 1.8463, respectively. It can be thus comprehended that ACO-based techniques are successful in enhancing the output image quality of the popular techniques. Thus, by using this experiment, we can understand that AI plays a significant role in image processing.

5.9 Conclusion

In this chapter, the focus is on the introduction and usage of AI techniques in image processing. Image processing is the method of improving image quality or deriving information from an image. The role of image processing encompasses many domains, like satellite imaging, medical imaging, precision agriculture, computer vision, etc. The ability of the machine, particularly the computer system, to mimic and apply logic in a way similar to human beings is known as AI. AI is important in many different types of application area to build automated devices to improve the quality of life. In image processing also, AI plays a significant role, like optimization of segmentation, building classifiers, etc. This chapter studied the effects of AI in popular edge detection and segmentation techniques on the basis on performance analysis. The result is the observation that if AI methods such as ACO are introduced in the image-processing techniques, it provides better and effective results. Further, studies can be done to find the effect of AI in image classification, preprocessing, restoration, etc.

REFERENCES

1. Artificial Intelligent overview. Available at: https://becominghuman.ai/where-is-artificial-intelligence-used-today-3fd076d15b68 [accessed April 12, 2019].
2. L. Sharma, P. Garg (Eds.), *From Visual Surveillance to Internet of Things*. New York: Chapman and Hall/CRC, 2020, https://doi.org/10.1201/9780429297922
3. L. Sharma, "Human Detection and Tracking Using Background Subtraction in Visual Surveillance", *Towards Smart World*. New York: Chapman and Hall/CRC, pp. 317–328, December 2020. https://doi.org/10.1201/9781003056751

4. Artificial Intelligence: How Does AI Work? Author B. J. Copeland, Independently Published, 2019.
5. L. Sharma (Ed.), *Towards Smart World*. New York: Chapman and Hall/CRC, 2021, https://doi.org/10.1201/9781003056751
6. L. Sharma, A. Singh, D. K. Yadav, "Fisher's Linear Discriminant Ratio based Threshold for Moving Human Detection in Thermal Video", *Infrared Physics and Technology*, Elsevier, March 2016.
7. L. Sharma, D. K. Yadav, S. K. Bharti, *"An Improved Method for Visual Surveillance using Background Subtraction Technique"*, *IEEE, 2nd International Conference on Signal Processing and Integrated Networks (SPIN-2015)*, Amity University, Noida, India, February 19–20, 2015.
8. Dileep Kumar Yadav, Lavanya Sharma, Sunil Kumar Bharti, *"Moving Object Detection in Real-Time Visual Surveillance using Background Subtraction Technique"*, *IEEE, 14th International Conference in Hybrid Intelligent Computing (HIS-2014)*, Gulf University for Science and Technology, Kuwait, December 14–16, 2014.
9. N. Abramson, D. Braverman, G. Sebestyen, Pattern recognition. *IEEE Transactions on Information Theory*, vol. *IT-9*, pp. 257–261, 1963.
10. A. Newell, H. Simon, The logic theory machine—a complex information processing system, *IRE Transactions on Information Theory*, vol. *IT-2*, pp. 61–79, 1956.
11. L. Sharma, P. K. Garg, "IoT and its applications", *From Visual Surveillance to Internet of Things*, Taylor & Francis Group, CRC Press, Vol. *1*, pp. 29.
12. L. Sharma, P. K. Garg, "Smart E-healthcare with Internet of Things: Current Trends Challenges, Solutions and Technologies", *From Visual Surveillance to Internet of Things*, CRC Press, Taylor & Francis Group, pp. 215–234, 2019.
13. L. Sharma, P. K. Garg, "A foresight on e-healthcare Trailblazers", *From Visual Surveillance to Internet of Things*, CRC Press, Taylor & Francis Group, pp. 235–244, 2019.
14. L. Sharma, S. Sengupta, B. Kumar, An improved technique for enhancement of satellite images. *Journal of Physics: Conference Series, 1714*, 012051, 2021.
15. Artificial Intelligent overview. Available at: https://sisu.ut.ee/imageprocessing/avaleht [accessed April 12, 2019].
16. Artificial Intelligent overview. Available at: https://www.apriorit.com/dev-blog/599-ai-for-image-processing [accessed April 12, 2019].
17. Gonazalez and woods, "Digital Image processing", pearson Eductaion international, third edition
18. S. Singh, L. Sharma, B. Kumar, A machine learning based predictive model for coronavirus pandemic scenario. *Journal of Physics: Conference Series, 1714*, 012023, 2021.
19. X. P. B. Artizzu, A. Ribeiro, A. Tellaeche, G. Pajares, C. F. Quintanilla, Improving weed pressure assessment using digital images from an experience-based reasoning approach, *Computers and Electronics in Agriculture*, *65*, pp. 176–185, 2009.
20. B. Sridhar, Cross-layered embedding of watermark on image for high authentication, *Pattern Recognition and Image Analysis*, vol. *29*, no. 1, pp. 194–199, 2019.
21. U. Walz, *Remote sensing and digital image processing*, Bastian, Olaf and Steinhardt, Uta editors. Kluwer Academic Publishers, 2002.
22. Artificial Intelligent basic. Available at: https://www.tutorialandexample.com/artificial-intelligence-tutorial/ [accessed April 12, 2019].
23. Artificial Intelligent algorithm. Available at: https://www.upgrad.com/blog/types-of-artificial-intelligence-algorithms/ [accessed April 12, 2019].
24. R. Faulk, An inductive approach to mechanical translation. *Communications of the ACM*, vol. *1*, pp. 647–655, 1964.
25. Xin Zhang, Wang Dahu, Application of artificial intelligence algorithms in image processing, *Journal of Visual Communication and Image Representation*, vol. *61*, pp. 42–49. 2019. ISSN 1047-3203, https://doi.org/10.1016/j.jvcir.2019.03.004.
26. John Stoitsis, Ioannis Valavanis, Stavroula G. Mougiakakou, Spyretta Golemati, Alexandra Nikita, Konstantina S. Nikita, Computer aided diagnosis based on medical image processing and artificial intelligence methods, *Nuclear Instruments and Methods in Physics Research Section A: Accelerators, Spectrometers, Detectors and Associated Equipment*, vol. *569*, no. 2, pp. 591–595, 2006. ISSN 0168-9002, https://doi.org/10.1016/j.nima.2006.08.134.
27. Bala Prabhakar, Rishi Kumar Singh, Khushwant S. Yadav, Artificial Intelligence (AI) impacting Diagnosis of Glaucoma and Understanding the Regulatory aspects of AI-based software as medical device, *Computerized Medical Imaging and Graphics*, 101818, 2020.

28. Manu Goyal, Thomas Knackstedt, Shaofeng Yan, Saeed Hassanpour, Artificial intelligence-based image classification methods for diagnosis of skin cancer: Challenges and opportunities, *Computers in Biology and Medicine*, vol. *127*, 2020.

29. Y. Zhong, L. Zhang, W. Gong, Unsupervised remote sensing image classification using an artificial immune network. *International Journal of Remote Sensing*, vol. *32*, no. 19, pp. 5461–5483, 2011. DOI: 10.1080/01431161.2010.502155

30. J. Chen, M. Zhang, B. Xu, J. Sun, A. S. Mujumdar, Artificial intelligence assisted technologies for controlling the drying of fruits and vegetables using physical fields: A review. *Trends in Food Science & Technology*, vol. *105*, pp. 251–260, 2020.

31. A. I. Mubarak, Mahmoud, H. Ren, Forest Fire Detection Using a Rule-Based Image Processing Algorithm and Temporal Variation. *Mathematical Problems in Engineering*, 2018 | Article ID 7612487 | https://doi.org/10.1155/2018/7612487

6

Deep Learning Applications on Very High-Resolution Aerial Imagery

Avinash Chouhan, Dibyajyoti Chutia, and P. L. N. Raju
North Eastern Space Applications Centre (NESAC), Government of India, Umiam, India

CONTENTS

6.1 Introduction

Artificial Intelligence (AI) is the intelligence demonstrated by machines. It is a field of study to develop computational systems with human-level analysis capabilities. It is also the development of computational systems that behave rationally. In 1950, Alan Turing introduced the Turing test to detect if a machine is intelligent enough. John McCarthy coined the term "Artificial Intelligence." Intelligence is composed of:

DOI: 10.1201/9781003140351-6

- Reasoning: It is the ability to make decisions based on observations and situations. It can be further divided into inductive reasoning and deductive reasoning. In inductive reasoning, a number of observations have to come upon general statements and theories. It is a bottom-up approach. Deductive reasoning worked on a top-down approach. It first assumes some general opinions and views, and those have been refined and redefined based on several observations.
- Learning: It is the ability to improve through experiences. Herbert Simon describes learning as a change in a system that enables it to do the same tasks more efficiently.
- Problem Solving: It is the ability to reach the solution based on the present situation using decision-making. It is to choose the best possible alternative from a complete set of other options.
- Perception and Linguistic Intelligence: It is the ability to acquire, process, and analyze sensory input. It is mostly involved in the reproduction of human perception ability using a computational system. Linguistic intelligence represents the ability to understand input language and produce possible output.

AI approaches can be divided into two categories: symbolic and non-symbolic [1]. Symbolic approaches are based on the human-readable representation of problems through symbols and solutions using rule- and logic-based approaches. Advantages of these are: the solution approach is more precise, and it requires less amount of data and computation power. Non-symbolic approaches do not rely on the symbolic representation of the problem. These require a large amount of data to learn. These follow established approaches to solve the problem without having a complete explanation of the solution. Deep learning, neural network, and genetic algorithms are examples of non-symbolic strategies.AI can be divided into three types:

- Artificial narrow intelligence: It is also known as weak AI. It is designed as a way to perform pre-defined limited tasks. It can perform very well in specific predefined tasks. Most of our AI applications are examples of weak AI.
- Artificial general intelligence: It is also known as strong AI. It is the ability of a computational system to show human-level intelligence in any task. Our AI system aims to transform from weak AI to strong AI.
- Artificial superintelligence: If the computational model shows ability better than any human cognitive ability, it is called artificial superintelligence. Currently, no computational model is able to achieve it.

6.2 Machine Learning

Machine learning is a type of approach that learns from data. It is a technique to understand the behavior based on experience gained from data without being explicitly programmed. Machine learning algorithms can be categorized into three types, as shown in Figure 6.1.

- Supervised Learning: It deals with paired data. Here we require input data and respective ground truth. Based on the passed data, ML algorithms will try to find out the mapping between input and output.
- Unsupervised Learning: It is applicable to a scenario where we do not have paired data. In this case, hidden patterns within the input data have been identified by algorithms. Algorithms tried to cluster the inputs into different groups based on parameters.
- Reinforcement Learning: It is an approach in which algorithms treat problems as a game. All rules of games need to be defined in the algorithm. The object of the algorithm is to solve the game. Reward and penalties have been defined based on right or wrong solutions.

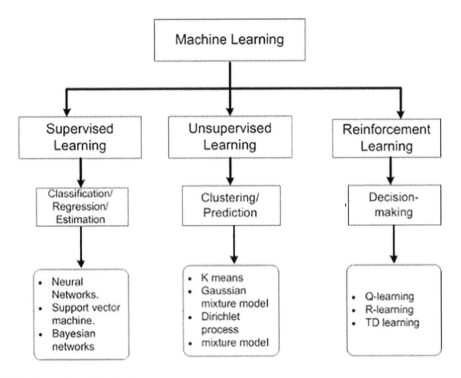

FIGURE 6.1 Types of machine learning.

6.3 Deep Learning

In deep learning (DL), the first event can be related to Walter Pitts and Warren McCulloch's experiments, where they have tried to create a computational model based on the workings of the human brain. Henry J. Kelley and Stuart Dreyfus worked on an initial computational model for backpropagation in the 1960s. Alexey Grigoryevich Ivakhnenko and Valentin Lapa introduced the hierarchical representation of neural networks with polynomial activation functions.

DL is a subclass of machine learning algorithms. It is an automatic feature of learning. It replaces feature engineering steps involved in other machine learning algorithms. It is motivated by the connectivity and working of the brain. It uses an artificial neural network (ANN) to do supervised, unsupervised, or reinforcement learning tasks. ANN is a network of interconnected neuron nodes. To represent the connection between neuron nodes, weights have been assigned. Positive weight means excitatory connection, and negative weight shows the inhibitory relationship between nodes. During the ANN training, the process learns these weights, and based on this, connection pattern changes.

In supervised deep learning, input has been passed through the neural networks to learn representation and predict the output. This expected output has been compared with ground truth. The gap between predicted output and ground-truth has been represented through the loss function. The ultimate goal of model training is to minimize the loss function using an optimization algorithm.

6.3.1 Comparison between ML and DL

In Figure 6.2, relation between AI, ML, and DL is shown. AI consists of ML and many other approaches. In spite of the fact that DL is a subclass of ML algorithms, there are major differences between the two. These differences are:

- Feature Learning: In traditional ML algorithms, we need to do explicit feature engineering. This feature extraction is a difficult task and requires domain expertise. This domain knowledge requirement is a limitation of traditional algorithms. DL replaced this with an automatic feature

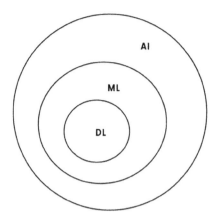

FIGURE 6.2 The relation between AI, ML, and DL.

extraction step built in. This enables users to directly utilize raw data in algorithms without rely-ing on domain experts.

- Solution Technique: In ML algorithms, mostly, we divide problems into different subproblems. Each subproblem is solved individually, and then they are brought together to form a solution. In DL algorithms, mostly, these steps are not explicitly created.
- Data Requirement: Differences in ML and DL algorithm's performance depend on data availabil-ity. With more data availability, DL algorithm's performance increases. DL algorithms heavily rely on large data sets. ML algorithms do not have these massive data availability requirements.
- Time Requirement: Time requirements can be split into two different categories of training time and test time. For training time, DL algorithms perform poorly. For test time, DL algorithms per-form better or similar to ML algorithms.
- Resource Dependency: DL algorithms are resource-intensive. For training of DL algorithms, graphical processing units (GPUs) are required. ML algorithms do not depend on GPUs and can be trained on central processing units (CPUs).
- Explainability: ML algorithms are well explainable because the entire work is well defined. DL algorithms are mostly a black box. It is very difficult to explain the output of DL algorithms from the input.

6.3.2 Types of Neural Network

- Perceptron: It is a fundamental model of neural networks. It has been proposed by Misky. It takes input and combines it with weights. It applied activation function on input and weights to produce output. Basic operation AND, OR, NOT can be implemented by it. The major limitation of it is that it can be used for linearly separable problems.
- Feedforward Neural Network: It is a type of ANN where input goes only in one direction from the input end to the output end. It may have one or more hidden layers in between. The number of lay-ers is based on the complexity of the function that needs to be implemented. It also combines inputs with weights and output passed to the activation function for output. It has static weights, as there is no backpropagation. It can be used in nonlinearly separable problems. It cannot be utilized in deep learning.
- Radial Basis Function (RBF) Neural Network: RBF neural network passes input to RBF neuron where it measures its similarity with available class prototypes from the dataset. Input is assigned to classes with the highest prototype similarity.

- Kohonen Self-Organizing Neural Network: It is an unsupervised learning model. It needs to be trained to create its own representation of input data. Input vectors that are near in high dimensions are mapped nearby in two-dimensional space. It can be used for dimensionality reduction.
- Recurrent Neural Network (RNN): RNN saves the output of the layer and feeds it back to the same layer. In this way, RNN neurons store a partial amount of information on previous steps. This stored information has been used for enhancing the output. In each step, useful information about the data has been stored in the state of RNNs. Different mechanisms are there to decide useful information.
- Convolutional Neural Network (CNN): CNN is a type of neural network where a fixed-size kernel is convolved over input to learn the weight of the kernel. The connectivity pattern in CNN between neurons is partial and learned during training. More details about CNN will be presented in the next section.
- Modular Neural Network [3]: It uses a number of neural networks to solve the problem. It workes on the divide-and-conquer approach. Problem is divided into smaller subproblems and passed to each neural network that acts as a module. In the final stage, output of each module is combined to produce final output.

6.3.3 Convolutional Neural Network (CNN)

CNN has been first proposed by Fukushima [4] in Neocognitron. It used a hierarchical network of many layers. Limited computation capability hinders its widespread use at that time. Lecun [5] used gradient-based learning on a multi-layer neural network for handwritten character recognition. Hinton [6, 7] proposed a fast training algorithm for deep belief networks and data dimensionality reduction using a neural network. Glorot [8] found logistic sigmoid activation unsuitable for neural network training with random initialization. Ranzato [9] used a max-pooling layer for sub-sampling. Krizhevsky [10] used a deep convolutional neural network in the ImageNet classification contest and produced better results than state-of-the-art.

CNN architecture is made up of a feature extractor and classifier. Feature extraction is performed by a sequence of layers as shown in Figure 6.3. Various layers used in CNN are:

- Convolutional Layer: It consists of learnable kernels. Kernels are a matrix of weights. These weights have been learning during the training process. Kernels convolve over input and produce feature maps. Convolution operation can be defined as

$$(M * K)(p) = \sum_{s+t=p} M(s)K(t) \qquad (6.1)$$

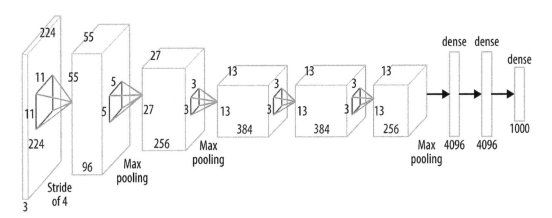

FIGURE 6.3 Different layers in convolutional neural network.

Here * is convolution operation. Sliding of kernels over input acts as a feature extractor. The amount of slide of the kernel has been decided by the stride parameter. Feature maps produced by kernels have been passed through the activation function. In CNN, nonlinear activation functions like sigmoid, rectified linear unit, hyperbolic tangent, max out, softmax have been used. Output dimension of convolution layer is calculated as

$$(H,W) = \begin{cases} \left(\left[\dfrac{n_h + 2p - f}{s} + +1\right], \left[\dfrac{n_w + 2p - f}{s} + 1\right]\right); & if \ stride > 0 \\ (n_h + 2p - f, n_w + 2p - f); & if \ stride = 0 \end{cases} \tag{6.2}$$

Here (H, W) is output feature map height and width, n_h is input height, n_w is input width, f is filter size, p is padding, and s is stride.

- Fully Connected Layer: In it, all input neurons are connected to every output neuron of the layer. It is mostly used as the last layer for classification tasks. It takes flatten output of previous convolution and pooling layers. Flattening is an operation to convert two-dimensional input to single-dimensional vector format. This flattened input has been passed through fully connected layers to capture global information required for classification.

- Pooling Layer: It is used to scale down the size of the input. It divides the input into different blocks. Based on the selected pooling type, the operation has been performed on blocks, and values have been returned. For each feature map, separate pooled maps have been generated. It makes network local translation invariance. It does not have any learnable parameter. Mostly max pooling and average pooling have been used. In average pooling, the average features of each block have been used in the pooled map. In max pooling, the maximum feature value of each block has been used.

6.3.4 Different CNN Architectures

- LeNet: It has been introduced by Lecun [5]. It consists of seven layers, excluding input. It takes the input of 32×32 pixels. It used two convolutional layers, two subsampling layers, and two fully connected layers. The first convolutional layer takes the input of 32×32×1 and, using 5×5 filters, produced an output of six feature maps of size 28×28. Input image has been normalized to −1 and 1.175. This makes mean input 0 and variance 1. The subsampling layer has a 2×2 receptive field and with no overlap. Figure 6.4 shows LeNet-5 architecture. Here Cx is a convolutional layer, Sx is a subsampling layer, and Fx is a fully connected layer. In this, x represents the location of the layer in the network.

- AlexNet [10]: It produced a state of art results in ImageNet Large-Scale Visual Recognition Challenge 2012. It has more number of filters per layer compared to LeNet. It used a parallel convolutional path to process input. Cross-connections were there between these paths. It used a Rectified Linear Unit (ReLU) in place of the hyperbolic tangent activation function. ReLU makes the network run faster. It used Local Response Normalization (LRN), which helps in faster convergence. LRN is defined as

$$b_{x,y}^{i} = \frac{a_{x,y}^{i}}{\left(k + \alpha \displaystyle\sum_{j=max\left(0, i-\frac{n}{2}\right)}^{min\left(N-1, i+\frac{n}{2}\right)} \left(a_{x,y}^{j}\right)^2\right)} \tag{6.3}$$

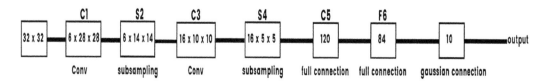

FIGURE 6.4   LeNet-5 Architecture.

Here a is the activity of the neuron, i is ith in the layer, $a_{x,y}^i$ represents the activity of kernel, N is a total number of kernels neuron calculated after applying kernel i at (x, y) location, and k, n, α, β are hyperparameters. Used value of these hyperparameters are $k = 2$, $n = 5$, $= 10^{-4}$, and $= 0.75$. It used overlapping pooling using a stride less than kernel size. Overlapping pooling makes the network difficult to overfit. It uses a dropout of 0.5 in the first two fully connected layers. Dropout is also used to reduce overfitting. It uses data augmentation of image translation, horizontal reflection, and altering the intensities of the RGB channels.

- AlexNet [10] architecture has been shown in [6]. Input size of 224×224×3 has been processed through 11×11×3 sizes 96 kernels and produce 96 feature maps of size 55 55. These feature maps have been passed through 3×3 overlapping pooling with stride two and produced 27×27×96 size feature maps. This pooled output has been passed through LRN. The same has been repeated for the other four convolutional layers with the increased number of feature maps. A total of eight layers with five convolutional layers and three fully connected layers have been used. The final layer produces an output of 1,000 neurons for 1,000 classes (Figure 6.5).

- VGGNet [11]: It came second in the ILSVRC 2014 challenge. It uses 16 convolutional layers. It uses a filter size of 3×3 only, but a large number of feature maps have been used. It uses ReLU activation after each convolutional layer. A large number of feature maps make it computationally expensive. In Figure 6.6 architecture and the various variant of VGGNet [11] are shown. In all variants input of size 224×224×3 has been passed. It used softmax activation in the final layer.

- GoogleNet [12]: It was the winner of the ILSVR 2014 challenge. It proposed inception blocks. In these blocks, kernel of different sizes have been used to have a variable receptive field. It consists of 22 layers. It has far fewer trainable parameters than VGGNet [11]. Inception blocks have been shown in Figure 6.7. Here input has been passed through the bottleneck layer of 1×1 before passing to large kernels. This helps in reducing the computational complexity of the network. Inception blocks help in capturing multi-context information.

ConvNet Configuration					
A	A-LRN	B	C	D	E
11 weight layers	11 weight layers	13 weight layers	16 weight layers	16 weight layers	19 weight layers
input (224 × 224 RGB image)					
conv3-64	conv3-64 **LRN**	conv3-64 **conv3-64**	conv3-64 conv3-64	conv3-64 conv3-64	conv3-64 conv3-64
maxpool					
conv3-128	conv3-128	conv3-128 **conv3-128**	conv3-128 conv3-128	conv3-128 conv3-128	conv3-128 conv3-128
maxpool					
conv3-256 conv3-256	conv3-256 conv3-256	conv3-256 conv3-256	conv3-256 conv3-256 **conv1-256**	conv3-256 conv3-256 **conv3-256**	conv3-256 conv3-256 conv3-256 **conv3-256**
maxpool					
conv3-512 conv3-512	conv3-512 conv3-512	conv3-512 conv3-512	conv3-512 conv3-512 **conv1-512**	conv3-512 conv3-512 **conv3-512**	conv3-512 conv3-512 conv3-512 **conv3-512**
maxpool					
conv3-512 conv3-512	conv3-512 conv3-512	conv3-512 conv3-512	conv3-512 conv3-512 **conv1-512**	conv3-512 conv3-512 **conv3-512**	conv3-512 conv3-512 conv3-512 **conv3-512**
maxpool					
FC-4096					
FC-4096					
FC-1000					
soft-max					

FIGURE 6.5 AlexNet architecture.

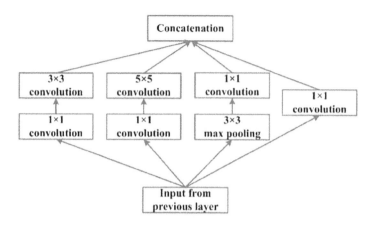

FIGURE 6.6   VGGNet architecture.

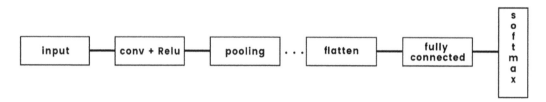

FIGURE 6.7   CNN architecture.

- ResNet [13]: It won the ILSVRC 2015 competition. It introduced a residual connection that enables the creation of a deeper neural network without the vanishing gradient problem. The residual connection can be defined as

$$R_i = H\left(R_{i-1}\right) + R_{i-1} \tag{6.4}$$

Here R_i is the output of ith layer, and H is an operation performed on the output of $(i-1)^{th}$ layer. Different variants of ResNet [13, 14] with the various number of layers have been proposed. Zagoruvko [15] introduced Wide Residual Networks. This increased the width of the network and reduced the depth of it.

6.3.5 Loss Functions and Optimization

The loss function represents the difference between the expected and actual output of the algorithm. It represents the ability of the algorithm to learn from the input. The high loss value represents the poorer performance of the algorithm. Loss function depends on the type and complexity of problems.

6.3.5.1 Loss Functions for Regression

- Mean Absolute Error (MAE) Loss: It is also known as L1 loss. It is the mean of absolute difference between ground-truth (y) and predicted output ($f(x)$). It is more robust to wrong ground-truth values than squared error.

$$MAE = \frac{\sum_i^N |y - f(x)|}{N} \tag{6.5}$$

- Mean Square Error (MSE) Loss: It is also known as L2 loss. It is the mean of the difference between the square of ground-truth (y) and predicted output ($f(x)$). It is useful when a large error value is undesirable.

$$MSE = \frac{\sum_i^N |y^2 - f(x)^2|}{N} \tag{6.6}$$

- Huber Loss (HL): It takes good properties of MAE and MSE. For smaller MAE, the loss value is calculated as MSE. For large errors, the loss value is de ned by MAE. The additional parameter has been used to decide which error to be used.

$$Huber\ Loss = \begin{cases} |y^2 - f(x)^2|, \ if \ |y - f(x)| < \delta \\ \delta|y - f(x)| - \dfrac{\delta^2}{2}, \ otherwise \end{cases} \tag{6.7}$$

6.3.5.2 Loss Functions for Classification

- Cross-Entropy Loss: Cross-entropy is defined as the difference between two probability distributions for a random variable. It showcases the similarity between the softmax of output distribution with one hot encoded input.

$$CE = -\sum_i t_i \log\left(f(s)_i\right) \tag{6.8}$$

Here t_i is a class vector, and $f(s)_i$ is softmax output for ith input.

- KL-Divergence: It is a measure of the difference between two probability distributions. In this loss, we use input ground-truth and predicted output as two probability distributions. It is mostly used in auto-encoders.

$$KL(p|q) = -\sum_i p_i \log\left(\frac{p_i}{q_i}\right) \tag{6.9}$$

Here p, q are two probability distributions.

- Hinge Loss: It is mostly used in SVM. For ground-truth y and predicted output $f(x)$, it is defined as

$$HL = max\left(0, 1 - yf(x)\right) \tag{6.10}$$

6.3.5.3 Optimization

In neural network training, its weights have been changed to find the global minimum of the loss function. These weight change has been decided by optimization function. For supervised learning, the goal of the optimization algorithm is to minimize the loss. For loss function L, N samples, ground-truth y, predicted output f(x;), this can be expressed as

$$min_\theta \frac{\sum_i^N L(y^i - f(x^i, \theta)}{N} + \lambda\|\theta^2\| \tag{6.11}$$

In the above equation, the second part represents regularization. It is controlled by a compromising factor. Regularization has been added to avoid overfitting. Sun [16] has discussed optimization algorithm categorization based on gradient information. Optimization algorithms can be categorized into three categories:

- First-order optimization: These methods are mostly based on gradient descent. It updates parameters in the reverse direction of the gradient to reach minimum loss. Some variant of gradient descents methods are

    Batch gradient descent (BGD): It used complete data in each training iteration. The gradient of the objective function is calculated for full data. It is very slow compared to other methods. A larger dataset can not be put in memory completely; hence it cannot use batch gradient descent. For each parameter θ, learning rate η, objective function $L(\theta,x,y)$, parameter updated using complete dataset as

$$\theta = \theta - \eta \frac{dL(\theta)}{d\theta} \qquad (6.12)$$

    For N samples and each input dimension D, each iteration requires O(ND) amount of computation complexity.

    Stochastic Gradient Descent (SGD): It used one random sample for the calculation of the gradient in each iteration. It makes SGD faster. It can be used for online training for larger datasets. It is having a computation complexity of O(D) for each iteration, where D is the input dimension. It showed fluctuation during convergence but perform better than SGD. It can be represented as

$$\theta = \theta - \eta \frac{dL(\theta,x^i,y^i)}{d\theta} \qquad (6.13)$$

    Here $L(\theta, x^i, y^i)$ is the loss function for randomly selected sample x^i and its ground truth y^i.

    Mini-batch gradient descent: It combined advantage of BGD and SGD. It used complete batch samples for gradient calculation. For a batch size of b, it can be represented as

$$\theta = \theta - \eta \frac{dL(\theta,x^{(i:i+b)},y^{(i:i+b)})}{d\theta} \qquad (6.14)$$

    Here $x^{(i:i+b)}$ is all samples and $y^{(i:i+b)}$ is all ground truth in batch. Gradient descent methods find difficulty on local minima and saddle points. It has a computational complexity of O(bD) for each iteration.

    Momentum [17] has been used to overcome these difficulties. Other methods as Adagrad [18], Adadelta [19], Adam [20] have been developed.

- Second-order optimization: Newton's method can be used for second-order optimization. It used the inverse of hassian matrix, which contains second-order partial derivatives. The computation of the inverse of hassian matrix is very computationally intensive. In practice, quasi-Newton methods have been used for second-order optimization. These methods approximate the inverse hassian matrix. L-BFGS [21] is an example of such an approach.

6.4 Deep Learning Applications in Remote Sensing

6.4.1 Image Scene Classification

Image scene classification is a problem to associate input image scenes with predefined classes. In remote sensing, image classification has important applications [22]. It can be utilized for urban planning [23, 24], environment monitoring [25–27], and vegetation mapping [28–30]. The earlier method relies on pixel level [31, 32] and object-based [31, 32] classification. CNN-based approaches [2, 33–35] have

TABLE 6.1

Remote Sensing Benchmark Datasets for Image Scene Classification

Dataset	Classes	Total Images	Image Size	Year
UC-Merced dataset [39]	21	2100	256×256 1	2010
WHU-RS19 [40]	7	1005	600×600	2012
Brazilian Coffee Scene [41]	2	2876	64×64	2015
SAT4/6 [42]	4/6	500000/405000	28×28	2015
SIRI-WHU [43]	12	2400	200×200	2016
EuroSAT [44]	10	27000	64×64	2019
BigEarthNet [45]	19	590326	120×120	2019

produced a state of art results in remote sensing scene classification. Advanced deep learning methods based on GANs [36] also emerged for image scene classification [37, 38, 46]. In Table 6.1, we have listed publicly available datasets for image scene classification in remote sensing. These datasets contain object ranges from 2 to 16 classes.

6.4.2 Object Detection

Object detection is a combination of classification and localization problems. It returned class probability of present predefined object class in input and spatial location of the same object. For that, the input is divided into multiple regions, and each part has been classified. In DL, object detection algorithms can be grouped into two categories, as shown in Figure 6.8. These categories are single-stage detector and multi-stage detector.

In remote sensing, Aksoy [47] has proposed Gaussian mixture models (GMMs) for heterogeneous compound structure detection. Bai [48] has used a ranking support vector machine (SVM) for object detection. Cheng utilized a multi-scale feature pyramid of the histogram of oriented gradients (HOG). Han [49] used a deep Boltzmann machine for feature encoding. Weakly supervised learning has been used for object detection.

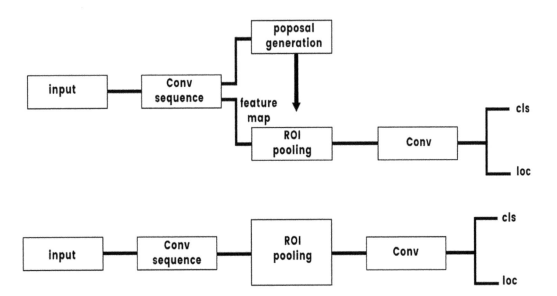

FIGURE 6.8 DL-based object detection algorithms approach.

In DL approaches, CNN-based feature extraction [50–52] is used for object detection and localization. Cheng [53] proposed rotation-invariant CNN (RICNN) for object detection. RICNN introduced a rotation-invariant layer. Deng [54] used region-based convolutional neural networks (R-CNNs) for fast vehicle detection on aerial images. Long [54] introduced unsupervised score-based bounding box regression for object detection. Zhong [55] utilized position-sensitive balancing for better region proposal generation. In Table 6.2, details of benchmark datasets for object detection have been provided.

6.4.3 Semantic Segmentation

The semantic segmentation objective is to classify each pixel into predefined classes. Long [63] has used a fully convolutional network for semantic segmentation. It used backbone to VGGNet [11]. Chen introduces deeplab [64], where they have used atorous convolution for semantic segmentation. ParseNet [65] used global pooling to capture global context. Zho [66] used pyramid pooling module. It pools features from different scales. Various benchmark dataset details have been included in Table 6.3.

6.4.4 Object Tracking

Szottka [70] uses particle-based methods for long-term vehicle tracking using airborne data. Fang [71] proposed Recurrent Autoregressive Network (RAN), which combined appearance and motion information for object tracking. Kim [72] used bilinear LSTM for multi-object tracking. Henschel [73] proposed to model the tracking problem as a weighted graph labeling problem. They fused two detectors in the tracking problem. In Table 6.4, details of the publicly available dataset for object tracking have been shown.

TABLE 6.2
Remote Sensing Benchmark Datasets for Object Detection

Dataset	Classes	Total Images	Instance	Image Width	Year
NWPU VHR-10 [56]	10	800	3,775	1,000	2014
VEDAI [57]	9	1,210	3,640	1,024	2015
UCAS-AOD [58]	2	910	6,020	1,280	2015
DLR3K Vehicle [59]	2	20	14,235	5,616	2015
HRSC2016 [60]	1	1,070	2,976	1,000	2016
DOTA [61]	15	2,806	188,282	800–4,000	2017
DIOR [62]	20	23,463	192,472	800	2018

TABLE 6.3
Remote Sensing Benchmark Datasets for Semantic Segmentation

Dataset	Classes	Total Images	Instance	Image Width	Year
INRIA [67]	10	800	3,775	1,000	2014
WHU [68]	9	1,210	3,640	1,024	2015
DeepGlobe [69]	2	910	6,020	1,280	2015
ISPRS 2D	2	20	14,235	5,616	2015

TABLE 6.4
Remote Sensing Benchmark Datasets for Aerial Object Tracking

Dataset	No of Frames	Tracking type	Spectrum
UTB [74]	15,000	Single object	Visible
UAV123 [75]	113,000	Single object	Visible
UAVDT [76]	80,000	Detection, Single and Multi objects	Visible
PTB-TIR [77]	30,000	Detection, Single and Multi objects	Thermal

6.5 Case Study

In this case study, we have taken the significant problem of road extraction from aerial imagery. Very high-resolution aerial imagery can be used for mapping road networks from time to time. Manual digitization is a very tedious task and not very suitable for aerial imagery, as data volume and data acquisition frequency are very high. Deep learning plays a significant role in automatic extraction tasks in such a scenario.

6.5.1 Objective

To extract the road network from very high-resolution aerial imagery using deep learning.

6.5.2 Dataset Preparation and Preprocessing

For VHR aerial imagery road class, there is no public data set available. To tackle this problem, we have manually created a training dataset. For this, we have used 200 images of aerial data from various survey locations. Each image has 4,000×3,000 pixels. We have used red, green, blue (RGB) bands of the images. These images have been manually annotated. Significant challenges in this step are:

- The selection of images for the dataset is a difficult task. We need to include as many as possible images with different locations, illumination conditions, geographic conditions. This is important to ensure generalized model development.
- We also need to ensure that selected images should have road features atleast 20% of the total area.

Our image size is massive, and it isn't easy to fit such a large image on GPU memory. GPU resources are mostly limited and scarce. After the selection of images and generation of ground-truth, we have divided these image pairs into patches of 128×128 pixels. A total of 7,000 random patches have been generated. From these patches, 4,200 patches have been used for training the neural network. A total 1,400 patches have been used for validation during training, and the remaining 1,400 patches have been used for testing.

6.5.3 Network Architecture Design

As there is no dataset available, no pretrained weight with road features on the aerial images is available either. To handle this, we have designed our own neural network architecture. We have created a neural network based on UNet [78] encoder-decoder architecture. UNet has proven very useful on small datasets and training from scratch without pretrained weight. Our network consists of an encoder and a decoder. We have used multi-scale encoder and decoder blocks. In each encoder block, we have used a sequence of convolution, batch normalization, ReLU, and downsampling layers. Max pooling as a downsampling layer has been used to reduce the size of the input and also produce multi-scale features. In decoder blocks, we have used upsampling layer and sequence of convolution, batch normalization, and ReLU layers. We have used skip connections between encoder and decoder to utilize features of earlier layers. Skip connection also resolves the vanishing gradient problem. It provides an alternate path for gradient flow. In Figure 6.9, network architecture based on UNet [78] is shown.

6.5.4 Hyperparameter Selection

In a neural network, there are various parameters that we need to define before training. There are static in nature and not learned during training. These are called *hyperparameters*. These parameters are learning rate, batch size, momentum, and number of epochs. There are other hyperparameters also, which are

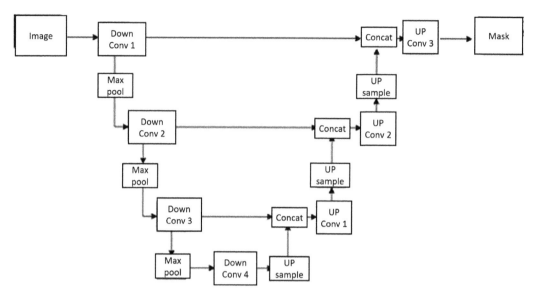

FIGURE 6.9 UNet architecture.

related to network architecture. These parameters are the number of the hidden layers, dropout, weight initialization, and activation function.

- Learning Rate: It determines the step size that needs to be moved during each iteration of training. It determines the speed at which the network can learn. A lower learning rate makes network take more time to learn, but it makes the network learn smoothly. A higher learning rate makes learning faster, but sometimes networks are unable to converge due to a large step size. We have used 0.0001 as the initial learning rate. This has been reduced after 50 epochs by a factor of half.

- Batch Size: It is the number of samples that the network uses in one iteration. After each iteration, the network updates its parameters. It also depends on the available GPU memory. We have used a batch size of 8.

- Momentum: It is a parameter with a value between 0 and 1. It is used to regulate the direction of learning toward a consistent gradient. It occurs by taking into consideration the running average of gradient value instead of the current value. We have used a value of 0.9.

- Number of Epochs: Epoch is defined as the number of iterations required to go through with all the training data. The number of epochs specifies how many times we have passed our complete data to the network for learning. We have set this parameter to 300.

- Dropout Rate: Dropout is a technique of randomly removing some neuron output during training. It increases the generalization power of the network. Dropout rate defines what percentage of neuron output from the previous layer needs to be ignored. We have used a dropout rate of 30%.

- Weight Initialization: It is the technique to initialize all weights of the network at the beginning of training. We have used uniform distribution for this.

- Activation Function: It is a nonlinear transformation applied to the output of layers. It introduces nonlinearity to the network. We have used a ReLU after each batch normalization layer. The sigmoid activation function has been used in the output layer of the network.

6.5.5 Inference Training

We have trained our network for 300 epochs on the Nvidia Quadro P4000 card. It has 8 GB of graphic memory. Our system has 16 GB of RAM. It takes around four days to complete training. We have used the Stochastic Gradient Descent (SGD) optimization algorithm.

6.5.6 Results

We have used Overall Accuracy (OA), mean intersection over union mIoU, to measure the performance of our model. Accuracy can be de ned as:

$$Accuracy = \frac{A + B}{A + B + C + D} \tag{6.15}$$

Here A is true positive, B is true negative, C is false positive, and D is false negative. Mean IoU is calculated as:

$$IoU = \frac{Area\ of\ Overlap}{Area\ of\ Union} = \frac{A}{A + B + C} \tag{6.16}$$

We have achieved 90.1% overall accuracy and a mean IoU of 74% for a testing dataset in our experiment. Comparative results of various methods have been shown in Table 6.5. In Figure 6.10, we have shown our output results.

TABLE 6.5

Output Comparison of Various Deep Semantic Segmentation Methods

	OA	mIoU
SegNet [79]	83.2	68.2
FCN [63]	88.5	71.1
UNet [78]	90.1	74

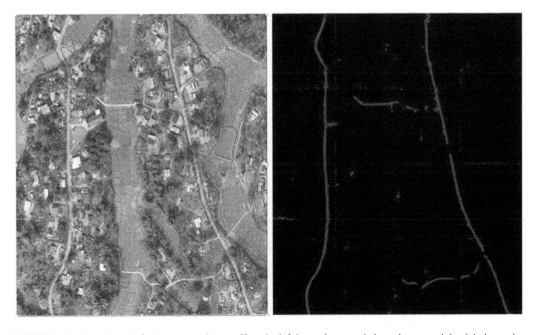

FIGURE 6.10 Mosaic result for the case study area. Here the left image is a mosaic input image, and the right image is a generated output road network map.

6.6 Challenges with DL on Remote Sensing Data

As mentioned by Ball [80], there are many challenges and opportunities in deep learning for remotely sensed data. These challenges are:

- Smaller Datasets: Large annotated dataset availability is a challenge in RS. DL algorithms rely heavily on the abundant availability of training data in large volumes. RS benchmark datasets mostly of few scenes size. It is difficult to produce a generalized solution from small datasets. Some benchmark dataset's performance results are nearly saturated using DL algorithms.

- Data augmentation has been used to generate more data from smaller datasets. Additional augmenting data have also been used in some cases to produce better results. New benchmark datasets have addressed this issue with bigger dataset size.

- Heterogeneous Data Source: RS data sources are heterogeneous in nature. These range from multispectral to synthetic aperture radar, LiDAR, aerial, multi-view, etc. Data properties, structure, and feature changes are based on the sensor of the satellite. Different DL algorithm approaches are required for different types of sensors.

One approach to tackle this problem is to train a different neural network for each data source type. Another solution is to train a single neural network with the fusion of multi-modal inputs from various sources and produce a generalized solution.

- DL Architecture and Hyperparameter Selection: DL architecture selection is a difficult task. The selection of different network components, number of layers, and sequence of different components in the network requires many experiments. Hyperparameter selection also affects model training and performance.

Various search algorithms have been used for the selection of hyperparameter selection of networks. Neural architecture search [81] has been used recently to generate network architecture automatically based on the problem.

- Issues with Transfer Learning: Applicability of transfer learning is an open question in RS because data properties change completely with the change in sensor type. Some successful attempts to utilize transfer learning have been mentioned in [82–84].

- Black Box Decoding: DL algorithms faced this issue. Solution explanation is very difficult in DL approaches.

- Resource Requirement: DL algorithms require huge GPU and memory resources. Without the availability of these, it is very difficult to train DL algorithms. Recently new research has been started on lite DL algorithms that can run on the system with minimal resources.

6.7 Conclusion

We have presented a brief discussion about DL techniques and applications in RS. The challenges and limitations of DL-based methods have been summarized. In remote sensing, the application of DL techniques is relatively new. As RS is always a data-intensive field, there is a lot of scope for DL algorithms development in RS data. We have also given details of publicly available datasets in RS. In the last section, we have included our case study for aerial road extraction using DL-based methods, and comparative results have been shown.

REFERENCES

1. Deep logic models. Integrating learning and reasoning with deep logic models. Available at: https://arxiv.org/abs/1901.04195 [accessed on 20 June 2021.

2. Kashif Sultan, Hazrat Ali, and Zhongshan Zhang. Big data perspective and challenges in next generation networks. *Future Internet, 10*:56, 2018.

3. G. Auda, M. Kamel, and H. Raafat. *Modular neural network architectures for classification.* In *Proceedings of International Conference on Neural Networks (ICNN'96), 2*:1279–1284, 1996.

4. Kunihiko Fukushima. Neocognitron: A hierarchical neural network capable of visual pattern recognition. *Neural Networks, 1*(2):119–130, 1988.

5. Y. Lecun, L. Bottou, Y. Bengio, and P. Haffner. Gradient-based learning applied to document recognition. *Proceedings of the IEEE, 86*(11):2278–2324, 1998.

6. Geoffrey E. Hinton, Simon Osindero, and Yee-Whye Teh. A fast learning algorithm for deep belief nets. *Neural Computation, 18*(7):1527–1554, 2006.

7. G. E. Hinton and R. R. Salakhutdinov. Reducing the dimensionality of data with neural networks. *Science, 313*(5786):504–507, 2006.

8. Xavier Glorot and Yoshua Bengio. *Understanding the difficulty of training deep feed-forward neural networks.* volume 9 of *Proceedings of Machine Learning Research*, pages 249–256, Chia Laguna Resort, Sardinia, Italy, 13–15 May 2010. JMLR Workshop and Conference Proceedings.

9. M. Ranzato, F. J. Huang, Y. Boureau, and Y. LeCun. *Unsupervised learning of invariant feature hierarchies with applications to object recognition.* In *2007 IEEE Conference on Computer Vision and Pattern Recognition*, pages 1–8, 2007.

10. Alex Krizhevsky, Ilya Sutskever, and Geoffrey E. Hinton. *Imagenet classification with deep convolutional neural networks.* NIPS'12, pages 1097–1105, Red Hook, NY, 2012. Curran Associates Inc.

11. Karen Simonyan and Andrew Zisserman. *Very deep convolutional networks for large-scale image recognition.* In *International Conference on Learning Representations*, 2015. https://arxiv.org/abs/1409.1556.

12. C. Szegedy, Wei Liu, Yangqing Jia, P. Sermanet, S. Reed, D. Anguelov, D. Erhan, V. Van-Houcke, and A. Rabinovich. *Going deeper with convolutions.* In *2015 IEEE Conference on Computer Vision and Pattern Recognition (CVPR)*, pages 1–9, 2015.

13. K. He, X. Zhang, S. Ren, and J. Sun. *Deep residual learning for image recognition.* In *2016 IEEE Conference on Computer Vision and Pattern Recognition (CVPR)*, pages 770–778, 2016.

14. Andreas Veit, Michael Wilber, and Serge Belongie. *Residual networks behave like ensem-bles of relatively shallow networks.* NIPS'16, page 550–558, Red Hook, NY, 2016. Curran Associates Inc.

15. Sergey Zagoruyko and Nikos Komodakis. Wide residual networks. In Edwin R. Hancock, Richard C. Wilson, and William A. P. Smith, editors, *Proceedings of the British Machine Vision Conference (BMVC)*, pages 87.1–87.12. BMVA Press, September 2016.

16. S. Sun, Z. Cao, H. Zhu, and J. Zhao. A survey of optimization methods from a machine learning perspective. *IEEE Transactions on Cybernetics, 50*(8):3668–3681, 2020.

17. Ning Qian. On the momentum term in gradient descent learning algorithms. *Neural Networks, 12*(1):145–151, 1999.

18. John Duchi, Elad Hazan, and Yoram Singer. Adaptive subgradient methods for online learning and stochastic optimization. *Journal of Machine Learning Research, 12*(null):2121–2159, 2011.

19. Matthew D. Zeiler. ADADELTA: an adaptive learning rate method. CoRR, abs/1212.5701, 2012.

20. Diederik P. Kingma and Jimmy Ba. *Adam: A method for stochastic optimization.* In Yoshua Bengio and Yann LeCun, editors, *3rd International Conference on Learning Rep-resentations, ICLR 2015*, San Diego, CA, May 7–9, 2015, Conference Track Proceedings, 2015.

21. Jorge Nocedal. Updating quasi newton matrices with limited storage. *Mathematics of Computation, 35*(151):951–958, 1980.

22. G. Cheng, X. Xie, J. Han, L. Guo, and G. Xia. Remote sensing image scene classification meets deep learning: Challenges, methods, benchmarks, and opportunities. *IEEE Journal of Selected Topics in Applied Earth Observations and Remote Sensing, 13*:3735–3756, 2020.

23. N. Longbotham, C. Chaapel, L. Bleiler, C. Padwick, W. J. Emery, and F. Pacifici. Very high resolution multiangle urban classification analysis. *IEEE Transactions on Geoscience and Remote Sensing, 50*(4):1155–1170, 2012.

24. Amin Tayyebi, Bryan Christopher Pijanowski, and Amir Hossein Tayyebi. An urban growth boundary model using neural networks, gis and radial parameterization: An ap-plication to Tehran, Iran. *Landscape and Urban Planning*, *100*(1):35–44, 2011.

25. T. Zhang and X. Huang. Monitoring of urban impervious surfaces using time series of high-resolution remote sensing images in rapidly urbanized areas: A case study of shenzhen. *IEEE Journal of Selected Topics in Applied Earth Observations and Remote Sensing*, *11*(8):2692–2708, 2018.

26. Xin Huang, Dawei Wen, Jiayi Li, and Rongjun Qin. Multi-level monitoring of subtle urban changes for the megacities of china using high-resolution multi-view satellite imagery. *Remote Sensing of Environment*, *196*:56–75, 2017.

27. Jun Li, Yanqiu Pei, Shaohua Zhao, Rulin Xiao, Xiao Sang, and Chengye Zhang. A review of remote sensing for environmental monitoring in China. *Remote Sensing*, *12*(7):1–25, 2020.

28. Xiaoxiao Li and Guofan Shao. Object-based urban vegetation mapping with high-resolution aerial photography as a single data source. *International Journal of Remote Sensing*, *34*(3):771–789, 2013.

29. Yichun Xie, Zongyao Sha, and Mei Yu. Remote sensing imagery in vegetation mapping: a review. *Journal of Plant Ecology*, *1*(1):9–23, 2008.

30. S.M. Hamylton, R.H. Morris, R.C. Carvalho, N. Roder, P. Barlow, K. Mills, and L. Wang. Evaluating techniques for mapping island vegetation from unmanned aerial vehicle (uav) images: Pixel classification, visual interpretation and machine learning approaches. *International Journal of Applied Earth Observation and Geoinformation*, *89*:102085, 2020.

31. T. Blaschke. Object based image analysis for remote sensing. *ISPRS Journal of Pho-togrammetry and Remote Sensing*, *65*(1):2–16, 2010.

32. Desheng Liu and Fan Xia. Assessing object-based classification: advantages and limita-tions. *Remote Sensing Letters*, *1*(4):187–194, 2010.

33. R. Minetto, M. Pamplona Segundo, and S. Sarkar. Hydra: An ensemble of convolutional neural networks for geospatial land classification. *IEEE Transactions on Geoscience and Remote Sensing*, *57*(9):6530–6541, 2019.

34. G. Cheng, C. Yang, X. Yao, L. Guo, and J. Han. When deep learning meets metric learning: Remote sensing image scene classification via learning discriminative cnns. *IEEE Transactions on Geoscience and Remote Sensing*, *56*(5):2811–2821, 2018.

35. Q. Wang, S. Liu, J. Chanussot, and X. Li. Scene classification with recurrent attention of vhr remote sensing images. *IEEE Transactions on Geoscience and Remote Sensing*, *57*(2):1155–1167, 2019.

36. Ian J. Goodfellow, Jean Pouget-Abadie, Mehdi Mirza, Bing Xu, David Warde-Farley, Sherjil Ozair, Aaron Courville, and Yoshua Bengio. *Generative adversarial nets. NIPS'14*, page 2672–2680, Cambridge, MA, 2014. MIT Press.

37. Y. Duan, X. Tao, M. Xu, C. Han, and J. Lu. *GAN-NL: Unsupervised representation learning for remote sensing image classification. In 2018 IEEE Global Conference on Signal and Information Processing (GlobalSIP)*, pages 375–379, 2018.

38. D. Lin, K. Fu, Y. Wang, G. Xu, and X. Sun. Marta gans: Unsupervised representation learning for remote sensing image classification. *IEEE Geoscience and Remote Sensing Letters*, *14*(11):2092–2096, 2017.

39. Yi Yang and Shawn Newsam. *Bag-of-visual-words and spatial extensions for land-use classification. GIS '10*, pages 270–279, New York, NY, 2010. Association for Computing Machinery.

40. Gui-Song Xia, Wen Yang, Julie Delon, Yann Gousseau, Hong Sun, and Henri Matre. Structural high-resolution satellite image indexing. International Archives of the Pho-togrammetry, Remote Sensing and Spatial Information Sciences – ISPRS Archives, 38, 07 2010.

41. O. A. B. Penatti, K. Nogueira, and J. A. dos Santos. *Do deep features generalize from everyday objects to remote sensing and aerial scenes domains? In 2015 IEEE Conference on Computer Vision and Pattern Recognition Workshops (CVPRW)*, pages 44–51, 2015.

42. Saikat Basu, Sangram Ganguly, Supratik Mukhopadhyay, Robert DiBiano, Manohar Karki, and Ramakrishna Nemani. *Deepsat – a learning framework for satellite imagery. Proceedings of the 23rd SIGSPATIAL International Conference on Advances in Geographic Information Systems*, November 2015, Article 37, pages 1–10.

43. Yan-Gang Zhao, Funing Zhong, and Min Zhang. Scene classification via latent dirichlet allocation using a hybrid generative/discriminative strategy for high spatial resolution remote sensing imagery. *Remote Sensing Letters*, *4*:07 2013.

44. P. Helber, B. Bischke, A. Dengel, and D. Borth. Eurosat: A novel dataset and deep learning benchmark for land use and land cover classification. *IEEE Journal of Selected Topics in Applied Earth Observations and Remote Sensing*, 12(7):2217–2226, 2019.

45. Gencer Sumbul, Marcela Charfuelan, Begum Demir, and Volker Markl. Bigearthnet: A large-scale benchmark archive for remote sensing image understanding. pages 5901–5904, 2019.

46. L. Jiao, F. Zhang, F. Liu, S. Yang, L. Li, Z. Feng, and R. Qu. A survey of deep learning-based object detection. *IEEE Access*, 7:128837–128868, 2019.

47. C. Ar and S. Aksoy. Detection of compound structures using a gaussian mixture model with spectral and spatial constraints. *IEEE Transactions on Geoscience and Remote Sensing*, 52(10):6627–6638, 2014.

48. X. Bai, H. Zhang, and J. Zhou. VHR object detection based on structural feature extraction and query expansion. *IEEE Transactions on Geoscience and Remote Sensing*, 52(10):6508–6520, 2014.

49. J. Han, D. Zhang, G. Cheng, L. Guo, and J. Ren. Object detection in optical remote sensing images based on weakly supervised learning and high-level feature learning. *IEEE Transactions on Geoscience and Remote Sensing*, 53(6):3325–3337, 2015.

50. Y. Long, Y. Gong, Z. Xiao, and Q. Liu. Accurate object localization in remote sensing images based on convolutional neural networks. *IEEE Transactions on Geoscience and Remote Sensing*, 55(5):2486–2498, 2017.

51. A. Salberg. *Detection of seals in remote sensing images using features extracted from deep convolutional neural networks*. In *2015 IEEE International Geoscience and Remote Sensing Symposium (IGARSS)*, pages 1893–1896, 2015.

52. I. Sevo and A. Avramovic. Convolutional neural network based automatic object detection on aerial images. *IEEE Geoscience and Remote Sensing Letters*, 13(5):740–744, 2016.

53. G. Cheng, P. Zhou, and J. Han. Learning rotation-invariant convolutional neural net-works for object detection in vhr optical remote sensing images. *IEEE Transactions on Geoscience and Remote Sensing*, 54(12):7405–7415, 2016.

54. Z. Deng, H. Sun, S. Zhou, J. Zhao, and H. Zou. Toward fast and accurate vehicle detection in aerial images using coupled region-based convolutional neural networks. *IEEE Journal of Selected Topics in Applied Earth Observations and Remote Sensing*, 10(8):3652–3664, 2017.

55. Yanfei Zhong, Xiaobing Han, and Liangpei Zhang. Multi-class geospatial object detec-tion based on a position-sensitive balancing framework for high spatial resolution remote sensing imagery. *ISPRS Journal of Photogrammetry and Remote Sensing*, 138:281–294, 2018.

56. Gong Cheng and Junwei Han. A survey on object detection in optical remote sensing images. *ISPRS Journal of Photogrammetry and Remote Sensing*, 117:11–28, 2016.

57. Sebastien Razakarivony and Frederic Jurie. Vehicle detection in aerial imagery: A small target detection benchmark. *Journal of Visual Communication and Image Representation*, 34:187–203, 2016.

58. H. Zhu, X. Chen, W. Dai, K. Fu, Q. Ye, and J. Jiao. *Orientation robust object detection in aerial images using deep convolutional neural network*. In *2015 IEEE International Conference on Image Processing (ICIP)*, pages 3735–3739, 2015.

59. K. Liu and G. Mattyus. Fast multiclass vehicle detection on aerial images. *IEEE Geo-science and Remote Sensing Letters*, 12(9):1938–1942, 2015.

60. Z. Liu, H. Wang, L. Weng, and Y. Yang. Ship rotated bounding box space for ship extraction from high-resolution optical satellite images with complex backgrounds. *IEEE Geoscience and Remote Sensing Letters*, 13(8):1074–1078, 2016.

61. G. Xia, X. Bai, J. Ding, Z. Zhu, S. Belongie, J. Luo, M. Datcu, M. Pelillo, and L. Zhang. *Dota: A large-scale dataset for object detection in aerial images*. In *2018 IEEE/CVF Conference on Computer Vision and Pattern Recognition*, pages 3974–3983, 2018.

62. Ke Li, Gang Wan, Gong Cheng, Liqiu Meng, and Junwei Han. Object detection in optical remote sens-ing images: A survey and a new benchmark. *ISPRS Journal of Photogram-metry and Remote Sensing*, 159:296–307, 2020.

63. J. Long, E. Shelhamer, and T. Darrell. *Fully convolutional networks for semantic segmen-tation*. In *2015 IEEE Conference on Computer Vision and Pattern Recognition (CVPR)*, pages 3431–3440, 2015.

64. L. Chen, G. Papandreou, I. Kokkinos, K. Murphy, and A. L. Yuille. Deeplab: Semantic image segmenta-tion with deep convolutional nets, atrous convolution, and fully connected crfs. *IEEE Transactions on Pattern Analysis and Machine Intelligence*, 40(4):834–848, 2018.

65. Wei Liu, Andrew Rabinovich, and Alexander C. Berg. ParseNet: Looking Wider to See Better. arXiv e-prints, page arXiv:1506.04579, June 2015.
66. H. Zhao, J. Shi, X. Qi, X. Wang, and J. Jia. *Pyramid scene parsing network*. In *2017 IEEE Conference on Computer Vision and Pattern Recognition (CVPR)*, pages 6230–6239, 2017.
67. E. Maggiori, Y. Tarabalka, G. Charpiat, and P. Alliez. *Can semantic labeling methods generalize to any city? The INRIA aerial image labeling benchmark*. In *2017 IEEE International Geoscience and Remote Sensing Symposium (IGARSS)*, pages 3226–3229, 2017.
68. S. Ji, S. Wei, and M. Lu. Fully convolutional networks for multisource building extraction from an open aerial and satellite imagery data set. *IEEE Transactions on Geoscience and Remote Sensing*, 57(1):574–586, 2019.
69. I. Demir, K. Koperski, D. Lindenbaum, G. Pang, J. Huang, S. Basu, F. Hughes, D. Tuia, and R. Raskar. *Deepglobe 2018: A challenge to parse the earth through satellite images*. In *2018 IEEE/CVF Conference on Computer Vision and Pattern Recognition Workshops (CVPRW)*, pages 172–179, 2018.
70. I. Szottka and M. Butenuth. *Tracking multiple vehicles in airborne image sequences of complex urban environments*. In *2011 Joint Urban Remote Sensing Event*, pages 13–16, 2011.
71. K. Fang, Y. Xiang, X. Li, and S. Savarese. *Recurrent autoregressive networks for online multi-object tracking*. In *2018 IEEE Winter Conference on Applications of Computer Vision (WACV)*, pages 466–475, 2018.
72. Chanho Kim, F. Li, and J. Rehg. *Multi-object tracking with neural gating using bilinear lstm*. In *ECCV*, 2018.
73. R. Henschel, L. Leal-Taixe, D. Cremers, and B. Rosenhahn. *Fusion of head and full-body detectors for multi-object tracking*. In *2018 IEEE/CVF Conference on Computer Vision and Pattern Recognition Workshops (CVPRW)*, pages 1509–150909, 2018.
74. Siyi Li and Dit-Yan Yeung. *Visual object tracking for unmanned aerial vehicles: A benchmark and new motion models*. In *AAAI*, 2017.
75. Matthias Mueller, Neil Smith, and Bernard Ghanem. A benchmark and simulator for UAV tracking. *European Conference on Computer Vision*, volume *9905*, pp. 445–461, 2016.
76. Hongyang Yu, Guorong Li, Weigang Zhang, Dawei Du, Qi Tian, and Nicu Sebe. The unmanned aerial vehicle benchmark: Object detection, tracking and baseline. *International Journal of Computer Vision*, *12*:201.
77. Q. Liu, Z. He, X. Li, and Y. Zheng. PTB-TIR: A thermal infrared pedestrian tracking benchmark. *IEEE Transactions on Multimedia*, 22(3):666–675, 2020.
78. Olaf Ronneberger, Philipp Fischer, and Thomas Brox. U-net: Convolutional networks for biomedicatl image segmentation. In: N. Navab, J. Hornegger, W. Wells, and A. Frangi (eds), *Medical Image Computing and Computer-Assisted Intervention – MICCAI 2015*. MICCAI 2015. Lecture Notes in Computer Science, volume *9351*. Cham, Switzerland: Springer. https://doi.org/10.1007/978-3-319-24574-4_28.
79. Vijay Badrinarayanan, Alex Kendall, and Roberto Cipolla. *Segnet: A deep convolutional encoder-decoder architecture for image segmentation*. *IEEE Transactions on Pattern Analysis and Machine Intelligence*, 2017:2481–2495.
80. John E. Ball, Derek T. Anderson, and Chee Seng Chan Sr. Comprehensive survey of deep learning in remote sensing: theories, tools, and challenges for the community. *Journal of Applied Remote Sensing*, *11*(4):1–54, 2017.
81. Thomas Elsken, Jan Hendrik Metzen, and Frank Hutter. Neural architecture search: A survey. *Journal of Machine Learning Research*, 20(55):1–21, 2019.
82. J. Yang, Y. Zhao, J. C. Chan, and C. Yi. *Hyperspectral image classification using two-channel deep convolutional neural network*. In *2016 IEEE International Geoscience and Remote Sensing Symposium (IGARSS)*, pages 5079–5082, 2016.
83. M. Iftene, Q. Liu, and Y. Wang. Very high resolution images classification by fine tuning deep convolutional neural networks. In Charles M. Falco and Xudong Jiang, editors, *Eighth International Conference on Digital Image Processing (ICDIP 2016)*, volume *10033*, pages 464–468. International Society for Optics and Photonics, SPIE, 2016.
84. Esam Othman, Yakoub Bazi, Naif Alajlan, Haikel Alhichri, and Farid Melgani. Using convolutional features and a sparse autoencoder for land-use scene classification. *International Journal of Remote Sensing*, *37*(10):2149–2167, 2016.

7

Improved Combinatorial Algorithms Test for Pairwise Testing Used for Testing Data Generation in Big Data Applications

Deepa Gupta and Lavanya Sharma
Amity Institute of Information Technology, Amity University, Noida, India

CONTENTS

7.1 Introduction

The most significant characteristic of big data is massive volumes of data. Industries use and deal with a huge amount of data, and the sizes of datasets often reach into terabytes. With this vast increase in data, processing speed suffers, and some datasets may take days or even weeks to be processed effectively. Therefore, smaller yet effective methods need to be found for testing while developing big data applications. Pairwise testing, also known as all-pairs testing, is a way to test the software using a combinatorial method, which allows all the possible different combinations of involved parameters to be tested. For example, for a small software item with 12 input fields and 12 adapted functionalities for every input field, there will be total 144 combinations number of possible inputs would be tested. Whenever the comprehensive testing approach is used, it is not possible to test all combinations [1–4]. The International Software Testing Qualifications Board defines pairwise testing as the black box testing which executes all the possible combinations of each pair of input parameters taken. Whenever we want the applications involving multiple parameters, pairwise testing is very helpful. For every pair of input parameters that have been taken by the user all distinct possible combinations is being tested [5–16].

A common type of big data applications is ETL: extraction, transformation, and load. Data is first extracted from various sources. These data may be in various structures, which are later converted into a common structural format and finally loaded for customers to view. For this process ETL applications

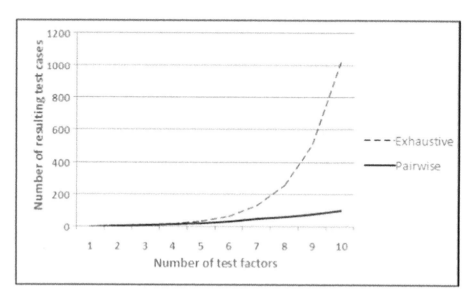

FIGURE 7.1   Enhanced lot of pairwise and entire tests along with the several of test levels.

are created, with the use of which input domain models (IDMs) are created, to which we will apply our improved algorithm for pairwise testing. It is difficult to write test cases for every non-trivial function. This is the same techniques as an equivalence partition in the boundary-value analysis [1]. This helps convert it to an oversized range to check the levels using a far smaller bunch with comparable defect-detection power. Whenever the program falls below test (SUT), this points to it being power-driven by variation of such factors, and complete testing once more becomes useless. Over the years, several com-binatorial methods have been developed to assist the testers with selection of subsets of input assortment that might be increasing the chance of detection of the defects. These methods include a random test, each-choice and base-option, randomly [17], and ultimately T-wise test. According to this intelligent test phase, all of them are equally important (Figure 7.1).

Pairwise is a Ruby-based tool for selecting a smaller number of test input combinations, using pair-wise generation rather than exhaustively testing all possible permutations.

This chapter introduces all pairs of tested element levels. In testing of a compiler, M and l are repre-sented with the use of orthogonal arrays [18]. Tatsumi, in his paper on a test suit framework network employed in Fujitsu Ltd [4], points out two standards of making the test arrays: (1) with all assortments coated precisely within the range of times the same (this is known as orthogonal arrays) or (2) each assortment coated at least once. According to shimokawa and the satoh one should accumulate and utilize the knowledge which is required in case of designing a test case [5]. Then extract test factors based on external specifications which will result in automatic generation of test case on the design experiment. According to Smith et al., Critical thinking and empirical analysis should be used while selecting the test cases which yields in more accurate results [3, 17–22].

7.2 Literature Review

Liu [23] proposed an effective method for performance testing in big data applications. In this paper the author provided test goal analysis, test design, and load design for applications of big data. The proposed framework by Li, Nan, et al. [24] used space partition testing to create a small dataset out of a large one for testing software that uses Hadoop-based big data technique. It also addressed three problems encoun-tered: huge time consumption while processing big data; validation of data at every transition point that is transferred and transformed to various services; and the issue of validating the data. Dipti and Alencar [25] focused on the present research scenario on how big data and its applications can be made using the software development project life cycle. Chandrasekaran et al. [26] used the software "weka," which is an open-source software for an experiment that applies combinatorial testing to five data-mining algorithms. Qi, Guanqiu, et al. [27] proposed a system that finds fault in a reciprocating system using machine learning techniques, which is used in the petroleum industry. Potential faults are identified and placed into two categories. One is used for testing purposes and the other is used for training purposes. The results clearly show that more accurate results are obtained with an accuracy of 80% and faults are easily diagnosed. Gupta et al. [28] also uses two algorithms, namely cuckoo algorithm and genetic algorithm, to generate small datasets from the big datasets without compromising quality. The method uses pairwise test case generation methodology.

7.3 Combinatorial Testing

Combinatorial testing is the way of testing in which not the whole data is tested, but instead only those parameters we consider needing to be tested. This type of testing is also called t-way testing – because t out of n parameters are tested.

Pairwise testing is a form of combinatorial testing, as here two out of n parameters are tested. These two parameters are considered a single input, which is why combinatorial testing reduces a lot of time and the number of tests that go into testing.

Combinatorial testing is very effective, as most of the bugs are spotted in pairwise testing itself. When we increase the number of parameters to be tested to three, we find the remaining bugs, and by the time we increase it to five, we have found almost every bug that is present in the software.

Here is an example of how this works. We are testing three parameters, and these parameters can have two values, 1 and 0. The test set that was created for these parameters for pairwise testing, which is a form of combinatorial testing, is given below.

Test parameter for two of three parameters $(p_1, p_2, p_3) = \{(0,0,0), (0,1,1), (1,0,1), (1,1,0)\}$

If this same test were to be performed by exhaustive testing where all the parameters are tested, then the test set would have been $2^3 = 8$; here we can see how pairwise testing has cut down the test set by half.

In real life, the number of parameters is much higher than three. One example is that for pairwise test set for 10 Boolean parameters, as few as 13 sets are required, whereas the exhaustive set requires $2^{10} = 1,024$ test sets.

7.3.1 Positives of Combinational Testing

The test set becomes finite and very small depending on the parameters we choose to be tested. If we choose more parameters to be tested, the chances of errors increase. One study by NASA showed that 67% of bugs are discovered when just one parameter is tested, 93% of bugs are discovered with two parameters tested, and 98% of bugs are discovered when three parameters are tested. Hence almost all the bugs can be fixed by performing two-way and three-way testing.

7.4 Applications of AI in Software Testing

Though AI has grown in many sectors of technology, when it comes to software testing, the field is still untouched by AI. Artificial intelligence can use the big data to select the parameters that have to be tested on [29, 30]. The big data, when cleaned, processed, and made uniform for software testing, can be used to select the edge cases and the relevant test cases where the software might break or give an unexpected output [31, 32]. Artificial intelligence will be used to generate test cases and parameterization, one of the best ways to create a test case that will be run across the software and then followed by pairwise testing, generating the results and helping remove the bugs [29–33].

The AI model will be used to understand where the software will fail, and it will finally remove the bugs by telling the developer where the bugs are likely to arise. AI will get the data from big data applications, which will be used to as a reference from the past mistakes in the similar applications bugs, preventing developers from creating the same bugs again [32, 33].

7.5 Big Data and Big Data Applications

Big data as a name is a reference to high-volume data that is ever expanding in our modern world. With the incoming Internet of Things (IoT) and increase in the number of clients, the volume of data has turned into higher orders of magnitude. With the volume of data being so huge, it becomes important to come up with efficient and reliable ways to process it. This data is being used for all sorts of things such as research, medical testing, and software testing.

One of the most common big data applications is ETL. The three functions comprising it – extract, transform, and load – perform specific roles when it comes to big data management.

   a. Extract: Selection, cleansing, scrubbing
   b. Transform: Splitting, joining, conversion
   c. Load: Initial load, incremental load, full refresh

7.5.1 Significance of Applications

The ETL applications are the first step in making sense of the big data. The raw data from where all the data is extracted is known as the source, and the final database where all the processed data is loaded is called the target. Even the data that is loaded in the target after being worked on by the ETL applications is of huge volume, so testing that data can still be very time consuming. Here we will discuss how the testing of this processed data can be done more efficiently.

Big data testing is done in two stages, the first stage being data preparation and second stage being data generation. The ETL applications make the data more sensible than the raw data, and now this processed data is used to create a data that will be required for testing.

7.5.2 Models of the Pairwise Testing

There are the three different principles designed by the PICT: the primary one is speed of generation; the second is simple use; and third is the core engine extensibility. PICT gives testers a lot of control over the way in which tests are generated, it raises the level of modeling abstraction, and makes the pairwise generation convenient and usable (see Figure 7.2). The tool generates a pairwise test array ($t = 2$) by

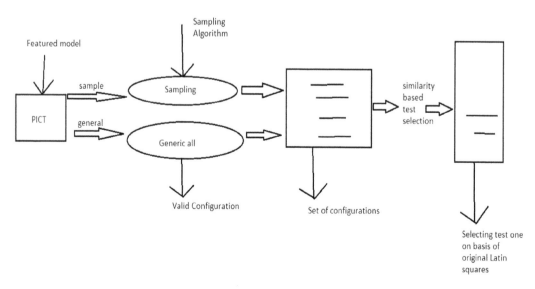

FIGURE 7.2 Improved Algorithm.

Task	AETG [14]	PairTest [21]	TConfig [24]	CTS [12]	Jenny [1]	DDA [8]	AllPairs [1]	PICT
3^4	9	9	9	9	11	?	9	9
3^{13}	15	17	15	15	18	18	17	18
$4^{15}3^{17}2^{29}$	41	34	40	39	38	35	34	37
$4^1 3^{39} 2^{35}$	28	26	30	29	28	27	26	27
2^{100}	10	15	14	10	16	15	14	15
10^{20}	180	212	231	210	193	201	197	210

DIAGRAM 7.1   PICT's generation efficiency compared with other tools.

```
Type:               Single, Spanned, Striped, Mirror, RAID-5
Size:               10, 100, 1000, 10000, 40000
Format method:      Quick, Slow
File system:        FAT, FAT32, NTFS
Cluster size:       512, 1024, 2048, 4096, 8192, 16384
Compression:        On, Off
```

DIAGRAM 7.2   Parameters for volume creation and formatting.

default. It employs a simple yet effective core generation algorithm which has separate preparation and generation phases. Diagram 7.2 shows the example of generation of test cases for volume creation and data formatting, with an easy model employed.

7.5.3 Test Case Generation Engine

There are two different test generation methods in PICT: the primary one is preparation and the second is generation. The test case generation engine was designed with usability, flexibility and speed in mind. The collection of all values to be coated is emulated during the interaction with the parameter structure. The tested device is easier to malfunction under more complex scenarios. Besides the coverage, complexity is another important factor which is covered by using the above-mentioned technique.

While using test generation engine percentage of test cases with higher value of complexity index is also increased. In the following if efficiency needs to be increased, then the process for the same is to consider a test suite with high complexity index which uses the combined effect of pairwise approach. The second approach used is to avoid the test suite scenarios that are impractical means factors remaining unchanged for a long time normally should be put into one scenario. Thus a compact test suite can be used, which is similar to the algorithmic program utilized approached used in AETG1 [18] using key variations using the PICT algorithm (Diagrams 7.1 and 7.2).

7.6 Input Domain Model

An input domain model is a dataset created out of the "target" data that we discussed earlier. The "source" data, which is the initial data, is loaded in "target," and they are mostly written in SQL, Hive and Pig. These input domain models are mostly used to apply combinatorial testing.

Input domain models are created by applying constraints on the dataset; these constraints, applied on the data, filter out the parameters that we will need. These constraints can be applied manually or by the software itself. These constraints can be classified broadly into three. The first constraint is applied by the model we are creating, basically according to the model, that suits our research or work. The second constraint comes from the user, the constraints that the user wants to apply. The third constraint comes from BIT-TAG, which stands for Combinatorial Big Data Test Data Generation – a type of software.

As mentioned earlier, data generation is the second step when it comes to big data testing. The specialty of BIT-TAG is that it will identify the parameter and later will create a domain model with those respective values. BIT-TAG can also select the characteristics of each parameter. The parameter is later used by the combinatorial algorithm to apply the t-way method, in which a selected number of parameters out of all the parameter is selected and testing is then performed. This prevents exhaustive testing: this input domain model provides a smaller dataset, and out of that dataset the algorithm selects an even smaller test set. In case of exhaustive testing, the test set is extremely large, wasting both resources and time.

7.7 Failure of Pairwise Testing

We present here the proposed work and survey of the mixed testing work, i.e., we show that pairwise information should be rewritten again. The loan arrange system (LAS) and the data management and the analysis (DMAS) are two kinds of software methods has been utilized along with the fault presented in Schroeder's study. Each method is tested with the use of n-way test sets, which is $n = 1, 3, 4$. After that the test sets of n-way of each kind are presenting with the use of the algorithm greedy, same one as mentioned by Cohn et al. [34].

Schroeder's study fault has been shown in this table: it was analyzing the fault basis for an n-way test set. The four-way category fault through several test suites the way it has been developed; five or more elements of data combination are needed to assure the fault is disclosed.

Pairwise testing fails if you do not select the proper values. In Table 7.1 the faults in the "not found" category are of a particular interest. There the one-way testing has been involved in pairwise testing, which is not found. Where the pairwise testing was not disclosed, also not disclosed were the n-way test sets. Still, if a full mixturial test set has been implemented on particular values, parameters are selected for exposing the fault because of the test data that is not selected.

Same as the variation of "not found" faults, this is the two-way faults that weren't exposed due to a specific collection of information values that have not been selected. In the communication of one or a pair of fields [21] are true divertingly, the cause of most field defects; interestingly, however, during the study, they were true both before and once the combine testing was implemented. This is not true for a pairwise test set, as it "protects across any incorrect implementation of the code involving any pair wise communication" [35]; the pairwise testing will solely conserve the interaction between the values of input used for testing. In this the selected values are usually and intensely small sets of the particular variation of potential values. The factor which result into the failure of the following approach is probability of any given combination of variables occurring in the field is not known.

In the MS Word choices example, the choice problem isn't a problem. In that case we tend to choose all the values for testing. The oracle problem remains one of the key challenges in software testing, for which little automated support has been developed so far. There is no other certain method to understand whether some better issue – say, gradual corruption of the document – isn't happening. A key research question is whether a more intrusive oracle placement is justified by its higher fault detection capability. So the solution to improvise the oracle technique is to initialize assertions in terms of mutation score and the possibility to detect real faults during improvement process.

TABLE 7.1

Classification and Evolution of the Fault

Type of fault	LAS	DMAS
1-way	32	29
3-way	6	12
4-way	9	1
4-way	6	3
Not detected	38	43

7.8 Improved Algorithm

The longer overwhelming and costly thought is software testing. In this example, there are n completely different input parameters with every parameter having multiple potential values, which makes complete testing that tests all attainable assortment practically impossible. Generating a better tests set can be done through positive software testing. The search-based test suite generation can be used to identify false positives and mutation testing which in turn identify false negatives in oracles represented in the form of program assertion. An approach used generates counter examples as test suites that demonstrate incompleteness and unsoundness, that will iteratively improve the assertion oracle. Finally, the process continues until it can generate new counterexamples and finishes with an improved result.

In this example, we tend to first assume the case wherever each parameter holds just two values. For example, all have been the same length $2k - 1$, and one has zero string. Let us sort out the pressure of the string quantity of 1's in it. Overall binary strings length k and $2k - 1$ have been gathered by S_{2k-1}. Note that

$$|S_{2k-1}| = \binom{2k-1}{k}$$

For example, here is the 10 strings of S_5 :

0	0	0	0	1	1	1	1	1	1
0	1	1	1	0	0	0	1	1	1
1	0	1	1	0	1	1	0	0	1
1	1	0	1	1	0	1	0	1	0
1	1	1	0	1	1	0	1	0	0

Take any of the two strings, for example, the first and last columns. Each of these collaborations (0, 1), (1, 0), and (1, 1) appear at least once. If one appends the 0 into the base of every string of the S_{2k-1}, then we assume that each of the four potential combinations (0, 0), (0, 1), and (1, 1) operates at least once.

ALGORITHM

The Input: group of the *n* parameter.
   Output: Test set.

1.   Smallest k output figure out as like follow

$$n \le \binom{2k-1}{k}$$

2.   from the S_{2k-1} choose any string n
3.   At the ending of all chosen string for the test set of *2k* achievement addition of zero.

End Algorithm

EXAMPLE 1

Let $n = 9$. Then $k = 3$, and the size of test set 6 is $f(0\ 0\ 0\ 0\ 1\ 1\ 1\ 1\ 1)$, $(0\ 1\ 1\ 1\ 0\ 0\ 0\ 1\ 1)$, $(1\ 0\ 1\ 1\ 0\ 1\ 1\ 0\ 0)$, $(1\ 1\ 0\ 1\ 1\ 0\ 1\ 0\ 1)$, $(1\ 1\ 1\ 0\ 1\ 1\ 0\ 1\ 0)$, $(0\ 0\ 0\ 0\ 0\ 0\ 0\ 0\ 0)g$. In that the set is achieved after choosing just first nine strings from the S_5.

We assume that each parameter holds larger than two values. This technique depends on reciprocally Orthogonal Latin Squares (MOLS). For pairwise testing in compiler styles the utilization of MOLS has been assumed within the past by M and l.

In the shape of the Latin square, of totally varied objects N is the array of N-copies (generally for the Latin letter Q, U, R :) So we can get the totally different objects into the any column and all the objects into the any row. Along with Latin square size, objects Q, U, R. ::: and Objects A, B, C, with: :: Orthogonal, which is supplemented by all X2 Q, q), (Q, u), :::, (U, q), (Q, q), :::. It shows that, if n is the principal, then for the construction of one will be 9 mols. For example, 2 mols of size are shown below three.

0 1 2	0 1 2
1 2 0	2 0 1
2 0 1	1 2 0

If the k (k_j 1)-values of the parameters, then k_j 2 mols of the size k_j 1 is needed. A set of k_j 2 mols of size k_j 1. Through the column and row, the two parameters are taking their values. They illuminate it. After considering the collection of test-like rows of matrix A, the existence of discussion matrix A has been created. Having the A matrix, A column, for example, to handle above testing, $A_1 = (0\ 0\ 1\ 1\ 2\ 2\ 2)\ T$, $A_2 = (0\ 1\ 2\ 0202\ 2)\ T$, $A_3 = (0\ 1\ 2\ 1\ 2\ 0\ 2\ 0\ 1)\ T$, and the $A_4 = (0\ 1\ 2\ 2\ 0\ 1\ 1\ 2\ 0)\ T$. The remaining parameters from the supermodel squares make their values ordered (KA_2) – from the adjacent squares they label 1, 2, 3 of 0, 1, 2 instead of rows and columns. Thus, while continuing for other entries, we are given a test set for another entries is 4-3 criteria: f (0), (0 1), (0 2 2), (1 2), (1 2) (1 2 0 1) (2 2 2) (2 1 2 2) (2 2 1 0) G. Can one build a test set, 16 3 with the use of test set for the valuable parameters, 4 3-honorary criteria? For example, 4 3-valuable criteria are required of two sizes of 3 mols, which are present above. Dual mols overlapping are obtained: (0; 0) (1; 1) (2; 2) (1; 2) (2; 0) (0; 1) (2; 1) (0; 2) (1; 0). The dark entry test configuration represents (2; 1; 0; 2): line label 2, column label 1, and entry (0, 2).

Follow the 18 £ 16 matrix 18 rows

$$\begin{pmatrix} A1A1A1A1A1A2.......A4 \\ A1A2A3A4A1.......A4 \end{pmatrix}$$

Here we define that total testing is the set conflict for the 16 3-value standards. It was easy to see that every pair of this collusion is known as a mixture of at least one time. To get set for any of these numbers, 3-norms of these concepts have been used. For example, we are considering 27 rows of 27 pounds 64 matrix: 64 £ 3 to get a test set for honorary norms.

$$A1A1A1A1A1A1..........A4$$

$$A1A1A1A1A2A2..........A4$$

$$A1A2A3A4A1A2..........A4$$

For motivating the test set of the n m-valued parameters, the first phase is L (m) MOLS is the size of the m. Full MOLS table can be found in ref [2].

7.9 Limitations of This System

Because of that for test suit generator, the constraint is a required feature. The test domain has some limitations, i.e., assortments that are impossible to be successfully performed around the given SUT context. Figure 7.3 shows volumes larger than 4 gigabytes (GB) that cannot be applied to the FAT classification system. Any test case that asks for volume and FAT larger the 4 GB can fail to work properly.

In this, we try to simplify any difficulties in analysis. There are some ways to handle constraints. The just method is to call the operator to manage the definition of parameter tests like the unwanted combinations not typically selected. This is achieved either by creating hybrid parameters or by separating parameter definitions onto dislocates subsets.

7.10 Conclusion

A huge amount of data is generated by big data applications, which requires ample amount of time for testing purposes while generating such applications. In this chapter, we have suggested that both pairwise testing and all-pairs testing represent an adequate test generating strategy. It indicates the size of set test generated and acquired for our technique and presents two different methods: pair-test and AETG. The most important observation is that in most cases our technique produces a better output compared to pair-test and AETG. Plus, there is no need to generate these test cases deeply and manually, freeing up time for product determining, choosing values and variables, and configuring and executing output. It is particularly effective because the pairwise bugs represent most combinatory bugs, based on the real dependencies of variables and product. Pairwise testing protects across pairwise faults or bugs while precipitously decreasing the number of tests necessary to perform. There is also no longer need to initiate this testing manually. This improved algorithm test provides an effective and efficient way to build smaller and more effective datasets for developing big data applications.

REFERENCES

1. K. Burr and W. Young, "*Combinatorial test techniques: Table-based automation, test generation, and test coverage,*" in *Proceedings of the International Conference on Software Testing, Analysis, and Review (STAR)*, San Diego, CA, 1998.
2. S. R. Dalal, A. Jain, N. Karunanithi, J. M. Leaton, C. Lott, G. C. Patton, and B. M. Horowitz, "*Model-Based Testing in Practice,*" in *Proceedings of the International Conference on Software Engineering*, Los Angeles, CA, pp. 285–294, 1999.
3. I. S. Dunietz, W. K. Ehrlich, B. D. Szablak, C. L. Mallows, and A. Iannino, "*Applying design of experiments to software testing,*" in *Proceedings of the International Conference on Software Engineering (ICSE 97)*, New York, pp. 205–215, 1997.
4. P. E. Ammann and A. J. Offutt, "*Using formal methods to derive test frames in category-partition testing,*" in *Ninth Annual Conference on Computer Assurance (COMPASS'94)*, Gaithersburg, MD, pp. 69–80, 1994.
5. D. M. Cohen, S. R. Dalal, J. Parelius, and G. C. Patton, "The combinatorial design approach to automatic test generation," *IEEE Software*, vol. *13*, no. 5, pp. 83–87, 1996.
6. C. Kaner, J. Bach, and B. Pettichord, *Lessons Learned in Software Testing: A Context Driven Approach*. New York: John Wiley & Sons, 2002.
7. S. R. Dalal, A. J. N. Karunanithi, J. M. L. Leaton, G. C. P. Patton, and B. M. Horowitz, "*Model-based testing in practice,*" in *Proceedings of the International Conference on Software Engineering (ICSE 99)*, New York, pp. 285–294, 1999.
8. M. Grindal, J. Offutt, and S. F. Andler. Combination testing strategies – A survey. GMU Technical Report, 2004.
9. S. Splaine and S. P. Jaskiel, *The Web Testing Handbook*. Orange Park, FL: STQE Publishing, 2001.
10. D. M. Cohen, S. R. Dalal, A. Kajla, and G. C. Patton, "*The Automatic Efficient Test Generator (AETG) System,*" in *Proceedings of the 5th Int'l Symposium on Software Reliability Engineering*, IEEE Computer Society Press, 1994, pp. 303–309.

11. R. Brownlie, J. Prowse, and M. Phadke, "Robust testing of AT&T PMX/star mail using OATS," *AT&T Technical Journal*, vol. *71*, no. 3, pp. 41–47, 1992.
12. S. Dalal, A. Jain, N. Karunanithi, J. Leaton, and C. Lott, "*Model-based testing of a highly programmable system*," in *Proceedings of the Nineth International Symposium on Software Reliability Engineering (ISSRE 98)*, Paderborn, Germany, 1998, pp. 174–178.
13. Q. Jiang, M. Hušková, S.G. Meintanis, and L. Zhu, "Asymptotics, finite-sample comparisons and applications for two-sample tests with functional data," *Journal of Multivariate Analysis*, vol. *170*, pp. 202–220, 2019.
14. S. López, A.A. Márquez, F.A. Márquez, and A. Peregrín, "Evolutionary design of linguistic fuzzy regression systems with adaptive defuzzification in big data environments," *Cognitive Computation*, pp. 1–12, 2019.
15. G. Aneiros, R. Cao, R. Fraiman, C. Genest, and P. Vieu, "Recent advances in functional data analysis and high-dimensional statistics," *Journal of Multivariate Analysis*, vol. *170*, pp. 3–9, 2019.
16. L. Sharma, S. Sengupta and B. Kumar, "An Improved Technique for Enhancement of Satellite Images," *Journal of Physics: Conference Series*, vol. *1714*, p. 012051, 2021.
17. D. M. Cohen, S. R. Dalal, M. L. Fredman, and G. C. Patton, "The AETG system: An approach to testing based on combinatorial design," *IEEE Transactions on Software Engineering*, 23(7), 1997.
18. J. Bach and P. Shroeder. "*Pair wise testing – A best practice that isn't*," in *Proceedings of the 22nd Pacific North West Software Quality Conference*, pp. 180–196, 2004.
19. C. J. Colbourn, M. B. Cohen, and R. C. Turban, "*A deterministic density algorithm for pair-wise interaction coverage*," in *Proceedings of the IASTED International Conference on Software Engineering*, 2004.
20. L. Copeland, *A Practitioner's Guide to Software Test Design*. Boston, MA: Artech House Publishers, 2003.
21. S. Dalal and C. L. Mallows, "Factor-Covering Designs for Testing Software," *Technometrics*, vol. *50*, no. 3, pp. 234–243, 1998.
22. Y. Lei and K. C. Tai, "*In-parameter-order: A test generation strategy for pair-wise testing*," in *Proceedings of the 3rd IEEE High-Assurance Systems Engineering Symposium*, 1998, pp. 254–261.
23. Z. Liu, "*Research of performance test technology for big data applications*," in *2014 IEEE International Conference on Information and Automation (ICIA)*. IEEE, 2014.
24. N. Li et al. "*A scalable big data test framework*," in *2015 IEEE 8th international conference on software testing, verification and validation (ICST)*. IEEE, 2015.
25. V. D. Kumar and P. Alencar. "*Software engineering for big data projects: Domains, methodologies and gaps*," in *2016 IEEE International Conference on Big Data (Big Data)*. IEEE, 2016.
26. J. Chandrasekaran et al. "*Applying combinatorial testing to data mining algorithms*," in *2017 IEEE International Conference on Software Testing, Verification and Validation Workshops (ICSTW)*. IEEE, 2017.
27. G. Qi et al. "Fault-diagnosis for reciprocating compressors using big data and machine learning," *Simulation Modelling Practice and Theory*, vol. *80*, pp. 104–127, 2018.
28. D. Gupta, A. Rana, and S. Tyagi. "A novel representative dataset generation approach for big data using hybrid cuckoo search," *International Journal of Advances in Soft Computing & Its Applications*, vol. *10*, no. 1, 2018.
29. S. Singh, L. Sharma and B. Kumar, "A machine learning based predictive model for coronavirus pandemic scenario," *Journal of Physics: Conference Series*, vol. *1714*, p. 012023, 2021.
30. L. Sharma, P. Garg, (Eds.). (2020). *From Visual Surveillance to Internet of Things*. New York: Chapman and Hall/CRC, https://doi.org/10.1201/9780429297922
31. L. Sharma, (Ed.). *Towards Smart World*. New York: Chapman and Hall/CRC, 2021. https://doi.org/10.1201/9781003056751
32. L. Sharma, "Human Detection and Tracking Using Background Subtraction in Visual Surveillance", *Towards Smart World*. New York: Chapman and Hall/CRC, https://doi.org/10.1201/9781003056751, pp. 317–328, December 2020.
33. T. Bouwmans, S. Javed, M. Sultana, and S. K. Jung, Deep neural network concepts for background subtraction: A systematic review and comparative evaluation. *Neural Networks*, vol. *117*, pp. 8–66, 2019.
34. K. Burroughs, A. Jain, and R. L. Erickson. "*Improved quality of protocol testing through techniques of experimental design*," in *Proceedings of the IEEE International Conference on Communications (Supercomm/ICC'94)*, May 1–5, New Orleans, Louisiana, pp. 745–752, 1994.
35. J. D. McGregor and D. A. Sykes, *A Practical Guide to Testing Object-Oriented Software*. Boston, MA: Addison-Wesley, 2001.

8

Potential Applications of Artificial Intelligence in Medical Imaging and Health Care Industry

T. Venkat Narayana Rao
Sreenidhi Institute of Science and Technology, Telangana, India

K. Sarvani
Narayana Junior College, Telangana, India

K. Spandana
Sreenidhi Institute of Science and Technology, Telangana, India

CONTENTS

8.1 Introduction

What makes a human unique? The most complex quality of a human being is the ability to think and make decisions in everyday life. Previously a robot could only perform mundane jobs that are programmed with a couple of fixed steps. But in a world filled with situations that require a capability to think, analyze, and arrive at a decision, a human is necessary. Although a robot cannot replace a human brain, a certain level of training can be done to enable intelligence in machines. This is possible through artificial intelligence

(AI) [1–3]. As the name suggests, it can mimic a human brain to a certain extent on training it through a couple of algorithms and datasets.

In the current technologically advanced time, the biomedical industry has a great potential of incorporating AI into its various practices [4, 5]. Biomedical technology uses various ways to diagnose a patient, such as diagnostic imaging, biomedical procedures, biochemical procedures, and so on.

8.1.1 Diagnostic Imaging

It is a biomedical practice by which abnormalities or diseases are diagnosed by retrieving images of the internal areas of the infected area [6]. For example, when a person suffers pain in any of their joints and the doctor needs a deeper knowledge on the injury, the doctor prescribes an inner scan of the particular area. There is a wide spectrum of scanning techniques that are used [7]. They are: X-ray, Computerized axial tomography (CAT), magnetic resonance imaging (MRI), sonography, and ultrasound.

8.1.1.1 Biomedical Practices

Observing the functioning of an organ besides its physical state is important to better understand some serious health conditions, and it is possible through biomedical procedures. These procedures diagnose the functioning of vital organs like the heart, brain, intestine, etc. Several techniques are used to achieve this, such as endoscopy, electrocardiography (ECG), and electroencephalography (EEG).

8.1.2 Biochemical Procedures

They are the diagnostic procedures that help in estimating the levels of chemicals in the body, such as hormones, antigens, antibodies, etc., hence providing more information to the doctor regarding the condition of the patient. The tests include: enzyme-linked immuno-sorbent assay (ELISA), swab test, urine analysis, and serum analysis.

8.2 Existing Structure and Design Issues

8.2.1 CAT Scanner

In a conventional X-ray machine, images or scans retrieved are two-dimensional, and as a result, they may seem unproductive for a doctor to analyze the condition of the patient. On the hunt for a better visual picture of the affected area, scientists came up with CAT and MRI scanning techniques [8]. These techniques enable a three-dimensional view of the affected area with X-ray beams and magnetic techniques, respectively.

Diving deeper into the CAT scanning, the procedure involves a computer that takes images generated by sending narrow X-ray beams through the patient's body from all angles. The images are then overlapped systematically, thereby arriving at a three-dimensional (3D) picture of the area under focus [9].

8.2.1.1 The Architecture

Figure 8.1 illustrates the basic architecture of a CAT scan. The machine has a motorized platform on which the patient rests. The donut-shaped machine has an X-ray tube that rotates around the patient clicking images in all directions by passing significant dosage of X-rays through the patient's body for better results. The retrieved images, called the tomographic images, are transferred into the computer that integrates the images to form a 3D view.

8.2.1.2 Design Issues

Although this technique helps in retrieving better-quality scans, it has a set of risks associated with it.

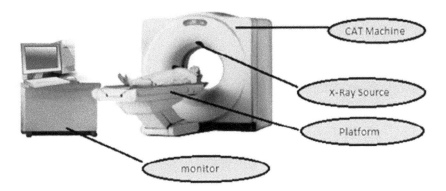

FIGURE 8.1   CAT machine existing architecture.

1. High dosage of radiation: The spiraling X-ray tube used in CT emits excessive amounts of radiation in an attempt to retrieve high-quality images. This issue has not been sufficiently covered until recently; it also raises the possibility of carcinogenic effects in the long run. Especially for children, who could have a sensitivity to radiation, CAT scans could be extremely dangerous and thus unnecessary [10].

2. Noise: To solve the issue of high dosage, low-dose CAT scans were examined, but the noise was inversely proportional to the dosage used. In other words, a decrease in noise required an increase in the X-ray dosage. Given the risks of high dosage, it seems almost impossible to scan a patient with a low dosage and avoid noise [11].

3. Cost: Given the high dosage and the equipment used, the CAT scans are bound to cost a fortune not only for the patients but also for the maintenance and construction.

8.2.2 Robotic Surgeons

Automation has entered medicine quite a while ago and is now making precise and low-risk surgeries a possibility with robotic surgeons. The procedures that require repetitive tasks like planting hair follicles require continuous and precise effort that could be mundane for a surgeon. Such tasks now opened doors to robots that perform accurate surgeries with precise depth and pace.

Moreover, even the most experienced doctors could tremble as a result of muscle fatigue while operating or encounter times of drifting focus resulting in several accidental movements. However, a surgical robot that is programmed to perform the surgery has no issues with staying focused or performing the same motion for prolonged time periods [12]. Still the robot maneuvers over the patient on the basis of instructions from the actual human doctor, who observes the procedure on screen. Additionally, the incisions made by a robot are often smaller and more precise than those made by a human, which results in fewer post-op complications and shorter recovery time for patients after surgery.

8.2.2.1 Design Issues

Today, the capability exists to train a robot and create a community of robots that can share their surgery data and learn as they perform just like a human doctor, as shown in Figure 8.2.

Given the success of a robotic surgeon, there is tremendous potential for advancements [10]. Currently, robots are not completely trusted by physicians because of the following reasons:

- Lack of additional knowledge that might only come through performing a number of surgeries in the past.
- With AI being a central part of surgeries, the abundance of data is a given for increased effectiveness, which means an increased likelihood of privacy of personal health data and other sensitive

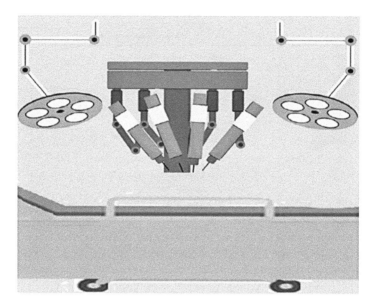

FIGURE 8.2   Robotic surgeries.

information of patients being compromised, touching up strongly on the issue of medical ethics [11]. This issue may be rectified in the near future by advancements in digital security that would allow to limit exposure during collection and processing of personal information.

8.3 Implementation of AI-based CAT Scanner

The AI-based CAT is the rightful solution to safer, faster, and cheaper scanning. It ticks all the boxes by solving every issue that the conventional CAT poses. With its deep learning capabilities, it could result in more detailed scans that could make it easy for the doctor to treat the patient.

First, when a traditional CAT scanner uses a low dose of X-ray, it results in noise, which is seen in the form of certain unknown random patterns appearing over the scan. However, with AI-enabled CAT, a low dosage of radiation can be used and yet the quality of the scan can be enhanced by reducing noise through deep learning, which integrates the images with neural network methods. Second, the speed is comparably high with AI-based systems. These systems also offer better structural fidelity, which is the quality with which the image shows the internal structure.

To optimize the cost and involve the doctor in the process, only the required area that doctor wishes to treat can be enhanced by increasing the quality of that particular area, thus resulting in the abstraction of data [13]. With the help of these neural networking algorithms, we not only arrive at high-quality scans with a low dosage, but we can also arrive at the phenotypic data of the scan by incorporating algorithms specific to the phenotype of the patient. In the existing systems, the machine compares the scan with a generalized phenotype that may or may not accurately judge the person's condition. But by training the system on various phenotypes, the system can accurately predict the defects based on what's normal for that person's phenotype and also give additional quantitative information, for instance, fat fraction from a mere liver scan.

In Figure 8.3, a flowchart of an algorithm is demonstrated, which identifies if the patient belongs to the COPD phenotype.

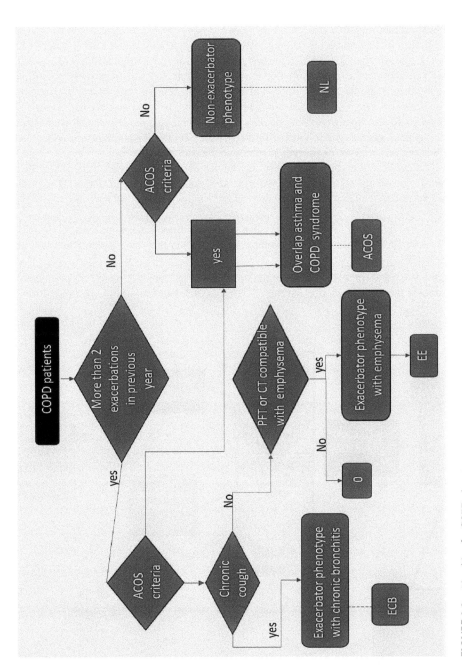

FIGURE 8.3   Algorithm for COPD phenotype.

8.4 Implementation of AI-based Robotic Surgeries

The existing robotic surgeons are programmed to operate through certain steps. However, they are not programmed for all scenarios. The AI and its complex deep learning algorithms can help robotics learn from previous surgeries and also improve their precision by learning from visual data for performing better and more accurate surgeries [14–16].

Some operations need extreme precision, and with one wrong step, they could risk a patient's life [17]. For instance, the pattern of cutting through the body for surgeries is essential, as it has to ensure that the surgeon doesn't cut through the important blood vessels. This kind of precision can only be ensured when the robot surgeon learns and has access to loads of previous data. Hence, several world-class companies are coming up with cloud networks for the surgical robots to upload their surgery data visuals and patterns. The cloud could also involve real-life doctors who could provide guidelines that could play a vital role in improving the surgeries that robots perform [18, 19]. With the immense data provided by the AI, the robot could precisely analyze its movements through the data points [1, 20].

AI can also be used for assessing the surgical instruments and surgical skill of the robot. To do this, we can introduce a tracking algorithm that performs segmentation of the instance. Through this, two instances are generated:

(1) As shown in Figure 8.4, a bounding box that is sent as input to tracking framework instance by instance keeps track of the surgical instruments, i.e., when they are changing and how they are operating [20]

(2) The mask is used to track the motion of the tips of these instruments, which are plotted on graphs to accurately measure the motion metrics [21]. Figure 8.5 demonstrates the plotting.

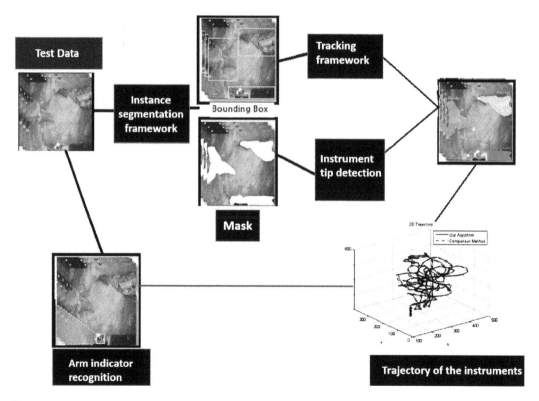

FIGURE 8.4 Working of surgical instruments.

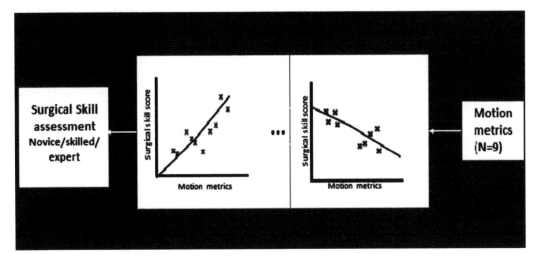

FIGURE 8.5 AI-based robotic surgeries.

8.5 Applications

8.5.1 Digital AI Doctors

Today, several AI chatbots take symptoms from the patients online and compare their medical background and their voice to the huge database of illnesses to accurately determine what the patient is suffering from. If the person is suffering from a serious illness, a referral can be made to see a live doctor, but if the person isn't suffering from any severe illnesses, this could save time and money associated with clinic or hospital visits. Several digital doctors have been created, such as Babylon in the UK [22].

8.5.2 Creating Drugs

As of the time of this writing, the world is still in the throes of the deadliest coronavirus epidemic in more than a hundred years – the COVID-19. It had taken almost a year of research and testing to come up with the most effective courses of treatment and, eventually, vaccines to prevent the disease. Assistance from AI pharmacy robots could help conduct testing faster and more effectively [23].

8.5.3 Mental Health

There are several AI chatbots today that recognize the behavioral patterns and pick up important data from the speech of the patient and make the correct determinations regarding a person's mental health. This could make it faster and easier for psychiatrists and psychologists to identify the scale of mental illness from which a person is suffering.

8.5.4 Medical Records

Storage and maintenance of medical data are essential, and with several artificial intelligence and data science techniques, better and more sustainable data storage can be possible.

8.5.5 Early and Accurate Cancer Detection

More often than not, manual cancer tests result in false positives or false negatives, leading to delayed or inaccurate treatment. With AI-aided cancer detection, the rate of false positives dropped by 11% and negatives by 5% [1–11]. With a scope of improving the AI-enabled cancer scanning, greater accuracy and earlier detection is a future possibility.

8.5.6 AI in Pregnancy Management

Electronic medical records collected throughout a pregnancy treatment are entered into the AI-driven EMR system, which notifies the doctor by giving valuable feedback and patterns to customize the treatment of each patient. Detection of cervical cancers and preterm risks of the pregnant mother, information on the health, age, etc. of the fetus are some of the many tasks AI can perform.

8.5.7 AI in Genomics

Human genome has 3 billion base pairs. These are analyzed through AI, which detects the variations in the genomes even at a single base locations. This helps in the diagnosis, treatment, and, in the future, prevention of genetic disorders. Gene therapy, i.e., replacement of mutated disease-causing genes with a functional gene, can be made possible through AI [24].

8.6 Conclusion

The AI has been a revolution that transformed our world on every scale. Health care is one of the most essential services of any country, and many surgical operations today take place with the involvement of AI. Efficiency, speed, and precision play an important role in any medical activity, be it surgeries, transplants, or even medical imaging. The scope of human error is always a risk in every medical activity, and the existing equipment is not always cost efficient and often risky as well. AI ticks all the right boxes and has the potential to transform the medical industry. AI-enabled image diagnosis techniques reduce the intensity of the X-rays to which a patient needs to be exposed, and give comparably more precise results than any of the conventional techniques like CAT, MRI, or X-ray. Surgeries require a high level of focus and precision, the lack of which, irrespective of how experienced a doctor is, lead to errors. Robotic surgeons have evolved significantly and are on the verge of welcoming AI to function more efficiently. Creating a cloud-based AI and deep learning–enabled robots that are trained over countless data and function on doctor's insights opens a door to endless possibilities in the health care industry.

REFERENCES

[1] R. Adams, "10 Powerful Examples of Artificial Intelligence In Use Today," 2019, [online]. Available: https://www.forbes.com/sites/robertadams/2017/01/10/10-powerful-examples-of-artificial-intelligence-in-use-today/#29b0c64420de.
[2] L. Sharma (Ed.). *Towards Smart World*. New York: Chapman and Hall/CRC, 2021. https://doi.org/10.1201/9781003056751
[3] L. Sharma and P. Garg (Eds.). *From Visual Surveillance to Internet of Things*. New York: Chapman and Hall/CRC, 2020. https://doi.org/10.1201/9780429297922
[4] S. Jha and E. J. Topol. "Adapting to Artificial Intelligence: Radiologists and pathologists as information specialists," in *JAMA* 2016; *316*: 2353–2354.
[5] F. Jiang, Y. Jiang, H. Zhi, et al. "Artificial intelligence in healthcare: Past, present and future," in *Stroke and Vascular Neurology*, 2017; e000101. doi:10.1136/svn-2017-000101
[6] C. Langlotz, B. Allen, and B. Erickson, et al. "*A roadmap for foundational research on artificial intelligence in medical imaging: From the 2018*," in *NIH/RSNA/ACR/The Academy Workshop*. Department of Radiology, 2019; 190613.
[7] A. Webb and G. C. Kagadis, "Introduction to biomedical imaging", in *Medical Physics*, 2003; *30*(8): 2267–2267.
[8] G. Freiherr, CAT and MRI Scans and their Significance, [online] Available: https://www.dicardiology.com/article/how-ai-can-unlock-data-ct-and-mri-scans

[9] P. Seebock, "Deep learning in medical image analysis," Master's thesis, Vienna University of Technology, Faculty of Informatics, 2015.

[10] A. Y. Letyagin et al., *"Artificial Intelligence for Imaging Diagnostics in Neurosurgery,"* in *2019 International Multi-Conference on Engineering, Computer and Information Sciences (SIBIRCON)*, Novosibirsk, Russia, 2019, pp. 0336–0337, doi:10.1109/SIBIRCON48586.2019.8958201.

[11] S. Dinakaran and P. Anitha, *"A review and study on AI in health care issues,"* in *International Journal of Scientific Research in Computer Science, Engineering and Information Technology*, 2018; 281–288. doi:10.32628/CSEIT183886.

[12] M. P. McBee, O. A. Awan, A. T. Colucci, et al., "Deep learning in radiology," in *Academic Radiology*, 2018; *25*(11): 1472–1480.

[13] A. S. Panayides et al., "AI in Medical Imaging Informatics: Current Challenges and Future Directions," in *IEEE Journal of Biomedical and Health Informatics*, vol. *24*, no. 7, pp. 1837–1857, July 2020, doi:10.1109/JBHI.2020.2991043.

[14] E. Köse, N. N. Öztürk, S. R. Karahan, et al., "Artificial intelligence in surgery," in *European Archives of Medical Research*, 2018; *34* (Suppl. 1): S4–S6.

[15] L. Sharma and P. K. Garg, *"Smart E-healthcare with Internet of Things: Current Trends Challenges, Solutions and Technologies"*, *From Visual Surveillance to Internet of Things*, CRC Press, Taylor & Francis Group, pp. 215–234, October 2019.

[16] L. Sharma and P. K. Garg, *"A foresight on e-healthcare Trailblazers"*, *From Visual Surveillance to Internet of Things*, CRC Press, Taylor & Francis Group, pp. 235–244, October 2019.

[17] W. Hsu, M. K. Markey M. D. Wang, et al. "Biomedical imaging informatics in the era of precision medicine: Progress challenges and opportunities," in *Journal of the American Medical Informatics Association*, 2013; *20*: 1010–1013.

[18] A. A. Shvets, A. Rakhlin, A. A. Kalinin, and V. I. Iglovikov, *"Automatic Instrument Segmentation in Robot-Assisted Surgery using Deep Learning,"* in *2018 17th IEEE International Conference on Machine Learning and Applications (ICMLA)*, Orlando, FL, 2018, pp. 624–628, doi:10.1109/ICMLA.2018.00100.

[19] S. Speidel, M. Delles, C. Gutt, and R. Dillmann, et al. "Tracking of instruments in minimally invasive surgery for surgical skill analysis," in *Medical Imaging and Augmented Reality*, Springer, Berlin, Heidelberg, pp. 148–155, 2006.

[20] Z. Pezzementi, S. Voros G. D. Hager, et al. *"Articulated object tracking by rendering consistent appearance parts"*, in *Robotics and Automation 2009. ICRA'09. IEEE International Conference on*, pp. 3940–3947, 2009.

[21] D. Lee, H. W. Yu, H. Kwon, H. J. Kong, K. E. Lee, H. C. Kim, et al., "Evaluation of surgical skills during robotic surgery by deep learning-based multiple surgical instrument tracking in training and actual operations," *Journal of Clinical Medicine*, 2020; *9*: 1964.

[22] K. Brian, et al. "10 promising AI applications in healthcare," 2018, [online] Available: https://hbr.org/2018/05/10-promising-ai-applications-in-health-care.

[23] Ch. Krishnaveni, S. Arvapalli, and J. Sharma, et al. "Artificial intelligence in pharma industry–A review," 2020. doi:10.21276/IJIPSR.2019.07.10.506.

[24] E. D'Agaro, "Artificial intelligence used in genome analysis studies," in *The EuroBiotech Journal*, 2018; 2. doi:10.2478/ebtj-2018-0012.

9

Virtual and Augmented Reality Mental Health Research and Applications

Priyanka Srivastava
Kohli Research Centre for Intelligent Systems, IIIT Hyderabad, India

CONTENTS

9.1 Introduction

Virtual reality (VR) is blurring the gaps between direct and indirect access to perceived reality. The experience of VR is like Truman Burbank's experience in *The Truman Show*, the motion picture in which the illusion created by the well-designed and well-programmed environment was indistinguishable from reality. Truman lived in this community (known as Sea Heaven) since birth, and he was utterly ignorant of living inside a TV show. His family, friends, neighbors, relatives, the weather, and any other object were part of a vast, carefully designed TV studio configured to resemble the real world. Truman's natural response to this grand ultimate illusion was not merely temporary, but persisted throughout his life until the age of twenty-nine when he finally realized his surrounding's falsehood.

What made Truman never question this artificial reality? The coupling between perception and action, the ability to predict environment properties, the physical and social reciprocation, and the unawareness of another world could be plausible explanations. A sense of being present and plausibility of actions occurring in the environment made him believe that Sea Haven was the real world. This film offers a stage for philosophical and empirical dialogue to understand reality. It lets us think about problems like how we come to know about reality. How do we know what we're perceiving and experiencing is indeed real? This quest had intrigued philosophers since at least the seventeenth century, when Descartes questioned one's belief and perception of reality.

The increasingly growing relevance and importance of VR in various fields, ranging from gaming entertainment, military training (Smith, 2010), clinical practices (Michaliszyn et al., 2010; Rizzo & Shilling, 2017), spatial cognition (Chrastil & Warren, 2012, 2013; Srivastava et al., 2019) to emotion (Li et al., 2017), are demanding VR technologies to step forward to realize an "ultimate display." This ultimate display is a conceptual device (Sutherland, 1965) that aims to resemble reality and blurs the boundaries between the physical and the computer-generated artificial world, not unlike Truman's world in the aforementioned film. More importantly, it envisions transcending the physical space and opens a window to examine the ability to perceive-think-and-act, a phenomenological experience of feeling and consciousness, by enabling controls that oppose the rules of physical reality and yet create a sense of presence (Sutherland, 1965).

The presence is a complex phenomenon and comprises identification and self-location in a spatial-and-temporal frame of reference. The "identification" refers to being present as self. The "self-location" in a temporal frame of reference refers to being present as self in a given time, now and then, the very moment. The self-location in space refers to the feeling of being present in the virtual environment as if it's a physical reality (Metzinger, 2018). The perceived reality is a spectrum that ranges from the real environment to the virtual environment (Jerald, 2015; Slater, 2009). The ability to visualize, manipulate, and interact with the virtual environment determines the degree to which we perceive and feel reality.

The technological advances in VR are making us realize the once unthinkable concepts, like a sense of agency and body ownership (Blanke & Metzinger, 2009), out-of-body experience (Braun et al., 2018), hallucination (Blanke & Metzinger, 2009), virtual hand illusion (Sanchez-Vives et al., 2010), PTSD treatment (Rizzo & Shilling, 2017), schizophrenia treatment (Freeman, 2008), treatment for phobia (Freeman et al., 2018), and depression (Falconer et al., 2016), to name a few. The VR allows us to design more personalized, ecologically valid, unobtrusive, tightly controlled experiments that are essential for clinical therapeutic settings and scientific observations. The ability to replicate the real-life scenarios and yet be able to conduct the study with randomized controlled trials envisions more quantitative and objective clinical practices and clinical research in the future.

At present, the diagnosis of any psychological disorder, including ones as common as depression, mainly relies on clinical examination and subjective evaluation of self-reported symptoms. There are no globally accepted or approved biomarkers or psychological/cognitive markers used as a part of the diagnostic criteria (DSM-V & ICD-10) for any psychological disorders (American Psychiatric Association, 2013). The psycho-therapeutic and pharmacological interventions and their impact on a patient's health follow-up share similar practice protocols as a diagnostic procedure.

Recently, a consortium study (Collins et al., 2011) focusing on identifying the grand challenges of mental, neurological, and substance use (MNS) disorders worldwide has reported 25 grand challenges for MNS disorders in 2011, with the help of researchers, advocates, and clinicians working in over 60 countries. Among those 25 grand challenges, the need to identify social and biological risk factors, develop biomarkers, develop a more ecological and evidence-based intervention, and create more culturally informed methods got the top positions.

The subjective reporting and subjective examination make the clinical observations difficult and limit the scope of addressing those 25 grand challenges of MNS disorder reported by Collin and colleagues (2011). The subjective reporting relies on patients' memory, thought-based self-observation or introspection, and imagination, and alienates them from realizing the mental health issues. It also makes patients aware of others' perceptions and judgment when they share their interoceptive experiences, thoughts, and feelings with a doctor. However, virtual reality offers them a more personalized, ecologically valid, and controlled experimental environment. In addition, patients do not have to rely on their memory, introspection, interoceptive feeling, thoughts, and behavior while interacting with a VR environment. VR enables unobtrusive and more objective observation. VR environments with a clinical focus can help psychological and psychiatric research, and can make the clinical practices more humane and may expedite the development of psycho-social-biological markers proposed by the National Institute of Mental Health, US (Collins et al., 2011; Schacter et al., 2009).

The VR-based clinical research would be crucial in identifying psycho-social-biological risk factors. It would help clinicians design early intervention programs to delay the onset of various psychological disorders, such as depression and anxiety. Also, it would offer better psycho-therapeutic methods for more effective treatment of ongoing psychological illnesses. Thus, VR holds a great promise for better intervention, monitoring, and treatment practices in the future.

The advances in VR technology make it available at affordable prices and make it accessible to ordinary people. The VR is in the process of revolutionary transition, another revolution like computers that eventually moved from specific labs to ordinary people's homes. The VR technology is going to change the game and would make mental health monitoring more accessible, easy, and comfortable for patients, similar to physical health monitoring devices like blood-pressure or sugar-testing devices.

This chapter discusses the state-of-the-art virtual reality technologies' clinical research. It is divided into discussions of two major VR technologies: head-mounted virtual reality and augmented reality. First, we will discuss the state-of-the-art VR clinical research and highlight the limitations. This section will also discuss future directions in VR clinical research. The augmented reality (AR) clinical research section will first discuss the state-of-the-art AR clinical research, and after that, we will discuss its limitations and future directions.

The mental health section will highlight the limitations of the currently ongoing clinical practices and discuss the shift in focus from a clinical interview-based examination to more objective, cognitive, neurological, and social perspectives of clinical observations. The virtual reality section will first discuss the concept of reality and then highlight current VR clinical research. Further, this section will discuss the limitations based on the observation methods used in these studies. It will highlight the need for correlation and causal studies focusing on cognitive, biological, and social perspectives of a given psychological disorder. This section represents a significant portion of this chapter and establishes a foundation for the augmented reality section. The AR section will first discuss the difference between VR and AR and then highlight state-of-the-art AR clinical research. The limitations and future directions in this section would highlight the references from VR-clinical research. The summary section will complete the critical components of each of these sections.

9.2 Mental Health

This section highlights the limitations of current clinical practices. Further, it discusses the new framework proposed by the US National Institute of Mental Health (NIMH) for the assessment and treatment known as biopsychosocial perspectives of psychological disorder.

Mental health is not just an absence of dysfunctional thoughts, behavior, and emotion that causes significant distress or impairment, but it is also a presence of subjective well-being (Keyes, 2006; Reddy, 2019). The World Health Organization (WHO) 2001 report stated that approximately 450 million people suffered from mental disorders at the beginning of the twenty-first century, indicating mental disorders as one of the leading causes of ill health and disability worldwide. A 2011 NIMH report informed that mental disorders constituted 13% of the global burden of disease, surpassing cardiovascular disease and cancer (Collins et al., 2011). The report listed depression is the third leading contributor to the global disease burden. The report projected that by 2020, 1.5 million people might die by suicide each year, and 15–30 million people would attempt it (Collins et al., 2011).

Conceptualizing mental disorders as abnormal psychological experiences and treating those experiences as illnesses had a journey of 200 years when we interpreted mental illnesses as God's punishment or a result of religious or supernatural forces (Schacter et al., 2009). It is worth acknowledging that mental health practices have come a long way, gaining widespread attention today and becoming a part of primary health care services and research. The traditional clinical observations involving the clinicians and the patients' subjective reports on symptoms of behavior thought and emotion have been widely criticized for framing the everyday rich complex functional settings within the Diagnostic and Statistical Manual of Mental Disorders (DSM)-based clinical assessment (Rizzo & Shilling, 2017; Schacter et al., 2009). Although the DSM provides a useful framework for classifying mental disorders, it fails to map the recent scientific findings on mental illnesses using biopsychosocial perspectives (Schacter et al., 2009). In biopsychosocial views, the biological factors focus on genetics and epigenetics, the biochemical imbalances and brain correlates of abnormalities; the psychological factors focus on maladaptive learning and coping mechanisms, cognitive biases, dysfunctional attitude, and interpersonal problems; and the social factors focus on poor socialization, stressful life experiences, culture, and social inequities (Schacter et al., 2009). Realizing the importance of psycho-social-biological factors, the NIMH has proposed a new framework for mental health and well-being research. It has recommended a shift of focus from DSM categories to more of biological, psychological/cognitive, and behavioral constructs as building blocks of mental disorders (Schacter et al., 2009; Collins et al., 2011).

9.3 Virtual Reality

This section discusses the concept of reality, and then the state-of-the-art virtual reality clinical research. Here you will learn the transition made from traditional clinical observations to virtual reality–based clinical observations. We will follow up with the discussion on limitations and future directions by highlighting the need for more objective and quantitative VR-based measures on cognitive, affective, and behavioral engagements. The section will focus on discussing the head-mounted display VR-based research.

Any interaction that combines a sense of control and prediction creates an illusion of reality. The sense of reality is subjective. The interaction with the environment, sensorimotor engagement, experience, language, culture, emotion, and development together shape a personal sense of reality (Borghi et al., 2018; Schacter et al., 2009). For example, let's consider the electromagnetic spectrum of which we humans could perceive only a fraction of rays ranging from 400 to 700 nm, called visible light. In contrast, pythons and bullfrog can see the infrared (IR) rays, a few birds can see ultraviolet (UV) rays, and the bluebottle butterfly can see the UV rays along with visible light. With the clinically blind population, alternate sensory modality gives a sense of reality distinct from that of sighted people. For instance, visually impaired people use echolocation techniques or audio signals using sound waves or air pressure compression to conceptualize spatial properties, whereas sighted people primarily rely on visual inputs and corresponding visual feedbacks.

Another example that challenges the definition of reality is the position of the sensory system and its impact on the interface with the environment and further development of the brain. Animals with eyes at the side of their heads, like rabbits or pigeons, are most aware of their 360° local area but lack distance awareness. In contrast, the animals with eyes at the front have access to distant information but require head rotation to acquire rear 180° visual inputs to complete a full 360° information. The animals with distant area awareness are more predatory in nature, whereas animals with local area awareness are usually prey and develop other senses and skills to save their lives. Using specific sensorimotor experiences shapes the individual's ability to interact with the environment and predict the other objects' course of action in the environment accordingly. The small subset of the world that an animal can detect is its "umwelt" (Schroer, 2019), and that is its ultimate reality.

From the above baseline perspective of reality, VR comprises a computer-generated artificial environment and allows an advanced form of human–machine interaction. This interaction goes beyond traditional keystroke-based interfaces and allows users to experience a multisensory, immersive, and dynamic virtual environment (Cipresso et al., 2018; Rizzo & Shilling, 2017). Immersion is one of the critical factors for creating an illusion of umwelts as being present in the artificial environment. Immersion is a technical concept that primarily focuses on creating a virtual environment that is inclusive, extensive, vivid, involves surrounding in its projection, and ensures an interactive and matched interface (Jerald, 2015; Slater & Wilbur, 1997).

Inclusiveness refers to the extent to which the external physical reality is blocked out from the virtual environment (Moreno et al., 2019). Extensiveness refers to the range of sensory modalities that could be presented to the user. Vividness refers to the richness of the stimuli that are projected onto viewer's sensory receptors, such as color, resolution, and frame rate. Surroundness refers to the extent to which the virtual environment is panoramic and covers 360° tracking (Jerald, 2015; Slater & Wilbur, 1997). Surroundness is what puts artificial reality on par with physical reality in the display.

Though these aspects are essential for make-believe reality, the sense of presence requires more than these realistic displays. It requires a feeling of interaction that could be controlled, predicted, and reciprocated. Matching and interactability enable mapping the control realism in a virtual environment (Jerald, 2015; Slater & Wilbur, 1997). Matching ensures the physical reciprocation by establishing a correspondence between body movement and the change in the display, such as turning head to the right 90° should change the field of view matching the viewer's orientation or visual flow or auditory display following Doppler's effect. The interactability allows control/action and prediction, and enables users to change the virtual world and influence the future course of events. In addition, VR requires a plot to create a social ambiance with the surroundings and matched interaction with the virtual interface.

The degree of immersion is determined by the extent to which a VR system achieves the sensorimotor contingencies (Slater, 2009). The sensorimotor contingency is a coupling between perception and action (Lobo et al., 2018) that follows the laws of control and regulate responses in correspondence with the local environmental conditions (Warren, 2006). It is important to note that immersion is critical for the feeling of "being there or present" or "sense of presence" known as place illusion (PI) (Slater, 2009), and is critical for the feeling that the depicted scenario is actually happening in the virtual environment, known as the plausibility of illusion (PSI) (Slater, 2009). Although immersion and presence have often been used interchangeably, they are different concepts. As described above, immersion is an objective technical description of any VR system. Immersion enables a rich multimodal 360° panoramic view and a multi-modal interaction in correspondence with the user's body/head movement in the given virtual environment. In comparison, the presence or sense of presence or place illusion is a subjective experience of one's interaction with the virtual environment (Slater and Wilbur, 1997; Slater, 2009; Srivastava et al., 2019).

High immersion allows capturing the high level of display fidelity (the extent of exactness with which the virtual stimuli resemble the real stimuli projection on to one's sensory system) and high level of inter-action fidelity (the extent of exactness with which the virtual objects are controlled resembling the real objects). In addition, high immersion is determined by the way it prevents the interaction with external physical reality. Though technological advancement helps achieve a high level of immersion, it does not ensure the user's experience with place illusion (McMahan et al., 2012). Studies have shown that the level of presence may not correspond to the level of immersion. The user may experience a similar level of presence despite interacting with a different level of immersive environment (Srivastava et al., 2019), and the user may report a different level of presence despite interacting with a similar level of immersive environment (Wilson & Soranzo, 2015). Studies have reported that a task and/or user's emotional state might also influence the experience of place illusion (Bowman & McMahan, 2007). It suggests that psychological fidelity (refers to the extent to which the virtual stimuli evoke a physiological or emotional response similar to the one might experience in real life) along with technical fidelity together creates a place illusion (See, Wilson & Soranzo, 2015).

When it comes to the concept of immersion and place illusion, it becomes important to acknowledge the difference between various types of VR systems as per their degree of immersion. The distinction between the VR systems is not commonly accepted because of the heterogeneous inclusion/exclusion criteria for defining the VR system. There are three major categories of VR systems that are available in the market today. It varies from non-immersive VR systems like a desktop to a highly immersive system like head-mounted display VR (HMD).

The non-immersive VR system is low on display and interaction fidelity. The inclusiveness, vividness, and surroundness are the key display components that create a difference between high and low display fidelity VR systems. The high compared to the low display fidelity VR system allows the user to experience a rich, three-dimensional, 360° multi-modal ambiance. The inclusiveness in the high display fidelity VR systems, like HMD oculus rift, restricts the users to interact with the external world, whereas, the low display fidelity VR system, such as 2D desktop VR, fails to restrict user's distraction from the external physical reality. The 2D desktop visual display is limited to 20–40° and lacks mapping the exactness of real 360° surrounding. It breaches the experience of surroundness while viewing the 2D dynamic virtual environment. Similarly, the high compared to low control fidelity allows users to control and predict the action as they would in real settings. The control fidelity is achieved by matching (refers to the reciprocal relation between body movement and the change in environment) and interactability (refers to the ability to manipulate the virtual objects and predict its future course of action as per the manipulation). However, it is challenging to map the control fidelity to the real settings in the non-immersive VR system such as the desktop (Slater, 2009; Srivastava et al., 2019; Moreno et al., 2019).

9.3.1 Virtual Reality Clinical Research

The highly immersive HMD VR appears as a potential solution for studying individuals with mental disorders using biopsychosocial perspectives. Before we proceed further, we should reflect on one of the most pertinent questions that this field of research should envision to address, namely how HMD VR is going to revolutionize the clinical research and practices.

The highly immersive VR devices, as described in the earlier section, offers more real-time gestalt perspective, in which the patient is asked to become aware of their thoughts, behavior, experiences, and feelings and asks them to own the responsibility for those responses (Schacter et al., 2009). Unlike traditional practices that rely on patients' ability to imagine and use memory, VR allows them to experience the simulated environment and interact with virtual objects. This facilitates hightened awareness of their thoughts, behavior, experiences, and feelings. VR allows clinicians to create simulations akin to real life, which may be difficult to create otherwise for various reasons. It enables tracking head and body movement corresponding to the dynamic changes in the virtual environment, and allows clinicians to control and manipulate the environmental settings as per the patient's profile. The VR clinically controlled observations allow clinicians to conduct repeated sessions with graded complexity and aids assessment of cognitive, emotional, and sensorimotor or psycho-motor responses (Garrett et al., 2018).

Unlike traditional clinical research and practices, the VR mental health research could encapsulate the following four major objectives while using more dynamic real-life simulations to get a holistic picture of a given mental disorder. These four objectives are: (a) symptoms assessment and description; (b) testable, putative causal explanations for the disorder; (c) predictions and identification of risk factors or biological/social/psychological markers; and (d) treatment methods to control or intervention to delay the onset of the disorder. For instance, we can describe the symptoms of depression as pervasive and persistent low mood, low self-esteem, and anhedonia (DSM-V) (American Psychiatric Association, 2013). These symptoms can be explained by single or multiple factors including biological, cognitive, psychological, learning, behavioral, social, and interpersonal relationship. The biopsychosocial causal understanding would lead to a better prediction of risk factors of depression or would help in developing biological and cognitive or psychological markers for depression, which further helps control the onset of depression or aid managing it using more effective psycho-therapeutic interventions (Collins et al., 2011; Schacter et al., 2009).

Overwhelmingly, most of the VR studies have focused on treatment, a few focused on assessment, and fewer investigated the causal explanations and worked on theoretical development of psychological disorders (see Freeman et al., 2017, 2018; Rizzo & Shilling, 2017). The VR clinical psycho-therapeutic treatment research has employed majorly three methods: exposure therapy (exposing patients to graded environment conditions and allow systematic confrontation of external or internal stimuli for sensory desensitization, habituation, and stress inoculation, e.g., PTSD, phobia) (Freeman et al., 2018; Michaliszyn et al., 2010; Valmaggia et al., 2016), distraction therapy (helps patient disengage their attention from the stressful stimuli and reorient their attention to something pleasant to reduce pain or discomfort, e.g., pain, emotional distress) (Hoffman et al., 2004; Niharika et al., 2018), and training (focusing on mindfulness, breathing exercises, relaxation techniques, social engagement, interpersonal skills, cognitive strategies, e.g., mindful meditation) (Chandrasiri et al., 2020; Rizzo & Shilling, 2017).

There are a few mental disorders that have been the primary focus of VR mental research, such as PTSD (Rizzo & Shilling, 2017), anxiety (Rothbaum et al., 2002; Freeman et al., 2018), psychosis (Freeman et al., 2003; Freeman, 2008), substance disorder (Freeman et al., 2017), and eating disorder (Riva, 2011). However, disorders like depression that are common and reported as one of the leading causes of suicidal death and attempts (Collins et al., 2011) had been studied rarely (Falconer et al., 2016). The last 20 years of VR clinical research findings are encouraging, and have showed a substantial transfer of virtual learning to the real world, indicating that HMD VR might not be just a potential humane alternative for clinical sessions; it may substitute the traditional clinical practices in many clinical cases.

It is important to note that VR immersion and sense of presence has been widely exercised in VR clinical research, which allows users/patients to experience the rich, inclusive 360° surroundings and use extensive sensorimotor contingent controls matching the experience of real perception through natural movement. However, when it comes to measures, the VR clinical studies couldn't utilize the full potential of VR objective measures. VR measures, such as head-tracking, can help obtain more objective and quantitative data, along with clinical examinations and subjective reporting.

VR clinical studies have mainly employed self-report and clinical observations. Studies using VR exposure therapy to treat anxiety (Rothbaum et al., 2000), PTSD (Difede & Hoffman, 2002), PTSD using Bravemind (Rizzo & Shilling, 2017), and acrophobia (Freeman et al., 2018) have used pre–post self-report questionnaires on anxiety and avoidance (Rothbaum et al., 2000), clinician-administered PTSD scale

(CAPS) (Difede & Hoffman, 2002), PTSD-checklist military version (PCL-M), Beck Anxiety Inventory (BAI), patient health questionnaire (PHQ) (Rizzo & Shilling, 2017), heights interpretation questionnaire (HIQ), acrophobia questionnaire, improving access to psychological therapy (IAPT) questionnaire, along with suicidal tendency and simulator sickness questionnaire (SSQ) (Freeman et al., 2018). Similar observations are made with VR distractions and VR training studies. Though these studies could succeed in transferring the virtual experience to the real world, these studies are still limited to exercise the latest framework of mental health and well-being assessment and treatment protocols using biopsychosocial perspectives.

9.3.2 Virtual Reality and Related Measures

With the advancement in HMD VR devices, today we can use HMD with embedded eye-tracking (e.g., FOVE, Tobii with eye-tracking, and HTCvive eye-pro) and we can synchronize HMD VR with physiological monitoring devices (Soujanya Kodavalla et al., 2019), as well as use the head-tracking measures (Li et al., 2017; Lin, 2017). Employing multiple measures like eye-tracking and physiological monitoring along with head-tracking data would help us evaluate cognitive processing, such as attention, by assessing their focus or scope of attention; psychological/cognitive strategies, such as approach strategy or avoidance strategy; memory; and fear and other emotional responses more quantitatively and objectively. The last 20 years of VR clinical research had laid the foundation for future research and created demand for more standardized VR settings, experiment protocols, and observation methods. The multiple measures in VR studies would improve the cognitive interpretations of a psychological disorder. This section will discuss the importance of cognitive, biological, and head-tracking measures in the context of clinical research.

9.3.2.1 Cognitive and Biological Measures

- *Cognitive Measures*

  Assessing cognitive engagement and the lack of it will be an important step for VR clinical research, as it will help understand patients' ability to perceive, attend, construct memory, make decisions, develop self and empathy, and regulate emotions. Studies investigating the relationship between depression and attention have shown biased attentional processing (Cotrena et al., 2016; DeJong et al., 2019) and poor attentional control (Epp et al., 2012; Kertzman et al., 2010). Studies evaluating the relation between anxiety and attention have shown biased attentional deployment in response to the threat stimuli (Cisler & Koster, 2010; Lisk et al., 2020). In addition, attentional bias modification techniques have been used to modulate attentional biases in such clinical cases and have shown encouraging results (Bar-Haim, 2010; Heeren et al., 2015). However, studies on psychosis have reported altered state of cognition, perception, self-hood, temporal reality, and altered sense of embodiment (see Pienkos et al., 2019).

  Unlike traditional psychiatric assessments and measures, studies investigating the cognitive processes underlying mental disorders have shown differences in eye movement while scanning the scene when patients suffer from anxiety (Lisk et al., 2020) or depression (Armstrong & Olatunji, 2012; Duque Vázquez, 2015; Sanchez et al., 2013), to name a few. Analyzing cognitive processing by employing eye-tracking measures to VR would not only benefit the understanding of cognitive mechanism of various mental disorders while interacting with dynamic complex real-life scenarios, but it will also support developing cognitive risk factors or cognitive markers. Identifying cognitive/psychological risk factors to biomarkers in physical health practices may help prevent or delay the onset of at least a few disorders like depression or anxiety.

- *Biological Measures*

  Studying cognitive engagement and cognitive mechanisms has tremendous potential for advancing the understanding of mental disorders. However, it would never be a complete story without the support of biological correlates. Let's look at the cases of clinical depression and anxiety.

  Studies investigating the brain correlates of depression show changes in amygdala activity with emotional stimuli when presented on the fMRI display screen (Sheline et al., 2001), indicating biased affective processing. A fMRI study on body perception showed a difference in insula activity compared to healthy control participant, suggesting that altered insula activity is related to MDD patients' abnormal "material me" and somato-vegetative abnormalities (Wiebking et al., 2010). Greater activity in the lateral prefrontal cortex and anterior cingulate in depressive brain while performing the n-back task (measures working memory load) suggests that more effort is required to achieve the same level of excellence as healthy adults (Harvey et al., 2005).

Studies investigating the biological correlates of rumination in case of depression have shown marked differences in default mode network (DMN) when compared with the healthy adult brain (Berman et al., 2011; Sheline et al., 2009). Comparing DMN between healthy adults and major depressive disorder patients (MDD) have shown increased activity and connectivity in the subgenual anterior cingulate cortex, and failure to reduce activity in the ventromedial prefrontal cortex, anterior cingulate, lateral parietal, and lateral temporal cortex. In a nutshell, it suggests that the inability to downregulate the DMN activities is responsible for depressive rumination (Berman et al., 2011; Sheline et al., 2001).

Studies investigating the brain correlates of anxiety disorder reported reduced activation at subgenual anterior cingulate cortex when participants viewed the fearful stimuli after completing their fear conditioning sessions (Britton et al., 2013). The anxious brain showed distinction in activation in the ventromedial prefrontal cortex with age, reporting more heightened activity in adolescents compared to old-age patients (Britton et al., 2013). Patients with anxiety disorders have shown a difference in DMN regions that are critical for emotional processing. The anxious brain has shown comparatively lesser deactivation in the medial prefrontal cortex and a greater deactivation in the posterior cingulate cortex responsive for an altered state of emotional processing (Zhao et al., 2007).

The VR clinical research can employ a pre–post design to measure the biological correlates and DMN of patients and healthy adults' brains to learn the adaptability and plasticity after a certain intervention. In the future, VR clinical research should also make use of classical cognitive tests and the corresponding brain correlate measures to show the impact of highly immersive VR interaction on the brain and behavioral responses. In a nutshell, VR clinical studies need to incorporate multiple measures, ranging from psycho-physical and psycho-physiological measures to the eye-tracking to the brain correlates, along with clinical evaluation and patient's self-reporting to develop a holistic and complex biopsychosocial understanding of a given mental disorder.

The following sections focus on the importance of studying cognitive engagement and highlight the use of VR head-tracking measures to understand the cognitive, behavioral, and emotional engagement in VR clinical research.

9.3.2.2 Cognitive, Behavioral, and Emotional Engagement

Cognitive engagement is a widely accepted concept in classroom learning or similar setup (Richardson & Newby, 2006; Rotgans & Schmidt, 2011). Cognitive engagement is critical to understand one's motivation to interact with the given virtual environment, as it determines the state of thought processes and feelings that are crucial for mental health and well-being during any point of interaction. In classroom settings, cognitive engagement is operationalized by measuring individual student's learning effort in completing their assignments, class attendance, participation in extracurricular activities, and involvement in classroom discussions (Richardson & Newby, 2006). When this engagement is viewed in the context of complex subject matter along with the involvement of another agent, instructional design, sense of agency, and autonomy ascribed to students' cognitive engagement (Cummins, 2014; Luck & d'Inverno, 1995), then it is operationalized as a measure of the complex interplay between an individual (such as ability to think, attend, and intent to learn) and situational context in which the learning or interaction takes place (Ben-Eliyahu et al., 2018; Rotgans & Schmidt, 2011).

Autonomy and a sense of agency are pivotal to cognitive engagement. Autonomy is defined as a state of exercising independence and self-determination (Cummins, 2014; Luck & d'Inverno, 1995; Rotgans & Schmidt, 2011). Agency is defined as the exercise of a capacity to act, and a sense of agency refers to the feeling of control over action (Moore, 2016). The sense of agency can be explained by the feedback-comparator model of motor control (Pacherie, 2007). The feedback-comparator model works in a loop of feed-forward and feedback inputs, in which the agent compares the predicted and the intended trajectory of movement, as well as the predicted and the actual trajectory of movements. The consistent comparison between the predicted and intended action and predicted and actual outcome helps fine-tune the bodily movements for effective interaction with an environment (Pacherie, 2007). It is apparent that for the feedback-comparator model, sensorimotor contingency is the key that plays a critical role in constructing immersion.

The sense of agency is closely associated with the experience of flow (Csikszentmihalyi & Csikzentmihaly, 1990; Nakamura & Csikszentmihalyi, 2014). Flow is a state of mind that leads to an experience of deep pleasure and absorption in task (Csikszentmihalyi & Csikzentmihaly, 1990). Experience of flow is associated with mental health and well-being, and has positive impact on therapeutic and rehabilitation sessions in fostering meaningful changes in a patient's everyday life (Riva et al., 2016). Studies (Vuorre & Metcalfe, 2016) have shown that it is the effortful flow compared to effortless flow that yields a sense of agency. This could be because of the difference in automatic versus controlled processing. The effortful attention processes require a conscious choice of predictions and comparisons as suggested by the feedback-compactor model to enable a better sense of agency.

A sense of agency is essential to almost every therapeutic session that seeks patients' active contribution (Braun et al., 2018; Moore, 2016). The therapeutic session aims to motivate patients to achieve a sense of agency, especially in the case of major depressive disorder (MDD) (Braun et al., 2018). The sense of agency helps develop a sense of efficacy, an individual's belief and confidence in their capacity to attain the desired goal. In case of MDD, the negative thoughts about self, the world, and the future are associated with a sense of agency and a sense of efficacy (Beck, 2008). Losing one's ability to predict and plan for the future, inability to believe in achieving the desired goal, and feeling of helplessness develop loss of the sense of agency and eventually loss of the sense of efficacy (Braun et al., 2018). A few other commonly discussed clinical disorders that have shown a varying degree of the sense of agency and sense of efficacy are schizophrenia (Braun et al., 2018; Moore, 2016), phobia (Braun et al., 2018), and delusions (Braun et al., 2018), indicating the importance of sense of agency in treating these patients well.

Behavioral engagement refers to more explicit action-oriented evaluation (Ben-Eliyahu et al., 2018), in which the students are observed reading aloud and engaging in peer and teacher–students discussions in a classroom setting. The emotional engagement refers to experiencing a pleasant or unpleasant and high or low aroused state of mind during learning (Ben-Eliyahu et al., 2018). To perform any task effectively and pursue the task with persisting motivation requires a correlation between cognitive, behavioral, and affective experiences. VR clinical studies can use these measures to meet the four goals of VR clinical research: the symptoms description, causal explanations for the disorders, prediction, and treatment. One of the primary ways VR could achieve measuring the engagement objectively is VR head-tracking.

9.3.2.3 Head-tracking, Emotion, and Attention

The ability to track body and head movements under HMD VR has the potential to measure cognitive, behavior, and affective engagement while interacting with the dynamic 360° surrounding. This section focuses on head-tracking measures. Head-tracking parameters include yaw, pitch, and roll movement. The side-to-side movement along the y-axis is defined as yaw, up-down movement along the x-axis as pitch, and tilt movement toward the shoulder along the z-axis as roll.

A study by Lin (2017) investigating fear responses in VR evaluated self-help, avoidance, and approach coping strategies. The self-help/self-talk entails screaming, swearing, or auto-suggestion to calm oneself; the approach strategy involves confrontation with the threat stimuli and active virtual exploration; the avoidance strategy involves mental disengagement (e.g., reorienting one's thought or attention to something else using distracting strategy) and/or physical disengagement (e.g., closing eyes or moving away from the stimuli) (Lin, 2017). The study showed that participants using avoidance strategy have reported closing their eyes or moving away from the stimulus by changing their head orientation on the three axes: y, x, and z (Lin, 2017). Interestingly, Lin (2017) reported that its plausibility illusion (PSI) compared to place illusion (PI) that contributed to the fear response, especially those who used avoidance strategy. Lin's (2017) study suggests measuring place illusion is not enough for understanding the impact of interaction with the virtual environment on mental health. It also indicates that participants with different cognitive strategies show variations in cognitive engagement (attention capture and active scanning of the environment versus restricting the scope of attentional focus), behavioral engagement (closing eyes or moving head and body away from the threatening stimuli), and emotional engagement (feeling negatively aroused compared to positively aroused) while interacting with the fearful stimuli under HMD VR.

Another study investigating the relationship between affective experience and head-tracking while exploring the 360° videos presented using HMD VR (Li et al., 2017) has shown a positive correlation between affective experience and head movement. Li et al. (2017) observed a higher standard deviation of yaw movement when participants reported experiencing a pleasant feeling compared to an unpleasant feeling. In addition, they observed a higher pitch movement when participants reported experiencing high compared to low arousal. A study from my group observed a similar pattern along with the higher angular speed of yaw movement when participants reported pleasant compared to unpleasant feeling, suggesting coverage of a wider region of exploration during pleasant experience than unpleasant experience (manuscript under submission).

Li et al. (2017) results could be explained by the broaden-and-build theory of emotion (Fredrickson, 2001, 2013). It can be suggested that pleasant compared to unpleasant state modulates the scope of attention and motivates observer/viewer to seek more information. The broader scope of attention and seeking more information can be supported by wider spatiotemporal virtual exploration (Srivastava & Srinivasan, 2010). The broadening aspect of the theory of positive emotion suggests that emotion modulates the scope of attention and allows either a wider or a narrower focus of attention, as per the individual emotional state. The broader scope of attention allows more distributed/defocused attention to acquire more information compared to a negative state of mind that narrows the focus of attention and restricts the scanning to a limited object/event property (Fenske & Eastwood, 2003; Srivastava & Srinivasan, 2010). The building aspect of the theory of emotion suggests that pleasant compared to unpleasant state increases motivation to seek more information and urges doing more act (Fredrickson, 2001, 2013). With VR experience, the building aspect can be assessed by more exploration during VR experience, and urges and motivation to do more tasks or activity in their everyday scenarios could be assessed as a post VR experience.

Attention and affective state can modulate the experience of flow and sense of agency, and may affect the individual's response to the VR stimulus. The positive state of mind increases the ownership and feeling of control over one's decision, judgment, or any other actions (Yoshie & Haggard, 2017). Therefore, it is important to use VR measures along with clinical observations and self-report to assess the cognitive, behavioral, and affective engagement to increase the sense of agency and experience of flow to achieve more goal-directed behavior.

9.4 Augmented Reality

This section focuses on the second emerging technology, augmented reality, as an aid for mental health research and clinical practices. We'll discuss the state-of-the-art augmentative research and its limitations.

The emerging AR devices, such as HoloLens and Magic Leap, are becoming another tool for studying mental health and well-being issues. Like VR, augmented reality allows users to interact with stimuli that for various reasons may not be possible or suitable for real-life scenarios. The AR can be considered a subpart of virtual technologies, in which the virtual objects are placed in the real world in real time. In AR settings, the user interacts with the virtual object while experiencing the real world in parallel. In contrast to virtual reality, in which the user is transported completely to the virtual artificial environment, augmented reality does not displace user from their present physical space and displays the virtual artificial objects within the view of the user's real world. Since the augmented reality builds the virtual elements overlaid on the user's existing perception of the actual world, it offers a few advantages over virtual reality. The purpose of augmenting virtual objects onto the user's real world is to enhance the experience and/or knowledge of their present world (Azuma et al., 2001; Baus & Bouchard, 2014; Cipresso et al., 2018). For any system to be considered AR, the system must have the following features (Azuma et al., 2001; Baus & Bouchard, 2014; Cipresso et al., 2018):

- The system should merge the real and virtual objects in real world or real environment.
- The system should run interactively and synchronize with real time.
- The system should register (align) the virtual and real objects with each other.

User experience of augmented reality is fundamentally different from virtual reality. Unlike VR, the AR does not detract the sense of presence from the real world and does not take the users to Alice's wonderland fantasy through the binocular VR glasses. Rather, AR demands user's sense of presence to be kept intact in the actual world and yet makes user feel that the virtual object with which they are interacting is a part of this real world. It is expected that the virtual object should reciprocate the user's intention and action in real time, and make the user feel as if the virtual object is part of the real world. The illusion that makes the user believe that the scenario being depicted is actually occurring in the environment is known as the plausibility illusion (Slater, 2009). This means that AR might require different measures to evaluate user experience than user experience from VR. Given that it's the interaction with the virtual object that is presented onto the user's real-world space is the key difference between virtual reality and augmented reality, the augmented reality demands more plausibility illusion than place illusion (Slater, 2009) to let user imitate the actual reaction as it would in real time. Place illusion (PI) is determined by the sensorimotor contingencies rendered by the virtual reality system, whereas, as Slater (2009) describes, "plausibility illusion (PSI) is determined by the extent to which the system can produce events that directly relate to the participant and the overall credibility of the scenario being depicted in comparison with expectations." It is the tight coordination between the real and virtual objects and their interaction with the user that creates an experience in AR. However, the realism (degree of convergence between the expected experience and actual experience in the virtual environment) and reality (extent to which the user experiences the hybrid environment like the real one) could still be pertinent to augmented reality.

9.4.1 Augmented Reality Clinical Research

AR, unlike VR, does not need a fully immersive, dynamic, rich 360° surrounding, but rather uses the real physical spaces and introduces an object/event into the real world from the user's point of view. The AR clinical research primarily focused on treatments, more specifically therapies for phobias of small animals, like cockroaches and spiders, and phobia of heights, i.e., acrophobia (Baus & Bouchard, 2014; Giglioli et al., 2015). The AR clinical research is greatly diversified in their sample size and approach. For the most part, the AR clinical research employed cockroaches as virtual stimuli embedded into the patients' actual world (see, for more details, Baus & Bouchard, 2014; Giglioli et al., 2015). Literature varies from reporting single patient's case study (Juan et al., 2004), to less than 10 patients study (Botella et al., 2010; Bretón-López et al., 2010), to more than 20 healthy adults or patients (Botella et al., 2016; Bretón-López et al., 2010; Wrzesien et al., 2011, 2013).

Regarding measures, AR clinical research appears to follow in the footsteps of VR clinical research. Like VR clinical research, these studies used questionnaires and asked their participants to self-report on anxiety (Botella et al., 2010; Bretón-López et al., 2010; Juan et al., 2004; Wrzesien et al., 2011, 2013), presence (Bretón-López et al., 2010), reality judgment (Bretón-López et al., 2010), avoidance behavior (Botella et al., 2010, 2016; Wrzesien et al., 2011, 2013), clinical severity scale (Botella et al., 2016), and expectation and satisfaction rating (Botella et al., 2016). These studies did not report any AR and associated measures that could have objectively informed about patients' cognitive withdrawal (e.g., focusing on the stimulus, not moving away or turning their head from the stimulus, particularly when it is there in their current physical space; recall of details of such interaction and stimulus) and physical withdrawal (e.g., contraction of muscles while approaching the fearful stimuli or the frequency with which the muscles have been contracted or latency in approaching the stimuli). The AR measures could also be used for assessing the approach strategies (e.g., increased frequency of touching the animal, modulation in muscle contractions, more sustained focusing on the stimulus, better recall of the surrounding than the anxious state) during and after AR intervention. In the future, these measures may also incorporate the physiological recording like heart rate or skin conductance, or change in their DMN as mentioned earlier regarding clinical VR research.

Similar to VR clinical research, introducing virtual element in the actual world was tightly controlled and was performed under the supervision of a clinician with graded complexity of varying size, number, and surrounding details (Baus & Bouchard, 2014; Giglioli et al., 2015). However, AR stands out in the number of sessions that were required to taper off anxiety in response to the phobic stimulus. Clinical AR research has followed the one-session treatment guidelines (Öst et al., 1991), which intensively exposes them to the phobic stimulus lasting up to three hours. The AR single session comprised of seeing, touching, and finally killing one or two concerning stimuli in progression. As mentioned earlier, the therapist chooses the number, size, and dynamics of it as the patient progresses in AR interaction. Patients report their discomfort and anxiety before and after the AR exposure therapy, of which the comparisons lead clinical AR research to assess their level of anxiety and discomfort in correspondence with the phobic stimulus (Baus & Bouchard, 2014; Botella et al., 2016; Giglioli et al., 2015). As mentioned in VR measure section, AR also needs to scale up the measurement techniques to allow for objective biopsychosocial perspectives.

9.5 Summary and Future Directions

This chapter focused on virtual reality and augmented reality clinical research. The chapter was divided into three major parts: description of mental health, VR, and AR (Sections 9.2, 9.3, and 9.4), the current state of mental health and well-being practices (Section 9.2), virtual reality and augmented reality clinical research (Sections 9.3.1 and 9.4.1), and the need for virtual and augmented reality measures, cognitive measures, and multiple measures ranging from eye-tracking to brain correlates.

The mental health section described the current psychiatric and clinical psychological practices followed by highlighting the limitations. It discussed the new framework for assessment and treatment of psychological disorder proposed by the NIMH. The proposed new framework has highlighted the need to shift the focus from the Diagnostic and Statistical Manual of Mental Disorders (DSM) categories to more on biological, cognitive, and behavioral constructs that are believed to be the building blocks of mental disorders. This section discussed the biopsychosocial perspectives of psychological disorders, rather than the DSM categories.

The chapter described the conceptualization of virtual reality and augmented reality, and highlighted the key components that are necessary to get user experience as real as possible while interacting with the virtual objects placed in a fully immersive virtual environment used in highly immersive head-mounted display VR systems, such HTCVive or Oculus Rift, or the virtual objects placed in their present real world as part of augmented reality systems such as HoloLens or Magic Leap. The highly immersive VR interface separates the user from the real world and demands user's attention to the virtual environment. Virtual reality creates an inclusive, extensive, vivid, 360° displays and allows users to control the object and predict its future course of action matched with the user's egocentric/allocentric point of view in a given virtual environment. On the other hand, augmented reality creates an extensive and vivid virtual object and uses matching and interactability concepts similar to VR devices to let the user experience as real an interaction as possible while being present in the real world. The chapter introduced two important key concepts, sense of presence/place illusion and plausibility illusion, that determine the user experience while interacting with the virtual object placed in a virtual environment (VR) or real environment (AR). Upon realizing the fundamental difference between the two systems, it was suggested that both systems should use different measures for assessing user experience.

The VR and AR clinical research sections discussed the research trends in the last 20 years. It highlighted the major psychological disorders that have been studied using both technologies. The VR clinical research could explore the assessment, explanation/theory development, and treatment aspect of psychological disorder. The therapies used in VR clinical research were primarily exposure, distraction, and training, whereas AR was limited to exposure therapy only. These sections have also highlighted the limitations of observations made in these studies. We discussed the limitations to realize the future avenues of such research and yet recognize the departure from traditional clinical practices to more advanced clinical research and practices in the future.

The last contribution of this chapter was to make recommendations for future VR and AR clinical research. The chapter recommended the use of multiple measures to study more holistic biopsychosocial perspectives of psychological disorders. The chapter suggested three measures: VR/AR measures, biological correlates, and cognitive measures, which should be incorporated in future VR/AR clinical research. The biological correlates may vary from physiological monitoring to brain correlates while being present in the VR or AR system by employing electroencephalography or event-related potential measures. In the future, the VR and AR studies should use fMRI techniques, if possible, to study participants' brain correlates before and after interventions. It was suggested that such research could also assess the default mode network (DMN) areas of the unhealthy brain compared to a healthy brain and evaluate the adaptability and plasticity of an individual brain. The chapter also discussed the need for measuring cognitive engagement and cognitive processing, especially in the case of studying mental health patients. The chapter reported a few studies, including the study from my own lab to demonstrate the use of VR head-tracking measures to understand cognitive engagement while interacting with the virtual environment.

REFERENCES

Armstrong, T., & Olatunji, B. O. (2012). Eye tracking of attention in the affective disorders: A meta-analytic review and synthesis. *Clinical Psychology Review, 32*(8), 704–723.

American Psychiatric Association. (2013). *Diagnostic and statistical manual of mental disorders (DSM-5®).* American Psychiatric Pub.

Azuma, R., Baillot, Y., Behringer, R., Feiner, S., Julier, S., & MacIntyre, B. (2001). Recent advances in augmented reality. *IEEE Computer Graphics and Applications, 21*(6), 34–47.

Bar-Haim, Y. (2010). Research review: Attention bias modification (ABM): A novel treatment for anxiety disorders. *Journal of Child Psychology and Psychiatry, 51*(8), 859–870.

Baus, O., & Bouchard, S. (2014). Moving from virtual reality exposure-based therapy to augmented reality exposure-based therapy: A review. *Frontiers in Human Neuroscience, 8*, 112.

Beck, A. T. (2008). The evolution of the cognitive model of depression and its neurobiological correlates. *American Journal of Psychiatry, 165*(8), 969–977.

Ben-Eliyahu, A., Moore, D., Dorph, R., & Schunn, C. D. (2018). Investigating the multidimensionality of engagement: Affective, behavioral, and cognitive engagement across science activities and contexts. *Contemporary Educational Psychology, 53*, 87–105.

Berman, M. G., Peltier, S., Nee, D. E., Kross, E., Deldin, P. J., & Jonides, J. (2011). Depression, rumination and the default network. *Social Cognitive and Affective Neuroscience, 6*(5), 548–555.

Blanke, O., & Metzinger, T. (2009). Full-body illusions and minimal phenomenal selfhood. *Trends in Cognitive Sciences, 13*(1), 7–13.

Borghi, A. M., Barca, L., Binkofski, F., & Tummolini, L. (2018). Varieties of abstract concepts: Development, use and representation in the brain. *Philosophical Transactions of the Royal Society B: Biological Sciences, 373*(1752), 20170121.

Botella, C., Bretón-López, J., Quero, S., Baños, R., & García-Palacios, A. (2010). Treating cockroach phobia with augmented reality. *Behavior Therapy, 41*(3), 401–413.

Botella, C., Pérez-Ara, M. Á., Bretón-López, J., Quero, S., García-Palacios, A., & Baños, R. M. (2016). In vivo versus augmented reality exposure in the treatment of small animal phobia: A randomized controlled trial. *PloS one, 11*(2), e0148237.

Bowman, D. A. and McMahan, R. P. (2007). Virtual Reality: How Much Immersion Is Enough? *Computer, 40*(7), 36–43. doi:10.1109/MC.2007.257

Braun, N., Debener, S., Spychala, N., Bongartz, E., Sörös, P., Müller, H. H., & Philipsen, A. (2018). The senses of agency and ownership: A review. *Frontiers in Psychology, 9*, 535.

Bretón-López, J., Quero, S., Botella, C., García-Palacios, A., Baños, R. M., & Alcaniz, M. (2010). An augmented reality system validation for the treatment of cockroach phobia. *Cyberpsychology, Behavior, and Social Networking, 13*(6), 705–710.

Britton, J. C., Grillon, C., Lissek, S., Norcross, M. A., Szuhany, K. L., Chen, G., … Pine, D. S. (2013). Response to learned threat: An fMRI study in adolescent and adult anxiety. *American Journal of Psychiatry, 170*(10), 1195–1204.

Chandrasiri, A., Collett, J., Fassbender, E., & De Foe, A. (2020). A virtual reality approach to mindfulness skills training. *Virtual Reality, 24*(1), 143–149.

Chrastil, E. R., & Warren, W. H. (2012). Active and passive contributions to spatial learning. *Psychonomic Bulletin & Review, 19*(1), 1–23.

Chrastil, E. R., & Warren, W. H. (2013). Active and passive spatial learning in human navigation: Acquisition of survey knowledge. *Journal of Experimental Psychology: Learning, Memory, and Cognition*, *39*(5), 1520.

Cipresso, P., Giglioli, I. A. C., Raya, M. A., & Riva, G. (2018). The past, present, and future of virtual and augmented reality research: A network and cluster analysis of the literature. *Frontiers in Psychology*, *9*, 2086.

Cisler, J. M., & Koster, E. H. (2010). Mechanisms of attentional biases towards threat in anxiety disorders: An integrative review. *Clinical Psychology Review*, *30*(2), 203–216.

Collins, P. Y., Patel, V., Joestl, S. S., March, D., Insel, T. R., Daar, A. S., … Glass, R. I. (2011). Grand challenges in global mental health. *Nature*, *475*(7354), 27–30.

Cotrena, C., Branco, L. D., Shansis, F. M., & Fonseca, R. P. (2016). Executive function impairments in depression and bipolar disorder: Association with functional impairment and quality of life. *Journal of Affective Disorders*, *190*, 744–753.

Csikszentmihalyi, M., & Csikzentmihaly, M. (1990). *Flow: The psychology of optimal experience* (Vol. *1990*). New York: Harper & Row.

Cummins, F. (2014). Agency is distinct from autonomy. *AVANT. Pismo Awangardy Filozoficzno-Naukowej*, (2), 98–112.

DeJong, H., Fox, E., & Stein, A. (2019). Does rumination mediate the relationship between attentional control and symptoms of depression? *Journal of Behavior Therapy and Experimental Psychiatry*, *63*, 28–35.

Difede, J., & Hoffman, H. G. (2002). Virtual reality exposure therapy for World Trade Center post-traumatic stress disorder: A case report. *Cyberpsychology & Behavior*, *5*(6), 529–535.

Duque, A., & Vázquez, C. (2015). Double attention bias for positive and negative emotional faces in clinical depression: Evidence from an eye-tracking study. *Journal of Behavior Therapy and Experimental Psychiatry*, *46*, 107–114.

Epp, A. M., Dobson, K. S., Dozois, D. J., & Frewen, P. A. (2012). A systematic meta-analysis of the Stroop task in depression. *Clinical Psychology Review*, *32*(4), 316–328.

Falconer, C. J., Rovira, A., King, J. A., Gilbert, P., Antley, A., Fearon, P., … Brewin, C. R. (2016). Embodying self-compassion within virtual reality and its effects on patients with depression. *BJPsych Open*, *2*(1), 74–80.

Fenske, M. J., & Eastwood, J. D. (2003). Modulation of focused attention by faces expressing emotion: Evidence from flanker tasks. *Emotion, 3*(4), 327–343. https://doi.org/10.1037/1528-3542.3.4.327

Fredrickson, B. L. (2001). The role of positive emotions in positive psychology: The broaden-and-build theory of positive emotions. *American Psychologist*, *56*(3), 218.

Fredrickson, B. L. (2013). Positive emotions broaden and build. In *Advances in experimental social psychology* (Vol. *47*, pp. 1–53). New York: Academic Press.

Freeman, D., Slater, M., Bebbington, P. E., Garety, P. A., Kuipers, E., Fowler, D., … Vinayagamoorthy, V. (2003). Can virtual reality be used to investigate persecutory ideation? *The Journal of Nervous and Mental Disease*, *191*(8), 509–514.

Freeman, D., Haselton, P., Freeman, J., Spanlang, B., Kishore, S., Albery, E., … Nickless, A. (2018). Automated psychological therapy using immersive virtual reality for treatment of fear of heights: A single-blind, parallel-group, randomised controlled trial. *The Lancet Psychiatry*, *5*(8), 625–632.

Freeman, D., Reeve, S., Robinson, A., Ehlers, A., Clark, D., Spanlang, B., & Slater, M. (2017). Virtual reality in the assessment, understanding, and treatment of mental health disorders. *Psychological Medicine*, *47*(14), 2393–2400.

Freeman, D. (2008). Studying and treating schizophrenia using virtual reality: A new paradigm. *Schizophrenia Bulletin*, *34*(4), 605–610.

Giglioli, I. A. C., Pallavicini, F., Pedroli, E., Serino, S., & Riva, G. (2015). Augmented reality: A brand new challenge for the assessment and treatment of psychological disorders. *Computational and Mathematical Methods in Medicine*, *2015*, Article ID 862942. doi:10.1155/2015/862942

Harvey, P. O., Fossati, P., Pochon, J. B., Levy, R., LeBastard, G., Lehéricy, S., & Dubois, B. (2005). Cognitive control and brain resources in major depression: An fMRI study using the n-back task. *Neuroimage*, *26*(3), 860–869.

Heeren, A., Mogoaşe, C., Philippot, P., & McNally, R. J. (2015). Attention bias modification for social anxiety: A systematic review and meta-analysis. *Clinical Psychology Review*, *40*, 76–90.

Hoffman, H. G., Sharar, S. R., Coda, B., Everett, J. J., Ciol, M., Richards, T., & Patterson, D. R. (2004). Manipulating presence influences the magnitude of virtual reality analgesia. *Pain*, *111*(1–2), 162–168.

Jerald, J. (2015). *The VR book: Human-centered design for virtual reality*. Morgan & Claypool.

Juan, M. C., Botella, C., Alcaniz, M., Banos, R., Carrion, C., Melero, M., & Lozano, J. A. (2004, November). *An augmented reality system for treating psychological disorders: Application to phobia to cockroaches.* In *Third IEEE and ACM international symposium on mixed and augmented reality* (pp. 256–257). IEEE.

Kertzman, S., Reznik, I., Hornik-Lurie, T., Weizman, A., Kotler, M., & Amital, D. (2010). Stroop performance in major depression: Selective attention impairment or psychomotor slowness? *Journal of Affective Disorders, 122*(1–2), 167–173.

Keyes, C. L. (2006). Subjective well-being in mental health and human development research worldwide: An introduction. *Social Indicators Research, 77*(1), 1–10.

Li, B. J., Bailenson, J. N., Pines, A., Greenleaf, W. J., & Williams, L. M. (2017). A public database of immersive VR videos with corresponding ratings of arousal, valence, and correlations between head movements and self-report measures. *Frontiers in Psychology, 8,* 2116.

Lin, J. H. T. (2017). Fear in virtual reality (VR): Fear elements, coping reactions, immediate and next-day fright responses toward a survival horror zombie virtual reality game. *Computers in Human Behavior, 72,* 350–361.

Lisk, S., Vaswani, A., Linetzky, M., Bar-Haim, Y., & Lau, J. Y. (2020). Systematic review and meta-analysis: Eye-tracking of attention to threat in child and adolescent anxiety. *Journal of the American Academy of Child & Adolescent Psychiatry, 59*(1), 88–99.

Lobo, L., Heras-Escribano, M., & Travieso, D. (2018). The history and philosophy of ecological psychology. *Frontiers in Psychology, 9,* 2228.

Luck, M., & d'Inverno, M. (1995, June). A formal framework for agency and autonomy. In *ICMAS* (Vol. 95, pp. 254–260).

McMahan, R. P., Bowman, D. A., Zielinski, D. J., & Brady, R. B. (2012). Evaluating display fidelity and interaction fidelity in a virtual reality game. *IEEE Transactions on Visualization and Computer Graphics, 18*(4), 626–633.

Metzinger, T. K. (2018). Why is virtual reality interesting for philosophers? *Frontiers in Robotics and AI, 5,* 101.

Michaliszyn, D., Marchand, A., Bouchard, S., Martel, M. O., & Poirier-Bisson, J. (2010). A randomized, controlled clinical trial of in virtuo and in vivo exposure for spider phobia. *Cyberpsychology, Behavior, and Social Networking, 13*(6), 689–695.

Moore, J. W. (2016). What is the sense of agency and why does it matter? *Frontiers in Psychology, 7,* 1272.

Moreno, A., Wall, K. J., Thangavelu, K., Craven, L., Ward, E., & Dissanayaka, N. N. (2019). A systematic review of the use of virtual reality and its effects on cognition in individuals with neurocognitive disorders. *Alzheimer's & Dementia: Translational Research & Clinical Interventions, 5,* 834–850.

Nakamura, J., & Csikszentmihalyi, M. (2014). The concept of flow. In *Flow and the foundations of positive psychology* (pp. 239–263). Dordrecht: Springer.

Niharika, P., Reddy, N. V., Srujana, P., Srikanth, K., Daneswari, V., & Geetha, K. S. (2018). Effects of distraction using virtual reality technology on pain perception and anxiety levels in children during pulp therapy of primary molars. *Journal of Indian Society of Pedodontics and Preventive Dentistry, 36*(4), 364.

Öst, L. G., Salkovskis, P. M., & Hellström, K. (1991). One-session therapist-directed exposure vs. self-exposure in the treatment of spider phobia. *Behavior Therapy, 22*(3), 407–422.

Pacherie, E. (2007). The sense of control and the sense of agency. *Psyche, 13*(1), 1–30.

Pienkos, E., Giersch, A., Hansen, M., Humpston, C., McCarthy-Jones, S., Mishara, A., and Thomas, N. (2019). Hallucinations beyond voices: A conceptual review of the phenomenology of altered perception in psychosis. *Schizophrenia Bulletin, 45*(Supplement_1), S67–S77.

Reddy, V. (2019). Mental health issues and challenges in India: A review. *International Journal of Social Sciences Management and Entrepreneurship (IJSSME), 3*(2).

Richardson, J. C., & Newby, T. (2006). The role of students' cognitive engagement in online learning. *American Journal of Distance Education, 20*(1), 23–37.

Riva, E., Freire, T., & Bassi, M. (2016). The flow experience in clinical settings: Applications in psychotherapy and mental health rehabilitation. In *Flow experience* (pp. 309–326). Cham: Springer.

Riva, G. (2011). The key to unlocking the virtual body: Virtual reality in the treatment of obesity and eating disorders. *Journal of Diabetes Science and Technology, 5*(2), 283–292.

Rizzo, A. S., & Shilling, R. (2017). Clinical virtual reality tools to advance the prevention, assessment, and treatment of PTSD. *European Journal of Psychotraumatology, 8*(sup5), 1414560.

Rotgans, J. I., & Schmidt, H. G. (2011). Cognitive engagement in the problem-based learning classroom. *Advances in Health Sciences Education, 16*(4), 465–479.

Rothbaum, B. O., Hodges, L., Smith, S., Lee, J. H., & Price, L. (2000). A controlled study of virtual reality exposure therapy for the fear of flying. *Journal of Consulting and Clinical Psychology*, 68(6), 1020.

Rothbaum, B. O., Hodges, L., Smith, S., Lee, J. H., & Price, L. (2002). A controlled study of virtual reality exposure therapy for the fear of flying. *Year Book of Psychiatry and Applied Mental Health*, 2002(1), 109–111.

Sanchez, A., Vazquez, C., Marker, C., LeMoult, J., & Joormann, J. (2013). Attentional disengagement predicts stress recovery in depression: An eye-tracking study. *Journal of Abnormal Psychology*, 122(2), 303.

Sanchez-Vives, M. V., Spanlang, B., Frisoli, A., Bergamasco, M., & Slater, M. (2010). Virtual hand illusion induced by visuomotor correlations. *PloS one*, 5(4), e10381.

Schacter, D. L., Gilbert, D. T., & Wegner, D. M. (2009). *Introducing psychology*. New York: Macmillan.

Schroer, S. A. (2019). Jakob von Uexküll: The concept of umwelt and its potentials for an anthropology beyond the human. *Ethnos*, 1–21.

Sheline, Y. I., Barch, D. M., Price, J. L., Rundle, M. M., Vaishnavi, S. N., Snyder, A. Z., ... Raichle, M. E. (2009). The default mode network and self-referential processes in depression. *Proceedings of the National Academy of Sciences*, 106(6), 1942–1947.

Sheline, Y. I., Barch, D. M., Donnelly, J. M., Ollinger, J. M., Snyder, A. Z., & Mintun, M. A. (2001). Increased amygdala response to masked emotional faces in depressed subjects resolve with antidepressant treatment: An fMRI study. *Biological Psychiatry*, 50(9), 651–658.

Slater, M., & Wilbur, S. (1997). A framework for immersive virtual environments (FIVE): Speculations on the role of presence in virtual environments. *Presence: Teleoperators & Virtual Environments*, 6(6), 603–616.

Slater, M. (2009). Place illusion and plausibility can lead to realistic behaviour in immersive virtual environments. *Philosophical Transactions of the Royal Society B: Biological Sciences*, 364(1535), 3549–3557.

Smith, R. (2010). The long history of gaming in military training. *Simulation & Gaming*, 41(1), 6–19.

Soujanya Kodavalla, S., Jai Bhagwan Goel, M., & Srivastava, P. (2019, November). *Indian virtual reality affective database with self-report measures and EDA*. In *25th ACM Symposium on Virtual Reality Software and Technology* (pp. 1–2).

Srivastava, P., & Srinivasan, N. (2010). Time course of visual attention with emotional faces. *Attention, Perception, & Psychophysics*, 72(2), 369–377.

Srivastava, P., Rimzhim, A., Vijay, P., Singh, S., & Chandra, S. (2019). Desktop VR is better than nonambulatory HMD VR for spatial learning. *Frontiers in Robotics and AI*, 6, 50.

Sutherland, I. E. (1965). The ultimate display. *Multimedia: From Wagner to Virtual Reality*, 1.

Valmaggia, L. R., Latif, L., Kempton, M. J., & Rus-Calafell, M. (2016). Virtual reality in the psychological treatment for mental health problems: A systematic review of recent evidence. *Psychiatry Research*, 236, 189–195.

Vuorre, M., & Metcalfe, J. (2016). The relation between the sense of agency and the experience of flow. *Consciousness and Cognition*, 43, 133–142.

Warren, W. H. (2006). The dynamics of perception and action. *Psychological Review*, 113(2), 358.

Wiebking, C., Bauer, A., De Greck, M., Duncan, N. W., Tempelmann, C., & Northoff, G. (2010). Abnormal body perception and neural activity in the insula in depression: An fMRI study of the depressed "material me". *The World Journal of Biological Psychiatry*, 11(3), 538–549.

Wilson, C. J., & Soranzo, A. (2015). The use of virtual reality in psychology: A case study in visual perception. *Computational and Mathematical Methods in Medicine*, 2015.

Wrzesien, M., Burkhardt, J. M., Alcañiz, M., & Botella, C. (2011, September). *How technology influences the therapeutic process: A comparative field evaluation of augmented reality and in vivo exposure therapy for phobia of small animals*. In *IFIP Conference on Human-Computer Interaction* (pp. 523–540). Berlin, Heidelberg: Springer.

Wrzesien, M., Alcañiz, M., Botella, C., Burkhardt, J. M., Bretón-López, J., Ortega, M., & Brotons, D. B. (2013). The therapeutic lamp: Treating small-animal phobias. *IEEE Computer Graphics and Applications*, 33(1), 80–86.

Yoshie, M., & Haggard, P. (2017). Effects of emotional valence on sense of agency require a predictive model. *Scientific Reports*, 7(1), 1–8.

Zhao, X. H., Wang, P. J., Li, C. B., Hu, Z. H., Xi, Q., Wu, W. Y., & Tang, X. W. (2007). Altered default mode network activity in patient with anxiety disorders: An fMRI study. *European Journal of Radiology*, 63(3), 373–378.

10

Solar Potential Estimation and Management Using IoT, Big Data, and Remote Sensing in a Cloud Computing Environment

Mudit Kapoor and Rahul Dev Garg
Indian Institute of Technology Roorkee, Roorkee, India

CONTENTS

10.1 Introduction

The Internet of Things (IoT) allows machines to communicate and exchange information between themselves without a minimum number of human mediators. The IoT plays an important role in transferring information and making decisions. It helps remote sensing by taking a recording of the place from a remote location and sending the data to the users for processing. The term "IoT" has been coined by Kevin Ashton in 1999 while presenting the work to Procter & Gamble (P&G) (Wikipedia, 2018; Zaeem Hosain, 2018). The IoT deals with the sensors and physical devices connected to the internet. If the device is not connected to the internet now, it will be connected to the internet in the future. The Machine-to-Machine (M2M) communication is the connection between the different types of machines such as desktop, scanner, mouse, etc. This M2M is a subset of the IoT. In both cases, M2M and IoT cannot replace each other. The IoT is the superset of the connected machines (Mohseninia, 2017). The information from the satellites, solar panel, and sensors gets transmitted to the analytics using the internet. The four key components of the IoT are people, things, data, and processes (Engineering & Board, 2018; Kumar & Kalavathi, 2018). The people here refer to the endpoints connected to the internet to share information such as, for example, remote sensing, health, and fitness sensors. Real-time photos can be input to the algorithm to extract the features Badenko et al. (2013a, 2013b). These extracted features have been utilized for estimating a usable area to install the SPV panel. Research questions that have been framed to test the applicability of this research include: Is it feasible to install the solar energy–generating equipment at this location? and How much solar potential is available using tilted GHI from pyranometer? Estimation of the best management practices for the solar plant maintenance is also investigated in this research.

DOI: 10.1201/9781003140351-10

10.2 Literature Review

Smart monitoring devices have been developed for solar plant management using IoT: sensors to provide data analytics, SPV control, and detect faults (Spanias, 2017). An integrated information system has been created for environmental monitoring and management by utilizing the latest technologies such as IoT, geoinformatics, big data, and cloud computing (Fang et al., 2014). The IoT has been used by many researchers for the solar potential assessment, management, and monitoring purposes with the application of remote sensing, GIS, big data, and cloud computing (Escolar et al., 2014; Hu et al., 2015; Markovic et al., 2013; Sharma, 2016). Aeris IoT services provide solutions for the maintenance of the solar power projects using both GSM and CDMA connectivity including 2G, 3G, and 4G LTE (Hermann et al., 2014; Litjens et al., 2018; Mohseninia, 2017; Mulder, 2014; Ranganadham, 2018; Saran et al., 2015). This helps in analyzing and processing real-time updates and data of the solar project. These types of services help in getting accurate and fast information from the M2M and IoT project. The online solar irradiance data have been used by many researchers to carry out the solar potential assessment, energy, and cost analysis. Researchers have used solar irradiance (NASA), and land use dataset (European Space Agency) to perform the solar irradiance analysis (Teluguntla et al., 2018; Yingzi & Yexia, 2019).

Lefevre et al. (2004) have derived the basic principle of Heliosat-II as describing in Equation 10.1. The equation has defined the cloud index, n, as the value of the irradiance detected by the sensor versus the irradiance value at the pixel at the ground. If there no difference, then it is related to the clearness of the atmosphere.

$$n^t(i,j) = \left[\rho^t(i,j) - \rho_g^t(i,j) \right] / \left[\rho_{cloud}^t - \rho_g^t(i,j) \right] \tag{10.1}$$

$\rho^t(i,j)$ is the reflectance, or apparent albedo, observed by the spaceborne sensor for the time t and the pixel (I,j): $\rho^t(i,j) = \dfrac{\pi L^t(i,j)}{I_{0met} \in (t) \, cost \, \theta_S(t,i,j)}$, where $L^t(i,j)$ is the observed radiance;

$\rho_{cloud}^t(i,j)$ is the apparent albedo over the brightest clouds, and $\rho_g^t(i,j)$ is the apparent albedo over the ground under clear skies.

Zhang and Grijalva (2016) have devised a data-driven methodology for the identification, verification, and estimation of residential PV installations using change point detection algorithm to select nonfunctional energy utilizations and verified using a permutation list with Spearman's rank coefficient. They have calculated the PV load using a local meteorological dataset. Spanias (2017) has developed a framework to manage solar projects using the IoT sensors. He demonstrated the framework to manage and provide mobile analytics, enable solar plant control, software code to detect and manage faults, optimize power and reduce inverter transients with the help of Kalman filter, signal processing, neural network, and digital image processing methods.

10.3 Study Area and Data Used

The study area selected for this study is Govind Ballabh Pant Institute of Engineering and Technology (GBPIET), Ghurdauri, located in Pauri Garhwal district of Uttarakhand, India (Figure 10.1). The selected study area is located on a hilly terrain. This study has been selected to assess the global horizontal irradiance (GHI) at an altitude higher than plane regions. This has helped in understanding the pollution and aerosols effect on the GHI. Pyranometer field survey has been performed on September 17, 2018, as shown in Figure 10.2.

The satellite data of the study area have been downloaded using Google Earth and Elshayal Smart GIS (Elshayal, 2018; Google Earth, 2018). The complete area has been downloaded in 40 parts. These 40 parts

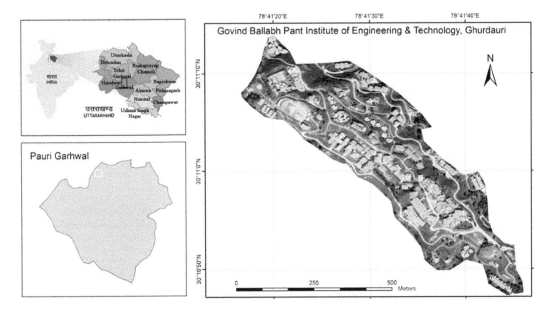

FIGURE 10.1   Study area selected for the proposed study.

FIGURE 10.2   Pyranometer data collection at GBPIET, Ghurdauri.

have been mosaicked using ArcMap to get the complete study for GIS analysis (Esri, 2018; Smith et al., 2018). Meteorological data of 29 years (1989 to 2017) have been utilized in assessing the solar potential of the study area.

10.4 Methodology

In this proposed methodology, latest technologies such as MapReduce and IoT have been implemented to predict the solar potential of the location. A feasibility study has been analyzed to carry out the applicability of the photovoltaic (PV) at the rooftops building. This study has been performed to integrate technologies such as cloud computing, big data, remote sensing, GIS, and programming to produce monthly and annual solar resource maps. Twenty-nine years' worth of meteorological data have been processed in a parallel computing environment using MapReduce to predict solar energy potential. Citrix XenCenter, a bare-metal hypervisor, has been used to create the cloud computing environment. The GHI data of few locations have been obtained from the satellite-derived methods such as the National Renewable Energy Laboratory (NREL) database and at local level using pyranometer. Other satellite-derived GHI

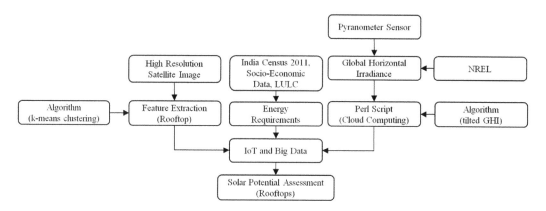

FIGURE 10.3 Proposed methodology for the study.

models, such as Copernicus Atmosphere Monitoring Service (CAMS) McClear, Simple Model of the Atmospheric Radiative Transfer of Sunshine (SMARTS), global solar atlas, and NASA Prediction of Worldwide Energy Resources (POWER), have also been assessed in this research work. The GHI has been obtained from NREL, CAMS McClear, SMART, and NASA POWER, and also using pyranometer with local environmental variables. The proposed methodology is shown as Figure 10.3.

The pyranometer data and satellite-derived NREL data have been utilized in this research to estimate the solar potential. The satellite-derived models, such as CAMS McClear, SMART, and NASA POWER, have been evaluated based on local environmental variables using pyranometer's data. The k-means clustering algorithm has been applied to high-resolution satellite images from Google Earth Pro to extract the building footprints for solar potential estimation. India Census 2011 and socioeconomic data have been applied to estimate the energy requirements of the institute (Badenko et al., 2013a; Xiong et al., 2017). These datasets have been best considered to be processed in the big data platform using Hadoop, and solar potential has been estimated.

10.5 Results and Discussions

The high-resolution satellite images have been extracted using Google Earth Pro and split into 43 segments. The k-means algorithm has been applied to individual segmented parts to extract the building footprints for square footage. A few results obtained using k-means clustering algorithm to extract building footprints are shown in Figure 10.4. The extracted footprints of all the buildings from satellite images are shown as Figure 10.5. The total building rooftop area of GBPIET, Ghurdauri is 34,462 m². Out of this total rooftop area, 70% i.e., 24,124 m² has been considered for solar potential estimation. The 30% area has been kept reserved for infrastructure and maintenance of solar photovoltaics.

The compute-intensive tasks of finding the optimal tilt angle and tilted GHI have been performed using Hadoop architecture in XenCenter cloud computing. The value of the optimal tilt angle obtained after several iterations and a validation using generated and predicted PV outputs at the rooftops of GBPIET, Ghurdauri is approximately 20.18° (Figure 10.6). At this tilt angle, PV panels are capable of producing maximum quantum of electricity from solar energy. A similar approach can be implemented at other locations too. The evaluation of this tilt angle has also been validated with the local GHI (pyranometer) and meteorological data.

The NREL-derived GHI values have been utilized to estimate the monthly and annual solar estimation at tilted solar photovoltaics. The values GHI on monthly/annual basis are shown in Table 10.1. The instant solar estimation using pyranometer has been calculated and found that it is capable of meeting full energy requirements as shown as Table 10.2. This shows that the estimate solar potential is enough to supplement the requirements of electricity demand of GBPIET, Ghurdauri.

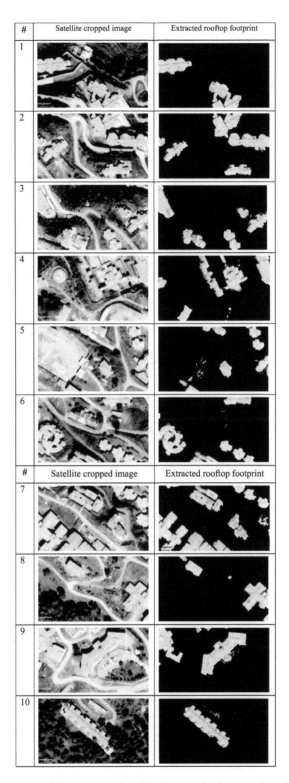

FIGURE 10.4 Few rooftops extracted from segmented satellite images using k-means clustering algorithm.

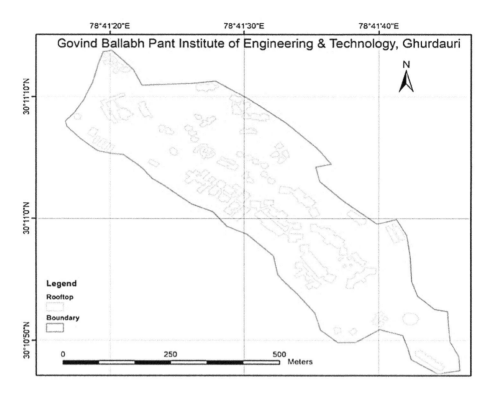

FIGURE 10.5   Rooftops of the study area using the semiautomatic feature and GIS analysis.

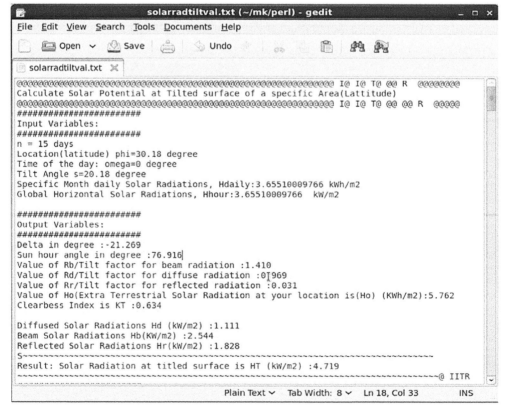

FIGURE 10.6   Snapshot of the tilted GHIs calculated using Perl script.

TABLE 10.1

Solar Irradiance Calculations at Variable Tilted Surface of GBPIET, Ghurdauri

Month/ Annual	GHI (kWh/m²/day)	Tilt angle (Degrees)	GHI at Tilted surface (kWh/m²/day)	Solar Potential calculated for Meerpur Rooftops, η*=57%, Area: 24,124 m² (70%) (MWh/day)
January	3.65	20.18	4.72	113.864
February	4.78	20.18	5.49	132.439
March	5.96	20.18	6.65	160.422
April	6.87	20.18	7.21	173.932
May	7.43	20.18	7.49	180.686
June	6.67	20.18	6.61	159.458
July	5.68	20.18	5.66	136.540
August	5.17	20.18	5.29	127.614
September	5.38	20.18	5.82	140.400
October	5.49	20.18	6.46	155.839
November	4.49	20.18	5.75	138.711
December	3.75	20.18	5.02	121.101
Annual	5.43	20.18	7.29	175.862

* D is Efficiency of the system

TABLE 10.2

Solar Potential Assessed using Pyranometer Data (instance) at GBPIET, Ghurdauri

Date/Time	GHI (h, Wh/m²)	Tilt Angle (s, degree)	Tilted GHI (h_T, Wh/m²)	Assessed Solar Potential* (SPA, MW)
17-09-18 /14:00	921	20.18	939	22.65

* η (efficiency) = 57%

The irradiance values range from 4 to 6 kWh/m²/day, 6–6.8 kWh/m²/day, 3.5–5.5 kWh/m²/day, and 6.9–7.1 kWh/m²/day using CAMS McClear, NASA POWER, SMART, and tilted GHI, respectively. The RMSE values obtained between the irradiances through tilted GHI and satellite derived models are 0.30 (CAMS McClear), 0.12 (NASA POWER), and 0.31 (SMARTS). The values obtained using CAMS McClear model uses real-time weather condition using web services, and provide irradiance values. The NASA POWER uses satellite-derived products and meteorological parameters to provide irradiance values.

The solar resource map obtained using NASA POWER model is more accurate to the ground-based measurements i.e., pyranometer. This has been validated with the reading obtained using pyranometer and tilted GHI. Therefore, it has been concluded from this research that NASA POWER model produces closer values to the tilted GHI, validated using pyranometer readings. The evaluation of these models has been performed using RMSE values obtained between the satellite-derived models (CAMS McClear, SMARTS, and NASA POWER) and tilted GHI approach. The RMSE value obtained for NASA POWER is 0.12. This shows a good correlation with developed tilted GHI approach.

10.6 Conclusion

The pyranometer sensor data and NREL-MNRE tilted GHI data have been used to carry out the solar potential assessment of the study area selected. Since the study area is located at the hilly terrain, the value of GHI is higher compared to the values of GHI at Roorkee (plain region). This clearly shows the effect of pollution and aerosol effect on the values of the GHI data. Solar potential assessment of the location has been performed using GHI values obtained from satellite-derived methods such as CAMS model, NREL, NASA atlas, and local GHI values using pyranometer. The semiautomatic feature extraction approach on high-resolution satellite dataset of WorldView-3 has been applied to extract the usable areas of the rooftops. Rooftop extraction was assisted by the approach of semiautomatic feature extraction. This semiautomatic approach of feature extraction extracted approximately 90% of rooftops. The NASA POWER model has produced best results in terms of ground-based validation with pyranomoter and tilted GHI approach. This has been evaluated using RMSE value.

The annual and monthly maps of GHI, tilted GHI, and solar potential are an excellent resource for the assessment of the individual and organizations for the feasibility of the small and large solar projects. The tilt angle of 20.18° is an essential parameter for the installations of the PV panels to harness the maximum quantum of solar energy. These maps are useful for the fast and quick analysis of the upcoming solar plants at the rooftops, water bodies, and land situated in Uttarakhand, India. These types of studies are needed for all the countries, and are particularly beneficial in areas with prolonged sun exposure.

REFERENCES

Badenko, V., Fedotov, A., & Vinogradov, K. (2013a). *Computational Science and Its Applications–ICCSA 2013* (Vol. *7974*). Springer International Publishing. https://doi.org/10.1007/978-3-642-39649-6

Badenko, V., Kurtener, D., Yakushev, V., Torbert, A., & Badenko, G. (2013b). Computational Science and Its Applications – ICCSA 2013, *7974*, 57–69. https://doi.org/10.1007/978-3-642-39649-6

Elshayal. (2018). Elshayal Smart GIS 18.022. Retrieved from https://freesmartgis.blogspot.com/

Engineering, C., & Board, D. (2018). *Role of IoT in Make in India.* (D. K. Tripathy, Ed.) (Vol. 2). Kolkata.

Escolar, S., Chessa, S., & Carretero, J. (2014). Energy management in solar cells powered wireless sensor networks for quality of service optimization. *Personal and Ubiquitous Computing*, *18*(2), 449–464. https://doi.org/10.1007/s00779-013-0663-1

Esri. (2018). ArcGIS 10.3.1 for Desktop quick start guide—Help I ArcGIS Desktop. Retrieved September 1, 2018, from http://desktop.arcgis.com/en/arcmap/10.3/get-started/quick-start-guides/arcgis-desktop-quick-start-guide.htm

Fang, S., Xu, L. D., Member, S., Zhu, Y., Ahati, J., Pei, H., … Liu, Z. (2014). An integrated system for regional environmental monitoring and management based on Internet of Things. *IEEE Transactions on Industrial Informatics*, *10*(2), 1596–1605. https://doi.org/10.1109/TII.2014.2302638

Google Earth. (2018). Google Earth. Retrieved January 30, 2018, from https://earth.google.com/web/

Hermann, S., Miketa, A., & Fichaux, N. (2014). *Estimating the Renewable Energy Potential in Africa.* International Renewable Energy Agency. https://www.irena.org/-/media/Files/IRENA/Agency/Publication/2014/IRENA_Africa_Resource_Potential_Aug2014.pdf

Hu, T., Zheng, M., Tan, J., Zhu, L., & Miao, W. (2015). Intelligent photovoltaic monitoring based on solar irradiance big data and wireless sensor networks. *Ad Hoc Networks*, *35*, 127–136. https://doi.org/10.1016/j.adhoc.2015.07.004

Kumar, K. R., & Kalavathi, M. S. (2018). Artificial intelligence based forecast models for predicting solar power generation. *Materials Today: Proceedings*, *5*(1), 796–802. https://doi.org/10.1016/j.matpr.2017.11.149

Lefevre, M., Albuisson, M., & Wald, L. (2004). Description of the software Heliosat-II for the conversion of images acquired by Meteosat satellites in the visible band into maps of solar radiation available at ground level. *HAL Archives-Ouvertes*, 1–44.

Litjens, G. B. M. A., Kausika, B. B., Worrell, E., & van Sark, W. G. J. H. M. (2018). A spatio-temporal city-scale assessment of residential photovoltaic power integration scenarios. *Solar Energy*, *174*(October), 1185–1197. https://doi.org/10.1016/j.solener.2018.09.055

Markovic, D. S., Zivkovic, D., Branovic, I., Popovic, R., & Cvetkovic, D. (2013). Smart power grid and cloud computing. *Renewable and Sustainable Energy Reviews*, *24*, 566–577. https://doi.org/10.1016/j.rser.2013.03.068

Mohseninia, M. (2017). Solar monitoring and the Internet of Things I PV Tech. Retrieved March 1, 2019, from https://www.pv-tech.org/news/solar-monitoring-and-the-internet-of-things

Mulder, F. M. (2014). Implications of diurnal and seasonal variations in renewable energy generation for large scale energy storage. *Journal of Renewable and Sustainable Energy*, *6*, 1–13. https://doi.org/10.1063/1.4874845

Ranganadham, M. V. S. (2018). *Energy Statistics*. Retrieved from http://mospi.nic.in/sites/default/files/publication_reports/Energy_Statistics_2018.pdf

Saran, S., Wate, P., Srivastav, S. K., & Krishna Murthy, Y. V. N. (2015). CityGML at semantic level for urban energy conservation strategies. *Annals of GIS, 21*(1), 27–41. https://doi.org/10.1080/19475683.2014.992370

Sharma, S. (2016). Expanded cloud plumes hiding Big Data ecosystem. *Future Generation Computer Systems, 59*, 63–92. https://doi.org/10.1016/j.future.2016.01.003

Smith, M. J. de, Goodchild, M. F., & Longley, P. A. (2018). *Geospatial Analysis*. Retrieved from www.spatialanalysisonline.com

Spanias, A. S. (2017). *Solar Energy Management as an Internet of Things (IoT) Application*. In *8th International Conference on Information, Intelligence, Systems & Applications (IISA)*. Larnaca, Cyprus: IEEE. https://doi.org/10.1109/IISA.2017.8316460

Teluguntla, P., Thenkabail, P., Oliphant, A., Xiong, J., Gumma, M. K., Congalton, R. G., … Huete, A. (2018). A 30-m landsat-derived cropland extent product of Australia and China using random forest machine learning algorithm on Google Earth Engine cloud computing platform. *ISPRS Journal of Photogrammetry and Remote Sensing, 144*, 325–340. https://doi.org/10.1016/j.isprsjprs.2018.07.017

Wikipedia. (2018). Kevin Ashton. Retrieved March 20, 2019, from https://en.wikipedia.org/wiki/Kevin_Ashton

Xiong, J., Thenkabail, P. S., Gumma, M. K., Teluguntla, P., Poehnelt, J., Congalton, R. G., … Thau, D. (2017). Automated cropland mapping of continental Africa using Google Earth Engine cloud computing. *ISPRS Journal of Photogrammetry and Remote Sensing, 126*, 225–244. https://doi.org/10.1016/j.isprsjprs.2017.01.019

Yingzi, L., & Yexia, H. (2019). Comparison and selection of solar radiation data for photovoltaic power generation project. *Journal of Electrical Engineering & Technology, 14*, 685–692. https://doi.org/10.1007/s42835-019-00110-3

Zaeem Hosain, S. (2018). *The Internet of Things for Business*. (S. Z. Hosain, Ed.) (3rd ed.). Aeris.

Zhang, X., & Grijalva, S. (2016). A data driven approach for detection and estimation of residential PV installations. *IEEE Transactions on Smart Grid, 3053*, 1–1. https://doi.org/10.1109/TSG.2016.2555906

Section III

Artificial Intelligence–Based Real-Time Applications

11

Object Detection under Hazy Environment for Real-Time Application

Dileep Yadav and Sneha Mishra
Galgotias University, Greater Noida, India

CONTENTS

11.1 Introduction

"Digital camera" is the term used for a device the role of which is to capture an image or record a sequence of frames without the use of an actual film on which to expose images. "Object detection" or "object tracking" is the term applicable exclusively to the digital camera. The branch of computer science that studies the phenomena associated with the creation and improvements of digital video recorders is called computer vision. The two terms – image processing and computer vision – are sometimes considered the same concepts and used interchangeably. In this chapter we will consider them one by one. Image processing focuses on processing the images by certain transformations like smoothening, increasing brightness or contrast, and detecting or highlighting the edges of an image. Image processing takes input as an image and reverts output as an image. On the other hand, computer vision is used to retrieve information or features from an image. Computer vision takes input as an image, but the output is the information regarding the image, extracted with the help of image-processing techniques. Thus, computer vision is dependent on the image-processing techniques. In many real-time scenarios where the computer is taking a place of human vision, the conditions are such in which human eyes fail to work.

Computer science has been attempting to equip machines with a visual system for a few decades now [1–3]. More than 150 years since the first photographic camera was invented, research is under way to make a machine see with better accuracy and attention to detail than humans can [4, 5]. Computer vision is increasingly taking on complex challenges by including machine learning, artificial intelligence, and deep learning to analyze more complex images with better accuracy.

DOI: 10.1201/9781003140351-11

There are some advanced areas of computer vision that need to be studied by the students and researchers, discussed in the later sections of this chapter. Object detection, discussed in Section 11.1, is a useful application for self-driving vehicles and a much studied area in computer vision. Video tracking, discussed in Section 11.2, is a useful application of surveillances that include facial recognition, activity recognition, etc. Section 11.3 describes the application areas of object detection and video tracking in real-time applications. Section 11.4 brings it to a conclusion and leaves some future scopes summarized in Section 11.5.

11.2 Object Detection

Object detection is a computer vision and image processing technique that allows detecting, locating, and identifying the objects in an image or video. Every object has its own feature, so classification of the object by analyzing the features of the object is also part of object detection [6–10]. The type of the object, whether it is a human, car, building, animal, etc., the total of the objects in the scene comes under features of the objects. Several methods of object detection are performed by several researchers. Basic background modeling, statistical background modeling, fuzzy background modeling, and background estimation are the types of background modeling. Object detection methods are also based on artificial intelligence and deep learning, but background modeling is the simplest and most efficient method used to detect the objects [7–12]. In machine learning the first necessity is to define the features using the techniques, such as support vector machine (SVM) for classification, while deep learning is based on the techniques, like CNN (convolutional neural network), RNN, Faster R-CNN, Retina-net, YOLO, that can perform end-to-end object detection without defining features that much specifically. In Figure 11.1, Zhou et al. describes one frame that is set as a background frame and another frame that is the current frame background subtraction, which gives the output as the object is detected. In Figure 11.2, moving object detection of the colored frames overcoming the various challenges like camouflage, lightening variations, and dynamic background is shown with the original as well as the detected frame.

FIGURE 11.1   Moving object detection by detecting contiguous outliers in the low-rank representation.

FIGURE 11.2 Moving object detection in colored video frames.

11.3 Video Tracking

Traffic surveillance system is the most important aspect of video tracking. Background modeling and foreground object detection play and important role in video surveillance system. Static camera or moving camera captures the videos, inherent changes like waving trees; unconditional weather, water surfaces, etc. may vary as background is not completely stationary. Optical flow algorithms calculate and analyze each pixel motion within two image sequences or frames. Edge detection algorithms detect the regions or edges to detect and classify vehicles in traffic. Figure 11.3 describes the moving object: in the first frame the foreground object is absent, but in the second frame there is a person standing in the image, so in third frame, after applying background subtraction algorithm, the third frame is blank because there is no foreground object, but in the fourth frame the object is detected by subtracting the background from the foreground [13–16].

FIGURE 11.3 Moving object detection.

11.4 Applications

Today's computer vision is playing a revolutionary role in many industries, like transport, medical sciences, agriculture, security, retail, banking, etc. It can be applied in various applications such as:

- **Indoor and outdoor surveillance**: For increased security of any facility or office space., the surveillance team, from a single control room, can visualize as many activities as the connected cameras can cover [16–18].
- **Target detection**: An essential task in video surveillance. Target detection is followed by object detection and tracking of moving objects, allowing efficient analysis of motion of the moving objects.
- **Traffic monitoring and analysis**: Traffic surveillance is the most attractive application area for computer vision. Traffic monitoring's task is to detect vehicles and to classify them. Traffic analysis produces results such as speeding vehicles, gap detection, stopped vehicles detection, and detection of vehicles coming from wrong direction.
- **Abandoned object detection**: Over the past few decades, abandoned object detection has become a wide area of research. Video surveillance identifies the objects that have been stationary for protracted amounts of time, which are then identified as an abandoned object. This application finds wide usage in monitoring of public places such as hospitals, railway stations, and airports, in which finding abandoned items presents a significant risk issue [19, 20].
- **Logo or packaging detection**: In packaging industries, the monitoring of the various objects that are being packed is required. A good example of importance of such monitoring is the packaging of medications to ensure the tamper-proof packaging hasn't been broken. This area needs high accuracy and efficient algorithms of object detection and tracking to maintain the product quality, because it is the direct and effective carrier for any brand [14]. In industries like cosmetics and food packaging, this is the way for a brand to make a good first impression on the consumer.
- **Defense**: Like with many other technological advancements, the military is at the forefront of application and usage of computer vision. Applications, to name just a few, include detection of enemy troops, target detection and tracking, automated firing sequences, and spying activities [20, 21].
- **Activity analysis**: The recognition of activities by regular observations of some particular actions [14, 19]. Computer vision provides activity recognition, which is an important application in industries such health care and security services.

11.5 Related Work

Computer vision is the field that detects objects in a sequence of frames and presents those in the form of a binary mask in each frame. Many vision-based applications such as traffic control, action recognition, human behavior analysis, industrial inspection, and intelligent surveillance are important issues for the researchers and industrialists today. Recently the drone technology has made a new appearance as an important application of machine vision, where the camera may move in a full 360-degree field. Even in stationary cameras, the background cannot be considered as static due to uncontrolled changes such as weather, lighting, etc. [6–8, 15, 22]. The increasing interest in moving object detection has forced researchers to develop robust methods to solve the problems related to camera and to overcome the challenges related to the same. In indoor scenes, the environment is not controlled due to some undesirable challenges of shadow, and sudden fluctuation in light by on-off is very obvious. Outdoor scenes have their own undesirable and uncontrollable challenges like moving trees, clouds, and varying lighting depending on the time of day [15–21, 23–27]. Considerable research in the field has addressed the presence and absence of the uncontrollable environments in the computer-observed scenes. The background subtraction technique, consisting of subtracting the current frame from the previous frames or background

TABLE 11.1

Authors' Contribution in Various Object Detection Techniques

Techniques	Contributions	Improvements
Fuzzy logic based	Zeng et al. [4] T. Bouwman [14]	• Suitable for infrared videos. • More robust in case of dynamic backgrounds such as waving vegetation.
Gaussian based	Wren et al. [5] Kim et al. [22] Lavanya et al. [15, 24] Toyama et al. [28] Zhao et al. [16]	• Better performance in the presence of gradual illumination changes and shadows. • Intrinsic and extrinsic improvements.
Kernel density estimation based	Elgammal et al. [6] Orten et al. [18] Tanaka et al. [25]	• Deals with multimodal backgrounds like waving trees and water rippling, other fast changes. • Many works adopted to reduce computation time.
Optical flow based	Sotirios et al. [29] Lucas et al. [30]	• Great work in object identification with high accuracy [16]. • Optical flow analysis may be performed on the series of images to determine a description of motion.
Subspace learning	Dileep et al. [21, 27] Tsai and Lai [9] Bucak et al. [31] Oliver [8]	• A noticeable improvement in presence of multimodal background. • An intrinsic improvement in presence of shadows.
Deep learning	Krishevesky et al. [26] Simonyan et al. [19] Tianming et al. [20]	• Background subtraction and convolution neural network combined for anomaly detection in much abnormal objects surveillance.

frames, detects a specific object within the scene. Table 11.1 describes the dedicated work done by various authors in the area of object detection, with the certain improvements they made.

Much advancement in the background modeling has been introduced over the years, such as low rank and sparse matrix representation in static camera, where the idea is that coherency is there between the sets of images so the low rank is formed and the sparse matrix contains the outliers. The goal of object tracking is different, as in tracking, the target is marked and then localized in the corresponding frames of the video sequences [8, 11]. Doing so effectively requires information such as histogram, color, texture, etc. to model similarities and update the characteristics to the next frame. The goal of the upcoming section is to find out the challenges that need to be overcome in the near future.

11.6 Challenges

Video surveillance is an active research area, but there are lots of difficulties in capturing a face whether the background is static or moving. Challenges may be related to bad weather conditions, light effects, and poor quality of equipment, among others. Challenges with object detection can arise in any image or video sequence [10, 11]. The crucial issues in video surveillance are given in the following sections.

11.6.1 Bootstrapping

By applying background modeling, if there is no frame available as a background, then it is difficult for the algorithms to detect the original object. Thus, it is very big challenge to model a background in such cases [10, 16, 23].

11.6.2 Camouflage

Camouflage is a condition when two objects of the same color are overlapping on each other in the frames and making it difficult to detect the actual object. In agriculture, for example, the detection of insects is difficult when the animals are of the same color as their leafy or flowery background. Many researchers in this area have worked to reduce this by learning representation of moving objects, but it is still a big challenge in the area of object detection and tracking [14–18, 23, 27, 29].

11.6.3 Illumination Variation

At different times of days, the lighting effects may differ due to variations in light sources, reflection from different sources of light, and blockage of light by other objects. This impacts the object's appearances, which may result in false-positive or false-negative detection. For these reasons lighting consequences are often challenging to handle [14, 15, 32, 33]. Many authors, as discussed in Section 11.2, have tried various approaches in such areas to lessen the challenges of lighting variations.

11.6.4 Foreground Aperture

In this condition the possibility of a false negative increases, as the foreground region has a uniform colored region and changes inside the foreground region are caused by various lighting effects and different directions and speeds of motion [10, 11, 30].

11.6.5 Motion in Background

Background may be of a dynamic nature in this condition. There are sequences of frames in which the background is changing that makes object detection and tracking a challenge. Sometimes the swaying trees, sprouting of water from the fountain or rain, moving flags, etc. detects the false-positive pixels [15, 20, 21, 27, 29, 30, 32, 33].

11.6.6 Occlusion

Occlusion is the condition in which some parts of an object are hidden behind the other object. Occlusion may be of two types: partial or complete, both of which are quite self-explanatory, but a reader can find more detailed definitions in refs. [24, 19–27]. For example, a pedestrian approaching an intersection of a tree-lined streets may be partially or completely occluded from the view of a traffic camera by those trees.

11.7 Conclusion

In real time, detection of moving objects, or target-based moving object detection, is a crucial area of research in computer vision. Object detection and tracking algorithms are facing various difficulties mentioned throughout the chapter. The resolution of such challenges requires good background modeling based on background subtraction, which has shown better performance in the real-time applications. Separation of the foreground from the background in the processed video frames will work well for surveillance. Here, the background subtraction method computes the difference between current frame and background model in order to extract information for many computer vision applications. This chapter provides rich information regarding object detection techniques and video tracking algorithms. The computer vision applications are wisely described with the crucial challenges that come in the way of object detection and tracking.

11.8 Future Scope

The existing methods for object detection and tracking with the combination of new technologies like IoT, deep learning, and cloud computing may form a better way to develop computer vision. Security and storage are a big concern that will surely be addressed in the near future, along with other challenges discussed in this chapter. Involvement of artificial intelligence and IoT in computer vision may provide a fruitful area of exploration.

REFERENCES

1. Shugang Zhang, Zhiqiang Wei, Jie Nie, Lei Huang, Shuang Wang, Zhen Li, "A Review on Human Activity Recognition Using Vision-Based Method", *Journal of Healthcare Engineering, 2017*:31, 2017. Article ID 3090343. https://doi.org/10.1155/2017/3090343.
2. N. McFarlane and C. Schofield. "Segmentation and tracking of piglets in images," *British Machine Vision and Applications, 41(12)*:187–193, 1995.
3. J. Zheng and Y. Wang. "Extracting roadway background image: A mode based approach," 2006. Transportation Research Board, TRB 2006.
4. J. Zeng, L. Xie, and Z. Liu. "Type-2 fuzzy Gaussian mixture models," *Pattern Recognition, 41(12)*:3636–3643, 2008.
5. C. Wren and A. Azarbayejani. "Pfinder: Real-time tracking of the human body," *IEEE Transactions on Pattern Analysis and Machine Intelligence, 19(7)*:780–785, 1997.
6. A. Elgammal and L. Davis. *"Non-parametric model for background subtraction,"* 6th European Conference on Computer Vision, ECCV 2000, pp. 751–767, 2000.
7. H. Lin, T. Liu, and J. Chuang. "A probabilistic SVM approach for background scene initialization." ICIP 2002, September 2002.
8. N. Oliver, B. Rosario, and A. Pentland. "A Bayesian computer vision system for modeling human interactions," *International Conference on Vision Systems, 22(8)*:841–843, 1999.
9. D. Tsai and C. Lai. "Independent component analysis based background subtraction for indoor surveillance," *IEEE Transactions on Image Processing, 18(1)*:158–167, 2009.
10. L. Sharma and P. Garg (Eds.). *From Visual Surveillance to Internet of Things.* New York: Chapman and Hall/CRC, 2020. https://doi.org/10.1201/9780429297922.
11. L. Sharma (Ed.). *Towards Smart World.* New York: Chapman and Hall/CRC, 2021. https://doi.org/10.1201/9781003056751.
12. S. Messelodi and C. Modena. "A Kalman filter based background updating algorithm robust to sharp illumination changes," *International Conference on Image Analysis and Processing, ICIAP 2005, 3617*:163–170, 2005.
13. R. Chang, T. Ghandi, and M. Trivedi. *"Vision modules for a multi sensory bridge monitoring approach,"* *IEEE Conference on Intelligent Transportation Systems*, ITS 2004, pp. 971–976, 2004.
14. M. Yazdi and T. Bouwmans. "New trends on moving object detection in video images captured by a moving camera: A survey," *Computer Science Review, 28*:157–177, 2018.
15. L. Sharma, "Human Detection and Tracking Using Background Subtraction in Visual Surveillance", *Towards Smart World.* New York: Chapman and Hall/CRC, https://doi.org/10.1201/9781003056751, pp. 317–328, December 2020.
16. M. Zhao, N. Li, and C. Chen. *"Robust automatic video object segmentation technique,"* *International Conference on Image Processing*, September 2002.
17. P. Pahalawatta, D. Depalov, T. Pappas, and A. Katsaggelos. *"Detection, classification, and collaborative tracking of multiple targets using video sensors,"* *International Workshop on Information Processing for Sensor Networks*, pp. 529–544, April 2003.
18. B. Orten, M. Soysal, and A. Alatan. "Person identification of surveillance video by combining Mpeg-7 experts." WIAMIS 2005, April 2005.
19. K. Simonyan and A. Zisserman. "Very deep convolutional networks for large-scale image recognition," arXiv:1409.1556, 2014.
20. T. Yu, J. Yang, and W. Lu. "Combining background subtraction and convolutional neural network for anomaly detection in pumping-unit surveillance," *Algorithms*, 12, May 2019. https://doi.org/10.3390/A12060115
21. L. Sharma, D. K. Yadav, S. K. Bharti, *"An improved method for visual surveillance using background subtraction technique,"* *IEEE, 2nd International Conference on Signal Processing and Integrated Networks (SPIN-2015)*, Amity University, Noida, India, February 19–20, 2015.
22. H. Kim, R. Sakamoto, I. Kitahara, T. Toriyama, and K. Kogure. "Background subtraction using generalized Gaussian family model," *IET Electronics Letters, 44(3)*:189–190, 2008.
23. A. Francois, G. Medioni. "Adaptive Color Background Modeling for Real –time Segmentation of Video Streams," *International Conference in Imaging Science, Systems and Technology*, pp. 227–232, June 1999.

24. L. Sharma, P. K. Garg, *"Block based adaptive learning rate for moving person detection in video surveil-lance,"* *From Visual Surveillance to Internet of Things*, CRC Press, Taylor & Francis Group, pp. 201–214, October 2019.

25. T. Tanaka, A. Shimada, D. Arita, and R. Taniguchi. *"Object segmentation under varying illumination based on combinational background modeling,"* *Proceeding of the 4th Joint Workshop on Machine Perception and Robotics*, 2008.

26. A. Krizhevsky, I. Sutskever, and G. E. Hinton. "ImageNet classification with deep convolutional neural networks," *Advances in Neural Information Processing Systems, 1*, 1097–1105, 2012.

27. D. K. Yadav, L. Sharma, and S. K. Bharti, *"Moving object detection in real-time visual surveillance using background subtraction technique,"* *IEEE, 14th International Conference in Hybrid Intelligent Computing (HIS-2014)*, Gulf University for Science and Technology, Kuwait, December 14–16, 2014.

28. K. Toyama and J. Krumm. "Wallflower: Principles and practice of background maintenance," *International Conference on Computer Vision ICCV 1999*, pp. 255–261, 1999.

29. S. Diamantas and K. Alexis. "Optical flow based background subtraction with a moving camera: application to autonomous driving." In: G. Bebis et al. (eds.), *Advances in Visual Computing, ISVC 2020*. Lecture notes in Computer Science, 12510. Cham, Switzerland: Springer, 2018.

30. B.D. Lucas and T. Kanade. *"An iterative image registration technique with an application to stereo vision,"* in *Proceedings of the 7th International Joint Conference on Artifitial Intelligence (IJCAI)*, pp. 674–679, August 24–28, 1981.

31. S. Bucak and B. Gunsel. "Incremental subspace learning and generating sparse representations via non-negative matrix factorization," *Pattern Recognition*, 42(5):788–797, 2009.

32. L. Sharma, A. Singh, and D. K. Yadav, *"Fisher's Linear Discriminant Ratio based Threshold for Moving Human Detection in Thermal Video,"* *Infrared Physics and Technology*, Elsevier, March 2016.

33. F. Schrof, D. Kalenichenko, and J. Philbin, *"FaceNet: a unifed embedding for face recognition and clustering,"* in *Proceedings of the IEEE Conference on Computer Vision and Pattern Recognition (CVPR '15)*, pp. 815–823, IEEE, Boston, MA, June 2015.

12

Real-Time Road Monitoring Using Deep Learning Algorithm Deployed on IoT Devices

Nilay Nishant
North Eastern Space Application Centre, Umiam, India

Ashish Maharjan
Sikkim Manipal Institute of Technology, Sikkim Manipal University, Majitar, India

Dibyajyoti Chutia and P. L. N. Raju
North Eastern Space Application Centre, Umiam, India

Ashis Pradhan
Sikkim Manipal Institute of Technology, Sikkim Manipal University, Majitar, India

CONTENTS

12.1 Introduction

Bad road conditions are more than an inconvenience to the general public; they cause passenger distress as well as vehicle damage and fatal accidents [1]. In India, over 9,300 people lost their lives and 25,000 people got injured due to the accidents caused by potholes alone in 3 years (2015, 2016, and 2017) [2]. Well-maintained roads can lessen the possibility of accidents. The Indian government has constructed 153,647 km of roads under various schemes between 2015 and 2017. Monitoring of these roads for cracks and potholes is an essential aspect of these projects. Potholes are one of the major causes of road-related accidents. The expansion and contraction of groundwater after the water has entered the ground under pavement form potholes [3]. Maintaining good road condition is imperative for drivers' safety as well as for the transportation and regulatory maintenance authorities. Road condition is assessed from some vital information such as ridability, surface and structural distress, and many other parameters. Many techniques and methods have been designed and developed to gather in situ information about the above-mentioned parameters and assess the quality of road [4, 5].

With the recent advancements in the areas of artificial intelligence (AI), many powerful tools that can learn semantic, high-level, deep features are being introduced to address real-world problems. Artificial neural network (ANN)-based systems utilizing deep learning technology have achieved significant growth in various fields [6–11] and demonstrated excellent performance in numerous applications. Deep

DOI: 10.1201/9781003140351-12

learning is a technique for assisting a computer system in accurately performing complex perception tasks [12]. Deep learning is made up of several layers of nonlinear processing units for feature extraction, with the output from the lower layer acting as the input to the subsequent layer. Many researchers have started applying deep learning in various applications. Li et al. [13] developed a deep learning algorithm to detect breast cancer by screening mammograms. A practical application of deep learning method was presented by Kłosowski [14] for language processing and modeling. Umme et al. [15] proposed a deep learning neural network composed of a set of convolutional neural networks (CNNs), rectified linear activation function (RELUs), and fully connected layers to quantify small dataset-based representation of a face. Vaidya et al. [16] proposed a deep neural network (DNN) to present an innovative method for offline handwritten character detection. Although much technological development has been reported in solving real-world problems, very limited efforts have been observed in monitoring the conditions of roads in real time. This chapter demonstrates the development of AI-based framework for pothole detection based on the state-of-the-art CNN, where we not only classify but also precisely locate and count the number of potholes detected. The algorithm tracks the potholes detected so as to avoid recording the same pothole more than once. The model is developed for real-time road condition monitoring and deployed into a Raspberry Pi single board computer for detecting the potholes in real-time. To accelerate the inference speed of the model, Edge TPU coprocessor, the Google Coral USB Accelerator along Raspberry Pi, was used. The Google Coral Dev Board and USB Accelerator are shown in Figure 12.1.

The Raspberry Pi 4 Model B single board computer is used for real-time monitoring of roads. The main features of the board are fast processing enabled by Broadcom BCM2711 following quad-core Cortex-A71 (ARM v8)-based architecture coupled with 4GB RAM and interfacing enabled by USB 3.0 ports (2 Nos.), Gigabit Ethernet, micro-HDMI ports (2Nos), and MicroSD card slot. Operating system is loaded in the MicroSD card along with data.

A three-dimensional (3D) reconstruction is a method of capturing the shape, appearance, and object depth in the real world [17]. A 3D scene reconstruction can be classified into stereo vision-based methods [18, 19], 3D laser scanner methods [20, 21], and visualization using other motion-sensing devices such as Microsoft Kinect [22, 23]. These methods require advanced, costly equipment for 3D visualization and have a high computational expense. The vibration-based methods use accelerometers for detecting the potholes. Mednis et al. [24], proposed a smartphone-based detection of road irregularities. Ghadge et al. [25] developed a Bumps Detection System (BDS) using K-means clustering and a random-forest classifier on training and testing data captured using accelerometer and GPS to evaluate road conditions and identify and detect locations of potholes. F. Seraj et al. [26] collected data using sensors such as accelerometer and gyroscope in a smartphone and classified road anomalies using a support vector machine (SVM).

In vision-based 2D, image or video data are used to detect the potholes. The methods for detecting potholes can be classified into two categories: traditional image-processing approach and CNN-based approach. The former relies on algorithms that extract low-level features, whereas the latter is a deep learning–based approach that tackles the problem by feature engineering. Dhiman et al. [17] proposed a single-frame stereo-vision-based method and multi-frame fusion-based method for pothole detection. Youngtae et al. proposed [27] a pothole detection algorithm using the black-box camera where candidate extraction extracts the pothole candidate regions. Oliveira et al. [28] applied a comprehensive set of image-processing

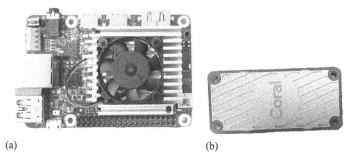

(a)   (b)

FIGURE 12.1   (a) Google Coral Dev Board; (b) Google Coral USB Accelerator.

algorithms for road crack detection. The current study focuses on a 2D computer vision-based technique to detect potholes from the video feed of a camera mounted on the windshield of the vehicle. Although pothole detection using a deep CNN has been studied for the past few years, its implementation in real-time monitoring remains a challenging task due to the computationally expensive nature of the algorithms. Information regarding the number of potholes is collected using CNN-based algorithms, ensuring that each pothole is accounted only once. The proposed system estimates the road quality index by computing the density of potholes in the road section measured every 5 minutes. We tested the model with images and videos before using it in real time for detecting the potholes. The experimental results confirm that our model performs sufficiently well for pothole detection with an accuracy of 89%.

The objective of our study is to automatically detect and track potholes and count them. The framework is designed to send the pothole count to the database server at every 5 minutes to compute road quality. The rest of this chapter is organized as follows. In Section 12.2, details of our dataset and the methods that have been proposed for pothole detection and road quality monitoring are presented. Experimental results and performance evaluation are reported in Section 12.3. Finally, the conclusion is provided in Section 12.4.

12.2 Methodology

A vehicle-mounted camera was used to collect images of streets of Umiam village in Meghalaya, India (Figure 12.2). The dataset consists of 1,374 road images along with images from [32], leading the total number a total of 20,318 images. The images were resized to 300×300 pixels before being fed into the

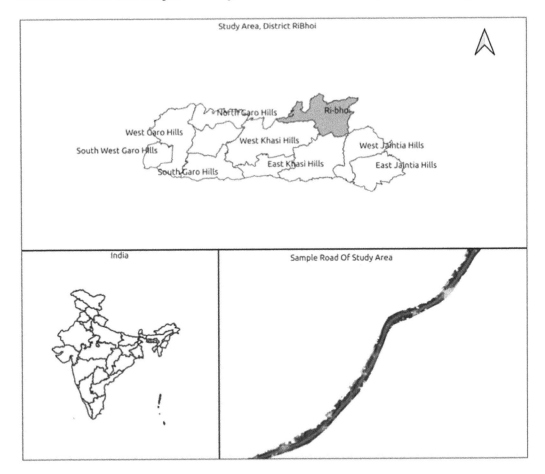

FIGURE 12.2 Study area: Umiam village of Meghalaya, India.

model. Images were annotated to prepare our dataset for the model. Additionally, out of the acquired sample of 20,318 images, 80% were used for training and 20% were used for testing.

12.2.1 Pothole Detection

Inspired by the results of CrackNet, we implemented object detection using Faster R-CNN and SSD models to detect potholes on the road. The models are trained to detect potholes of different shapes and sizes, as well as under different illumination conditions. Faster RCNN and SSD MobileNet models return the bounding box of each pothole. Also, Deep SORT-based tracking algorithm was adopted to avoid counting multiple instances of the same pothole. The Faster R-CNN [33] has two networks for detection. It uses selective search to generate region proposals. Initially, Faster R-CNN model takes an input image with the class of object to be detected along with its bounding box. The image is then extracted using a deep ConvNet, and the final convolution layer of the feature map is used to input the region proposal network (RPN) for prediction. Then feature maps are used to extract the features of the proposed regions by application of RPN. The softmax function predicts the object categories to achieve object classification and boundary regression objectives. The SSD [34] architecture is specially tailored for mobile and resource-constrained environments where the depth-wise convolutions replace the regular convolutions in a bounding box predictor. It is much more computationally efficient. The MobileNetV2 is proposed based on SSDLite framework [35].

Frames from a video of the road collected at different illumination conditions were used for validating the model's accuracy. The initial learning rate was 0.004 with learning decay of 0.9 every 10,000 iterations. We trained the network using RMSProp optimizer and a 0.9 learning momentum. The network was trained for over 9,000 iterations with a batch size of 24. In order to conserve and maintain uniform dimension of the image, zero padding is applied to the top and bottom of an image to preserve an aspect ratio, i.e., 300×300 pixels. The dataset is classified into two classes: one for pothole and one for background. Anchors are created by sliding windows. Since the sliding-window process is convolutional, a GPU can handle it easily. Any anchor with an IoU greater than 0.5 is considered a match. The loss function used is confidence loss along with localization loss. Smooth L1 loss between the predicted and the ground truth is used for localization, and softmax loss over several classes' confidences is used for confidence loss. Non-maximum suppression is used to filter multiple boxes per object that may be matched during prediction. We developed a prototype dashboard application to capture and visualize the real-time pothole detection with counting details in the spatial domain. The experimental result achieved from the prototype work presented here is quite encouraging and can monitor the quality of roads more effectively than the previously used methods.

12.2.2 Tracking Potholes

To ensure that only one instance of each pothole is reported to calculate the road quality index, each pothole is tracked all through its identification until it is visible in the video frame. The current study utilizes Deep SORT (Simple Real-time Tracker) [36] tracking framework to track the potholes. Deep SORT is an object-tracking framework based on SORT, which estimates the existing track in the current frame using the Kalman filter. The state contains a total of eight variables, four of which are bounding box positions and the remaining four are the velocity of each of the coordinates. In this way, when each frame arrives, the position of a current track is calculated based on its previous position [37]. An appearance descriptor obtains the appearance information of detections and tracks. CNN is used to train the appearance descriptor on large-scale re-identification dataset.

With the approximate location of the original tracks and appearance descriptors, the new detection results can be correlated with the existing tracks in each subsequent frame. A threshold is used to screen out all detections with a confidence level below it. The squared Mahalanobis distance incorporates the uncertainties from the Kalman filter. When a new detection is connected with an existing track, it is applied to the track, and the associated track's age is reset to zero. When the new detections fail to interact with the existing tracks in the frame, tentative tracks are formed. If the new detections are successfully linked, the track is modified as verified tracks, otherwise the tentative tracks are removed immediately.

Each frame's unassociated ages are increased by one if the current track fails to correlate with the new detections. If the unassociated age crosses the maximum age threshold, the track is deleted.

12.2.3 Deployment in IoT

We implemented a convolutional neural network based on the Faster R-CNN and SSD MobileNet framework. NVIDIA Quadro P400 was used to effectively handle speed and memory requirements for model training and inference. We then tested the trained model on a sample video containing potholes to evaluate its performance before deploying it in a Raspberry Pi for real-time pothole detection. The Raspberry Pi usually supports models having higher inference speed and low processing power. For deployment of lightweight deep learning models on resource-constrained edge computers, the model is converted to TensorFlow lite. They provide high-performance on-device inference for any TensorFlow models so that the models can have faster performance in real-time applications [38]. Since the Raspberry Pi has the low processing power, TensorFlow Lite utilizes TensorFlow models that have been converted into an optimized format. Then, the pothole detection model, webcam, and the GPS tracker are deployed into the Raspberry Pi for detecting potholes in real time. After that, the model is tested using static images and sample videos. Finally, the Raspberry Pi, webcam, and GPS tracker are mounted into a vehicle for real-time pothole detection. The runtime speed of the model is boosted using Google Coral USB Accelerator. Then, the number of potholes detected is transferred wirelessly to the database server every 5 minutes. In this way, the database stores the pothole count. A provision is also implemented to share the GPS coordinates based on a GPS tracker connected to an IoT device. Then, the pothole detection system calculates the road quality index by retrieving the potholes detected from the database server and computes the index by calculating the density of potholes at every source-destination instance pushed at the database server. Finally, it classifies the roads based on the road quality index and displays a road map on the web server accordingly.

Faster R-CNN and SSD MobileNet models were successful in detecting potholes in real time. The parameters considered were accuracy and speed. To ensure that the model reports only a single instance of all detected potholes, we implemented a Deep SORT-based tracking algorithm on top of the models. Figure 12.3 shows the workflow of the proposed methodology. Initially, the images required for training and testing the neural networks are captured using a camera mounted on a vehicle. The images are then resized into 300×300 pixels and labeled using the labeling method. We trained on 80% of all image data and tested and validated on the remaining 20%. Then we trained two models: Faster R-CNN and SSD MobileNet using the dataset. Here we implemented the models using Python 3.6 and trained the deep neural network using TensorFlow and OpenCV for image processing. We evaluated the trained models by comparing their output losses, accuracy rates, and speed. As a result, we finally chose SSD MobileNet as the basic framework for our method.

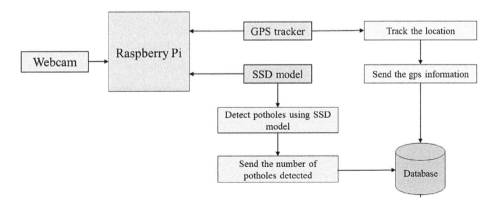

FIGURE 12.3 Proposed schematic methodology.

12.3 Results and Discussion

Loss is one of the widely accepted metrics for estimating the efficacy of the model. In general, a neural network with lower loss is much better [39]. We compared the total loss for two object detection models for their accuracy and loss results. We observed that Faster R-CNN has a lower loss compared to SSD MobileNet. We used the images collected from the vehicle-mounted camera as training data for the model. We computed the performance of pothole detection model using three performance indicators: recall, precision, and accuracy. The optimum model was selected based on the performance metrics of the models.

$$Precision = \frac{TruePositive}{TruePositive + FalsePositive}, \qquad (12.1)$$

$$Recall = \frac{TruePositive}{TruePositive + FalseNegative}, \qquad (12.2)$$

$$F1 = \frac{2 * precision * recall}{precision + recall}, \qquad (12.3)$$

Recall refers to the percentage of the actual number of potholes that were correctly classified by the model. Precision may refer to the percentage of potholes classified correctly with corresponding to a total number of potholes in the dataset; precision and recall are dependent on true positive (TP), false positive (FP), and false negative (FN). Overall performance of the detection model is represented by F1 score, which represents overall performance of the detection model. Intersection over Union is the index that determines accurate detection potholes and is calculated as,

$$IoU = \frac{AreaofOverlap}{Areaof \cup nion}, \qquad (12.4)$$

where "Area of overlap" defines the intersection area between the ground truth and the predicted bounding box from detected result, and area of union defines the union area between the two bounding boxes [40]. In the case of object detection, IoUs over 0.5 is considered precise detection [10]. The performance of the model is depicted in Table 12.1. It indicates that the combination of SSD with MobileNetV2 provides accuracy comparable to Faster R-CNN. We tested the models at different illumination conditions (see Figure 12.4). We observed that Faster R-CNN detected more potholes than did SSD MobileNetV2, and at the same time Faster R-CNN misclassified more potholes than did SSD MobileNetV2. The experimental results show that the SSD MobileNetV2 model performs well when there are shadows as well as when there are no shadows whereas its performance seems to decrease at dawn. The Faster R-CNN model detected almost all the potholes at dawn, but SSD MobileNetV2 failed to detect some potholes

TABLE 12.1

Comparison of the models

Models	TP	FP	FN	Precision	Recall	f-score
Faster R-CNN (without shadows)	40	2	2	95.23%	95.23%	95.23%
SSD + MobileNetV2 (without shadows)	35	7	-	83.33%	100%	90.90%
Faster R-CNN (with shadows)	34	4	2	89.47%	94.44%	91.89%
SSD + MobileNetV2 (with shadows)	35	4	1	89.74%	97.22%	93.33%
Faster R-CNN (dawn/dusk)	37	5	-	88.09%	100%	93.67%
SSD + MobileNetV2 (dawn/dusk)	30	10	1	75.00%	96.77%	84.51%

FIGURE 12.4 Pothole detection under different illumination conditions for the study area.

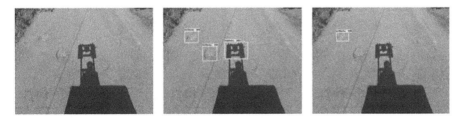

FIGURE 12.5 Shadow misclassified as a pothole using Faster R-CNN in the second image (from left) and pothole detected using SSD MobileNetV2 in the third image (from left).

at dawn. Figure 12.5 shows a shadow misclassified as a pothole using Faster R-CNN model. This result shows that even if the accuracy of the faster R-CNN model is high, the false detection rate of the model is high compared to that of SSD MobileNetV2. Also, the accuracy of SSD MobileNetV2 model is low compared to Faster R-CNN, since it was able to detect only one pothole in the image. Although the SSD MobileNetV2 model detected only one pothole in the image, the model didn't misclassify a shadow as a pothole. The overall accuracy of Faster R-CNN and SSD MobileNetV2 is 93.59% and 89.58%, respectively (Table 12.1). The tracking algorithm implemented in the proposed method is shown in Figure 12.6.

The real-time object detection strongly depends on the amount of computation time. Due to the lightweight nature of a microcomputer like Raspberry Pi and mobile devices used, computing capacity is limited. Thus, it is essential to select the right detection framework that achieves satisfying speed and accuracy on a single board computer. We tested the model using a 30-second video to evaluate its performance on Raspberry Pi in terms of accuracy and real-time detection speed. Here, we tested two

FIGURE 12.6   Detected potholes using SSD MobileNetV2 represented by images annotated with bounding boxes.

TABLE 12.2

Comparison of real-time speed

Detector System	Real-Time Speed (FPS)
Raspberry Pi without Google Coral USB Accelerator	3
Raspberry Pi with Google Coral USB Accelerator	7

models for classification and localization of the potholes: Faster R-CNN and SSD with mobilenetV2. We observed that the Faster R-CNN model is multiple times slower due to its large inference graph. So, SSD with MobileNet was selected for our framework because its lightweight nature and fast real-time speed make it the most suitable for Raspberry Pi. Since the inference speed for detecting the potholes in real time was not sufficient, we added the Google Coral USB Accelerator into the Raspberry Pi to boost up the image-processing speed of the model. Table 12.2 presents the real-time speed of the model before and after adding the Google Coral USB Accelerator.

12.4 Conclusion

We proposed a framework for automatic real-time detection of potholes, which also classifies roads based on the number of potholes detected via the IoT platform. Road images from the streets of Umiam, Meghalaya, India were captured using a dashboard-mounted on the vehicle and augmented by a publicly available data source. These images were then resized, labeled, and released as a training dataset. We trained and evaluated the pothole detection model using our dataset. The pothole detection model is fed into Raspberry Pi along with the Google Coral USB Accelerator to increase the real-time speed. The real-time inspection results show that the proposed pothole detection system achieves an accuracy of 89.58% at seven frames per second. We successfully trained and implemented a variety of models for real-time detection of potholes using Raspberry Pi. We also compared different CNN models to identify which fit most in Raspberry Pi for real-time pothole detection. We used precision, recall, and f-score to evaluate the model. The evaluation result showed that deep learning models are viable for deployment on mobile devices in terms of inference time and accuracy. This study shows that the trained models are feasible to use in a real-time environment on devices with limited computational capability while maintaining high accuracy and also reliable for the end-user. This framework can help in low-cost road condition maintenance. Further, this model can be improved by including more training datasets to improve performance. Also, it can be extended to the transportation system and extend the parameters to predict the repairing cost of the pothole and figure out which areas require urgent repair work.

ACKNOWLEDGMENTS

We would like to acknowledge the contribution of the UAV team at NESAC for their support in data collection.

REFERENCES

1. Bhatt Umang, Mani Shouvik, Xi Edgar, and J. Kolter, *"Intelligent Pothole Detection and Road Condition Assessment,"* *Data for Good Exchange 2017*, New York, September 2017.
2. Kamaljit Kaur Sandhu. (2018, July 24), "Over 9300 deaths, 25000 injured in 3 years due to potholes," *India Today,* Available online: https://www.indiatoday.in/india/story/over-9300-deaths-25000-injured-in-3-years-due-to-potholes-1294147-2018-07-24.
3. "Diving into the UK's pothole problem," *In Day Insure*, Available online: https://www.dayinsure.com/news/diving-into-the-uks-pothole-problem, September 30, 2019.
4. Kim Taehyeong and S. Ryu, "Review and analysis of pothole detection methods," *Journal of Emerging Trends in Computing and Information Sciences*, 2014, vol. *5*, pp. 603–608.
5. Akagic Amila, Buza Emir, and Omanovic Samir, *"Pothole Detection: An Efficient Vision Based Method Using RGB Color Space Image Segmentation,"* *2017 40th International Convention on Information and Communication Technology, Electronics and Microelectronics (MIPRO)*, IEEE, 2017, pp. 1104–1109 [doi:10.23919/MIPRO.2017.7973589].
6. Dingan Liao, Hu Lu, Xingpei Xu, and Quansheng Gao, *"Image Segmentation Based on Deep Learning Features,"* *2019 Eleventh International Conference on Advanced Computational Intelligence (ICACI)*, IEEE, 2019, pp. 296–301 [doi:10.1109/ICACI.2019.8778464].
7. Voulodimos Athanasios, Doulamis Nikolaos, Doulamis Anastasios, and Protopapadakis Eftychios, "Deep Learning for Computer Vision: A Brief Review," *Computational Intelligence and Neuroscience*, vol. *2018*, 2018, pp. 1–13 [doi:10.1155/2018/7068349].
8. Zhao Zhong-Qiu, Zheng Peng, Xu Shou-Tao, and Wu Xindong, "Object Detection with Deep Learning: A Review," *IEEE Transactions on Neural Networks and Learning Systems*, vol. *30*, no. 11, 2019, pp. 3212–3232 [doi:10.1109/TNNLS.2018.2876865].
9. Nassif Ali, Shahin Ismail, Attili Imtinan, Azzeh Mohammad, and Shaalan Khaled, "Speech Recognition Using Deep Neural Networks: A Systematic Review," *IEEE Access*, vol. *7*, 2019, pp. 19143–19165 [doi:10.1109/ACCESS.2019.2896880].
10. Vargas Manuel, Lima Beatriz, and Evsukoff Alexandre, *"Deep Learning for Stock Market Prediction from Financial News Articles,"* *2017 IEEE International Conference on Computational Intelligence and Virtual Environments for Measurement Systems and Applications (CIVEMSA)*, IEEE, 2017, pp. 60–65 [doi:10.1109/CIVEMSA.2017.7995302].
11. Chen Qi, Wang Wei, Wu Fangyu, De Suparna, Wang Ruili, Zhang Bailing, and Huang Xin, "A Survey on an Emerging Area: Deep Learning for Smart City Data," *IEEE Transactions on Emerging Topics in Computational Intelligence*, vol. *3*, no. 5, 2019, pp. 392–410 [doi:10.1109/TETCI.2019.2907718].
12. Dargan Shaveta, Kumar Munish, Maruthi Rohit Ayyagari, and Kumar Gulshan, "A Survey of Deep Learning and Its Applications: A New Paradigm to Machine Learning," *Archives of Computational Methods in Engineering*, June 2019 [doi:10.1007/s11831-019-09344-w].
13. Shen Li, Margolies Laurie, Rothstein Joseph, Fluder Eugene, McBride Russell, and Sieh Weiva, "Deep Learning to Improve Breast Cancer Detection on Screening Mammography," *Scientific Reports*, vol. *9*, no. 1, 2019, p. 12495 [doi:10.1038/s41598-019-48995-4].
14. Kłosowski Piotr, *"Deep Learning for Natural Language Processing and Language Modelling,"* *2018 Signal Processing: Algorithms, Architectures, Arrangements, and Applications (SPA)*, IEEE, 2018, pp. 223–228 [doi:10.23919/SPA.2018.8563389].
15. Aiman Umme and Vishwakarma Virendra, *"Face Recognition Using Modified Deep Learning Neural Network,"* *2017 8th International Conference on Computing, Communication and Networking Technologies (ICCCNT)*, IEEE, 2017, pp. 1–5 [doi:10.1109/ICCCNT.2017.8203981].
16. Vaidya Rohan, Trivedi Darshan, Satra Sagar, and Pimpale Mrunalini, *"Handwritten Character Recognition Using Deep-Learning,"* *2018 Second International Conference on Inventive Communication and Computational Technologies (ICICCT)*, IEEE, 2018, pp. 772–775 [doi:10.1109/ICICCT.2018.8473291].

17. Dhiman Amita and Klette Reinhard, *"Pothole Detection Using Computer Vision and Learning,"* *IEEE Transactions on Intelligent Transportation Systems*, 2019, pp. 1–15 [doi:10.1109/TITS.2019.2931297].
18. Wang Kelvin, *"Challenges and Feasibility for Comprehensive Automated Survey of Pavement Conditions,"* *Applications of Advanced Technologies in Transportation Engineering (2004)*, American Society of Civil Engineers, 2004, pp. 531–536 [doi:10.1061/40730(144)99].
19. Hou Zhiqiong, Wang Kelvin, and Gong Weiguo, *"Experimentation of 3D Pavement Imaging through Stereovision,"* *International Conference on Transportation Engineering 2007*, American Society of Civil Engineers, 2007, pp. 376–381 [doi:10.1061/40932(246)62].
20. Chang Kuan-Tsung, J. Chang, and Liu Jin-King, *"Detection of Pavement Distress using 3D Laser Scanning Technology,"* *In Proceedings of International Conference on Computing in Civil Engineering 2005*, pp. 1–11 [doi:10.1061/40794(179)103].
21. Li Qingguang, Yao Ming, Yao Xun, and Xu Bugao, *"A Real-Time 3D Scanning System for Pavement Distortion Inspection,"* *Measurement Science and Technology*, vol. *21*, no. 1, 2010, p. 015702 [doi:10.1088/0957-0233/21/1/015702].
22. Joubert Damien, Tyatyantsi Ayanda, J. Mphahlehle, and V. Manchidi, *"Pothole Tagging System,"* *In Proceedings of the 4th Robotics and Mechatronics Conference of South Africa*, CSIR International Conference Centre, Pretoria, November 2011, pp. 23–25.
23. Moazzam Imran, Kamal Khurram, Mathavan Senthan, Ahmed Syed, and Rahman Mujib, *"Metrology and visualization of potholes using the Microsoft Kinect sensor,"* *In Proceedings of the 16th International IEEE Annual Conference on Intelligent Transportation Systems*, 2013 [doi:10.1109/ITSC.2013.6728408].
24. Mednis Artis, Strazdins Girts, Zviedris Reinholds, Kanonirs Georgijs, and Selavo Leo, *"Real time pothole detection using Android smartphones with accelerometers,"* *2011 International Conference on Distributed Computing in Sensor Systems and Workshops (DCOSS)*, Barcelona, 2011, pp. 1–6 [doi:10.1109/DCOSS.2011.5982206].
25. Ghadge Manjusha, Pandey Dheeraj, and Kalbande Dhananjay, "Machine Learning Approach for Predicting Bumps on Road," *2015 International Conference on Applied and Theoretical Computing and Communication Technology (ICATccT)*, IEEE, 2015, pp. 481–485 doi:10.1109/ICATCCT.2015.7456932.
26. Seraj Fatjon, van der Zwaag, Berend Jan, Dilo Arta, Luarasi Tamara, and Havinga Paul, *"RoADS: A Road Pavement Monitoring System for Anomaly Detection Using Smart Phones,"* *International Workshop on Modeling Social Media International Workshop on Mining Ubiquitous and Social Environments International Workshop on Machine Learning for Urban Sensor Data*, SenseML 2014, Berlin: Springer, 2014. pp. 1–16.
27. Jo Youngtae and Ryu Seungki, "Pothole Detection System Using a Black-Box Camera," *Sensors*, vol. *15*, no. 11, 2015, pp. 29316–29331 [doi:10.3390/s151129316].
28. Oliveira Henrique and Correia Paulo, *"CrackIT; An Image Processing Toolbox for Crack Detection and Characterization."* *2014 IEEE International Conference on Image Processing (ICIP)*, IEEE, 2014, pp. 798–802 [doi:10.1109/ICIP.2014.7025160].
29. Zhang Lei, Yang Fan, Zhang Yimin, and Zhu Ying, *"Road Crack Detection Using Deep Convolutional Neural Network,"* *2016 IEEE International Conference on Image Processing (ICIP)*, IEEE, 2016, pp. 3708–3712 [doi:10.1109/ICIP.2016.7533052].
30. Hu Yong and Zhao Chun-Xia, "A Novel LBP Based Methods for Pavement Crack Detection," *Journal of Pattern Recognition Research*, vol. *5*, no. 1, 2010, pp. 140–147 [doi:10.13176/11.167].
31. Zou Qin, Cao Yu, Li Qingquan, Mao Qingzhou, and Wang Shoupeng, "CrackTree: Automatic Crack Detection from Pavement Images," *Pattern Recognition Letters*, vol. *33*, no. 3, 2012, pp. 227–238 [doi:10.1016/j.patrec.2011.11.004].
32. Maeda Hiroya, Sekimoto Yoshihide, Seto Toshikazu, Kashiyama Takehiro, and Omata Hiroshi, "Road Damage Detection and Classification Using Deep Neural Networks with Smartphone Images," *Computer-Aided Civil and Infrastructure Engineering*, vol. *33*, no. 12, 2018, pp. 1127–1141 [doi:10.1111/mice.12387].
33. Ren Shaoqing, He Kaiming, Girshick Ross, and Sun Jian, "Faster R-CNN: Towards Real-Time Object Detection with Region Proposal Networks," *IEEE Transactions on Pattern Analysis and Machine Intelligence*, vol. *39*, no. 6, 2017, pp. 1137–1149 [doi:10.1109/TPAMI.2016.2577031].
34. Liu Wei, Anguelov Dragomir, Erhan Dumitru, Szegedy Christian, Reed Scott, Fu Cheng-Yang, and Berg Alexander, "SSD: Single Shot MultiBox Detector," *European Conference on Computer Vision*, vol. *9905*, 2016, pp. 21–37 [doi:10.1007/978-3-319-46448-0_2].

35. Sandler Mark, Howard Andrew, Zhu Menglong, Zhmoginov Andrey, and Chen Liang-Chieh, *"MobileNetV2: Inverted Residuals and Linear Bottlenecks," 2018 IEEE/CVF Conference on Computer Vision and Pattern Recognition*, IEEE, 2018, pp. 4510–4520 [doi:10.1109/CVPR.2018.00474].

36. Wojke Nicolai, Bewley Alex, and Paulus Dietrich, *"Simple Online and Realtime Tracking with a Deep Association Metric," 2017 IEEE International Conference on Image Processing (ICIP)*, IEEE, 2017, pp. 3645–3649 [doi:10.1109/ICIP.2017.8296962].

37. K. Host, M. Ivasic-Kos, and M. Pobar, *"Tracking Handball Players with the DeepSORT Algorithm," In Proceedings of the 9th International Conference on Pattern Recognition Applications and Methods - Volume 1*, ICPRAM, 2020, ISBN 978-989-758-397-1, pp. 593–599. [doi:10.5220/0009177605930599]

38. "Deploy machine learning models on mobile and IoT devices," Available online: https://www.tensorflow.org/lite.

39. Murad A. Qurishee, "Low-cost deep learning UAV and Raspberry Pi solution to real-time pavement condition assessment," Degree of Master of Science, The University of Tennessee at Chattanooga Chattanooga, Tennessee, May 2019.

40. Rezatofighi Hamid, Tsoi Nathan, Gwak JunYoung, Sadeghian Amir, Reid Ian, and Savarese Silvio, *"Generalized intersection over union: A metric and a loss for bounding box regression." 2019 IEEE/CVF Conference on Computer Vision and Pattern Recognition (CVPR)*, 2019, pp. 658–666.

13

AI-Based Real-Time Application: Pattern Recognition Automatic License Plate and Vehicle Number Detection Using Image Processing and Deep Learning (with OpenCV)

Varshini Balaji and Vallidevi Krishnamurthy
SSN College of Engineering, Anna University, Chennai, India

CONTENTS

13.1 Introduction

Pattern recognition often involves analyzing 2D images and recognizing them based on a relevant area of interest. It is a fast-developing field that forms the core of many disciplines such as image processing, computer vision, and text and document analysis. The core goal of building object/text recognition systems is to extract meaningful features from high-dimensional patterns such as images and videos. A special topic of interest in pattern recognition is OCR, and a common use case is in the ANPR systems. It focuses on image transformations and representations prior to comparison process. This chapter highlights the challenges faced in implementing an efficient vehicle number detection system. The ANPR systems help police and local authorities identify and track suspicious cars and implement a regulated parking system. It has applications in many private businesses, as it aids real-time monitoring of vehicles entering and exiting static structures. This provides them secure control over vehicular access to their car parks and premises. It can satisfy the daily demands for security and surveillance on corporate and school/college premises and help police officials in tracking organized criminal suspects. The system is intended for surveillance purposes within the college campus to store all the vehicle information along with timestamp of entry and exit for future reference Agarwal (2014), Andrew et al. (2020), Bushkovskyi

(2020), Chand, Gupta, and Kavati (2020), Ghadage and Khedkar (2019), Sharma, Singh, and Yadav (2016). The security system monitors the front gates of the campus and is designed to run for 24 hours that notifies the security in charge on the entry of unauthorized vehicles. It also maintains an image log with the time stamp for the vehicles that enter and exit the campus Rao and Reddy (2011), Samra (2016). A database of authorized vehicles is maintained. This chapter also discusses the functions, specifications, design constraints, requirements (functional and nonfunctional), and other factors of the entry gate monitoring system implemented for real-time surveillance of the college campus.

13.2 Literature Survey

Use of deep learning for sequential patter recognition is described in detail. It introduces elements of pattern recognition and concepts of deep learning and neural networks. The use of probabilistic models and Markov models and deep architectures developed for sequential pattern recognition are discussed in detail (Safari & Kleinsteuber, 2013). Main ideas of unsupervised feature learning and deep learning including implementation of deep learning algorithms are elaborated in this tutorial contributed by professors at Stanford (Andrew et al., 2020). Effective solutions for ANPR system are proposed by TitanHz, which uses the latest technologies utilizing OCR to automatically read license plate characters using both stationary and mobile cameras in (ANPR systems and its solutions). Applications of ANPR systems are listed in Dhote (2020). Optimization of loss functions for classification and localization of license plates during nighttime vehicle detection is explored in Jiang, Qin, Zhang, and Zheng (2011). Image-processing techniques for noise reduction, smoothing, and segmentation through bottom-up and top-down approach toward pattern recognition are examined. Tasks performed by kernel approaches for building object recognition systems used for surveillance are analyzed in (Zoppis, Mauri, and Dondi (2018). The components of statistical recognition systems for designing systems to identify complex patterns are discussed in Golden (2002). Three basic approaches of pattern recognition, namely statistical, structural, and neural pattern recognition, are analyzed, and neural network–based approach to forming complex decision regions for pattern recognition is examined in Rao and Reddy (2011). Deep residual learning for pattern recognition to ease the training of deeper networks using residual learning frameworks is discussed in He, Zhang, Ren, and Sun (2016). Experimental results that demonstrate high accuracy and precision in pattern recognition are presented in the paper on deep learning for sequence pattern recognition (Gao, Zhang, & Wei 2018). Genetic algorithms for plate localization and number recognition to support fast evaluation and convergence in case of inclined plates and missing symbols are discussed in Samra (2016). A technique to detect license plate symbols without using any information about the plate's outer shape and internal colors using a genetic algorithm with a flexible fitness function is introduced in Gowshalya Shri and Arulprakash (2014). The basics of pattern recognition and its various applications are mentioned in Prasad (2020). A detailed research of the use of optical character recognition in license plate detection has been studied in Roy (2007). A system leveraging a Mask R-CNN framework for vehicle detection has been discussed in Patel, Shah, and Patel (2013). An overview of a neural approach toward pattern recognition is presented in Sawon (2020).

13.3 Various Applications

Pattern recognition supports people in the performance of tasks related to ensuring security, including access to premises and devices, detection of unusual changers especially in the field of medicine and geology, and diagnosing technical conditions of devices Gowshalya Shri and Arulprakash (2014), Ghadage and Khedkar (2019). It is the fastest-developing area due to high demand for such solutions in different domains. In spite of its short history, it has found applications in many areas of human activity (Sharma, 2021; Sharma & Garg, 2020a). Various applications of pattern recognition and uses specific to ANPR systems are discussed below.

13.3.1 Pattern Recognition and Its Application

Applications are based on the approach followed to perform pattern recognition. There are three common approaches followed:

a. Statistical approach: In this approach, historical data is gathered and analyzed based on observations. It uses supervised machine learning. It identifies whether a particular item belongs to class or label.

b. Syntactical or structural approach: It is used to define a more complex relationship between elements, like parts of speech. It uses semi-supervised machine learning. It relies on sub-patterns called primitives, like words.

The above two approaches are based on direct computation using math and statistics-related techniques.

c. Neural approach: The ANPR system implemented uses the neural approach for pattern recognition. Neurons are basic units of brain cells, and together these neurons create networks to control specific tasks. An artificial neural network simulates the working function of a biological neural network of human brains (Sharma & Garg, 2020b; Fogel, Owens, & Walsh 1965; Makkar & Sharma, 2019; Giraldo & Bouwmans, 2020). In this system, all neurons are well connected where input unit receives various forms of information based on an internal weighting system and the neural network attempts to predict the required output.

It goes through a training phase where it learns to recognize patterns in data. Back propagation, time delay neural nets, and recurrent neural networks are some commonly found models. Generally, feedforward neural networks are used for pattern recognition. Performance of the network can be improved by providing feedback to reconstruct input patterns and reduce error by comparing the results with the actual value and adjusting weights of layers accordingly, which can be done by implementing a back-propagation algorithm.

Some of the applications are:

Computer Vision: Extracting important features from images and videos and interpreting the same to gain a high-level understanding is termed "computer vision." This has many applications in real life, like biomedical imaging, video surveillance, intelligent transport systems, etc. (Sharma & Garg, 2020a; Fogel, Owens & Walsh, 1965; Makkar & Sharma, 2019; Giraldo & Bouwmans, 2020; Bouwmans, Javed, Sultana & Jung, 2019).

Speech Recognition: All the virtual assistants commonly used today, such as Alexa, Siri, Google Assistant, etc., are applications of speech recognition (Gao, Zhang & Wei, 2018; Prasad TVSNVC, 2020, Sharma, 2021). Pattern recognition is crucial in these applications.

Fingerprint Identification: Many recognition approaches exist to perform fingerprint identification. Biometric technologies rely heavily on pattern recognition.

Medical Diagnosis: Pattern recognition forms the foundation of expert medical diagnosis. Medical diagnosis refers to a process of determining the disease or condition of a person by analyzing the symptoms based on past medical cases (Sharma, 2020; Giraldo & Bouwmans, 2020; Bouwmans, Javed, Sultana & Jung, 2019). It has a vast scope, from breast cancer detection to COVID-19 prediction algorithms, with a high range of accuracy.

13.3.2 ANPR System and Its Application

Automatic number plate recognition (ANPR) has a wide range of applications, since the license number is the most widely accepted, human-readable identifier of motor vehicles. The ANPR provides automated access of the content of the number plate for computer systems managing databases and processing information of vehicle movements. It has a number of applications:

Parking: Parking management through ticketless parking fee management, parking automation, vehicle location guidance, and car theft prevention from parking lots can be done with the help of ANPR systems.

Law Enforcement: A common use of ANPR systems is for law enforcement purposes. That is, monitoring the vehicle movement can aid in identification of stolen cars and tracing vehicles violating traffic signal rules and speed limits.

Access Control: It refers to a mechanism for limiting the access to areas and users based on their identity and membership details in various groups. This can be achieved via license plate number detection, as personal identity can be verified by mapping it to the user's corresponding vehicle number, which is most often unique to every user.

Highway Road Tolling: Road tolling is a process of charging drivers for accessing particular regions of road infrastructure. Tolling is needed for funding construction and maintenance of highways, motorways, bridges, and roads. The ANPR systems can be used for automating toll fee management by reducing the toll account balance by detecting the vehicle number and deducting money from the account associated with the vehicle. FASTag, an electronic toll collection system in India, is an example of automated fee payment system that employs radio frequency identification (RFID) technology, which also relies on pattern detection but using a different approach. A similar system can be implemented using ANPR technology.

Journey Time Measurement: It is a widely used feature of ANPR systems. Using time stamps of the vehicle collected at entry and exit points of a region can be used in estimating the average speed of travel, analyze traffic, and accordingly plan efficient routes to reduce cost and time of travel.

Border Control: It is a state-coordinated effort established to regulate operational control of country's state borders with the central idea of increasing nation's security against terrorism, illegal border traffic, smuggling, and other criminal activities.

13.4 Research Methods

The ANPR performed in controlled lighting conditions with predictable license plate types can use basic image-processing techniques. More advanced ANPR systems utilize dedicated object detectors, such as HOG + Linear SVM, Faster R-CNN, SSDs, and YOLO, to localize license plates in images. State-of-the-art ANPR software utilizes recurrent neural networks (RNNs) and long short-term memory networks (LSTMs) to aid in improving OCR of the text from the license plates. More advanced ANPR systems use specialized neural network architectures to preprocess and clean images before OCR, thereby improving the ANPR accuracy.

ANPR system implemented using tradition image-processing techniques using OpenCV and improving the accuracy of the same by constructing a deep learning model is discussed in the following sections.

13.4.1 Image Processing Using OpenCV

13.4.1.1 Data Collection Module

This system aims to accomplish license plate number detection without using any deep learning model due to the lack of preexisting large and relevant dataset covering all possibilities to perform the necessary training. It uses built-in libraries and packages supported by Python, such as OpenCV, Tesseract, PyTesseract for OCR, etc., to execute the necessary steps. This method can be split into individual modules:

The first module locates potential license plate candidates. This is done by setting values for the necessary parameters, such as aspect ratio range (by setting minimum and maximum aspect ratio) and typical rectangular dimension values of the plate, to clean up the borders of the image. It uses a Blackhat morphological operation that performs operations on a binary image based on image shape. It takes two inputs, the original image and the structuring element, to decide the nature of operation.

There are different kinds of operators, such as erosion and dilation, with variants for each such as open, closing, and gradient operations. This step reveals dark characters (letters, digits, and symbols) against a light background. The next step is to find regions in the image that are light, which might contain the license plate characters. Small square kernels are used to apply a closing operation to fill small holes and help identify large structures in the image. Binary threshold is performed on the input image using Otsu's method to reveal the light regions in the image that may contain license plate characters. OpenCV functions are used to implement the closing and threshold operation as a part of the preprocessing step (Figure 13.1).

The Scharr gradient is further applied to the resultant image to detect the edges and emphasize the boundaries of characters in the license plate. This involves computing the Scharr gradient magnitude representation in the x-direction of the generated Blackhat image. The resulting intensity values are then scaled back to the range [0, 255]. Gaussian blur is applied to the gradient magnitude image. This is followed by an application of a closing operation and binary threshold using Otsu's method (Figure 13.2).

At the end of this module is a contiguous white region where the license plate characters are located. This, however, is less accurate, as it provides cluttered results while locating large white regions (Figure 13.3).

To reject the false positives, a series of erosion and dilation operations are performed in an attempt to de-noise the threshold image. This will reduce a lot of noise from the previous result and provide more accurate results. This results in light regions of the image (Figure 13.4).

The next step starts with identification of all contours and reverse-sorting them based on the pixel area and retaining the large contours. The operation returns the pruned and sorted list of contours.

To summarize the above process, input image is converted to grayscale and traditional image-processing techniques with emphasis on morphological operations are performed. This module will output a selection of candidate contours that might contain the license plate (Figure 13.5).

FIGURE 13.1 OpenCV is used to perform closing and threshold operations as a preprocessing pipeline step for automatic license/number plate recognition (ANPR) with Python.

FIGURE 13.2 Applying Scharr's algorithm in the x-direction emphasizes the edges in our Blackhat image as another ANPR image-processing pipeline step.

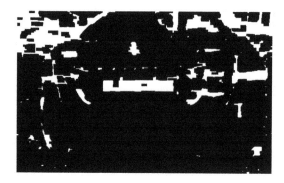

FIGURE 13.3   Blurring, closing, and thresholding operations using OpenCV and Python result in a contiguous white region on top of the license plate/number plate characters.

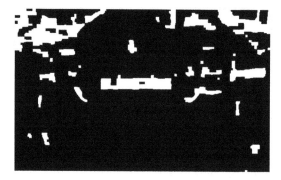

FIGURE 13.4   Erosions and dilations with OpenCV and Python clean up the threshold image, making it easier to find license plate characters for the ANPR system.

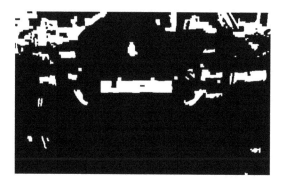

FIGURE 13.5   After a series of image-processing pipeline steps for ANPR/ALPR performed with OpenCV and Python, a clearly displayed region with the license plate characters is one of the larger contours.

13.4.1.2 Pruning License Plate

In this section we define a method to locate a license plate from the candidate regions. This includes three input parameters: grayscale image, contour candidates returned by the previous module, and a Boolean value to indicate if the system should eliminate any contours that touch the edge of the image. Using these input values, candidate contours are filtered further to finally locate the license plate region in the image.

The Method loops, a place where variables need to store license plate contour and license plate region of interest, are initialized with some value. This loop aims to isolate the contour that contains the license plate and extract the region of interest of the license plate. This step aims to determine the bounding box rectangle of the contour. The aspect ratio of contour bounding box is computed to help ensure the proper rectangular shape of the license plate. Here, the aspect ratio represents the relationship between the height and width of the rectangle. The correct license plate contour and region of interest values are set when the aspect ratio is acceptable and lies within the range of minimum and maximum aspect ratio.

13.4.1.2.1 Optical Character Recognition (OCR)

The OCR is performed on the license plate region detected in the previous step. This is done by defining Tesseract ANPR options including an OCR character whitelist and a page segmentation mode. Using OpenCV's morphological operations and contour processing to license plate is identified and a clean image is generated to feed as input to the next module, to send through the Tesseract OCR engine. The page segmentation method is implemented by setting the mode of operation to treat the image as a single line of text. It is a setting that indicates a layout analysis of images and documents. Whitelist is a listing of characters considered by Tesseract.

The OCR method takes a three-channel color image with license plate tag, PSM mode of operation, and a flag variable verifying the need to clean-up contours touching the borders of the license plate. By setting PyTesseract options OCR is performed (Figure 13.6).

To summarize, ANPR system implementation can be briefed using the following steps:

1. Load an input image from disk
2. Find the license plate in the input image
3. OCR the license plate
4. Display the ANPR result to our screen and store it in a database in real time for further processing

The ANPR system results are improved by clearing the border (clearing foreground pixels that touch the borders of the license plate) method. Results are obtained from the ANPR system using OpenCV (Figure 13.7).

The image in the figure displays several successful results obtained from the ANPR algorithm for number detection using Python, OpenCV, and Tesseract OCR to implement each module. However, the system is very sensitive to some conditions and has many incorrect predictions – in this case, allowing characters to touch the edges of the image result in noisy input to Tesseract OCR, thus leading to less accuracy. This is mainly because of the number of assumptions made while building the system (Figure 13.8).

13.4.1.3 Edge Cases and Assumptions

- European and international plates are often longer and not as tall as US license plates. In this chapter we're not considering US license plates.
- Sometimes, motorcycles and large dumpster trucks mount their plates sideways.
- Some countries and regions allow for multi-line plates with a near 1:1 aspect ratio.
- Not all cars have a similar contrast in the color of license plate, so a lighter background with darker text is assumed.

FIGURE 13.6 The results of our Python and OpenCV-based ANPR localization pipeline. This sample is very suitable to pass on for OCR with Tesseract.

FIGURE 13.7   Correct vehicle number prediction.

Implementing an ANPR system is challenging because of the following limitations:

- A vast diversity and assortment of license plate types across states and countries.
- Dynamic lighting conditions including reflections, shadows, and blurring
- Fast-moving vehicles
- Obstructions

13.4.1.4 Technologies and Their Definitions

The intention of the proposed system is to use some of the built-in packages to deploy some traditional image-processing techniques with an emphasis on morphological operations to identify contour with the license plate. This section is for the readers who do not have a clear idea about the definitions/purpose of certain libraries and technologies.

(a) *Optical Character Recognition*: OCR is the electronic or mechanical conversion of images of typed, handwritten, or printed text into machine-encoded text, whether from a scanned document, a photo of a document, a scene photo (for example, the text on signs and billboards in a landscape photo), or from subtitle text superimposed on an image. It is a common method of digitizing printed texts so that they can be electronically edited, searched, stored more compactly, displayed online, and used in machine processes such as cognitive computing, machine translation, (extracted) text-to-speech, and key data and text mining. The OCR is a field of research in pattern recognition, artificial intelligence, and computer vision.

FIGURE 13.8 Incorrect vehicle number predictions.

(b) *OpenCV*: OpenCV is a Python library that is designed to solve computer vision problems. It is a library of programming functions mainly aimed at real-time computer vision. It supports highly improved deep learning modules. It also supports several deep learning frameworks, including Caffe, TensorFlow, and Torch/PyTorch. It supports APIs for using pre-trained deep learning models that are compatible with many languages such as C++, API, and Python.

(c) *Tesseract*: Tesseract is an open-source optical character recognition engine and command line program. It is released under the Apache license. In the latest version of Tesseract there is a greater focus on line recognition; however, it still supports the legacy Tesseract OCR engine that recognizes character patterns. It can be used directly or (for programmers) using an API to extract printed text from images. It supports a wide variety of languages. Tesseract does not have a built-in GUI, but there are several available from third-party vendors.

(d) *Neural Networks*: Neural networks use neurons to transmit data in the form of input values and output values through connections. They are modeled loosely on neural pathways in human brain to learn useful features.

(e) *Convolutional Neural Network*: It is a layered architecture where each network node (connection point) has the capability to process input and forward output to other nodes in the network.

13.4.2 ANPPR Using Deep Learning

Tesseract OCR works effectively only with neatly cleaned and preprocessed input images. Advanced deep learning models are preferred mainly in uncontrolled environment. However, training a deep learning model requires large amounts of training data and takes countless hours to annotate thousands of images from the dataset.

The large and robust ANPR datasets for training/testing are difficult to obtain, for the following reasons:

1. These datasets contain sensitive, personal information, including time and location of a vehicle and its driver.
2. ANPR companies and government entities closely guard these datasets as proprietary information.

Due to lack of a large relevant dataset, only traditional image-processing techniques are often deployed for ANPR system without a deep learning model. The first step in building an ANPR system is usually to collect data and amass enough example plates under various conditions. In this case, a large set of images of different kinds of vehicles with different plate dimensions under varying lighting conditions and color features must be captures or gathered from some other source. To improve OCR using similar character images while training, a dataset including all commonly appearing alphanumeric characters in license plates with multiple examples must be collected.

To implement a fully functional system, it is necessary to integrate the real-time CCTV camera footage into the ANPR system, generating image frames if a moving vehicle is detected. The captured image is further processed before giving input to the ANPR modules. This is then fed into the trained deep learning model to locate the license plate and recognize the text present on the license plate.

The steps involved in the overall execution are listed in sequence below:

Image acquisition: Capture input frames of a moving vehicle through a video camera.

Plate detection: Deep learning–based object detection algorithm used to identify objects in an image or video. An object detection model is trained on lots of number plates to detect the location of the number plate. The location of a bounding box that contains number plate and a score indicates the confidence on whether the detection is correct.

Image preprocessing phases are:

1. Converting RGB to grayscale image: Processing of an RGB image is complex and time consuming, so the color image is converted to grayscale.
2. Image enhancement: Adaptive histogram equalization is to enhance contrast of the image (gray color image). Several histograms each for a distinct region in the image are constructed. This is advantageous because ordinarily, there is only one histogram is for an entire image. In addition to this, media filtering is done to remove noise in the image. This is followed by binarization where the gray image is converted into an image having pure black and pure white pixel values.

Segmentation: In this step, an image that is strongly in consonance with objects or areas contained in the captured image is compartmentalized.

Edge detection: Edge is a boundary between two regions with relatively distinct gray-level properties. It detects discontinuities in intensity values. The basic step in recognition of the plate is to detect its shape (rectangle); thus the edge of a rectangular plate is detected. Using the Sobel operator, the edges in the image are highlighted. This in turn reduces the amount of data in the image and processes the required data for further use.

Morphological image processing: A structuring element used to create output of the same size using dilation and by adding pixels to the boundary of the object to increase the thickness of the edges. Using the shrinking operation thins the image to eliminate the irrelevant parts.

Threshold: In this method, two distinct levels are awarded to pixels that are above and below the selected threshold value. To separate the object from a background image, it is converted in binary form. Gray-level threshold is a simple process. The value of the threshold is selected and compared with the pixel of the image. It also transforms the input image into an output binary image that is being segmented. In global threshold, the histogram of the image is partitioned using a single threshold value. Threshold means the volume of gray level falling between the baseline boundary that lies among the pixels found in the foreground and background.

Segmentation: Character segmentation is a bridge between a number plate extraction and character recognition. In this, different characters on a number plate area are segmented. Various reasons such as lighting variance, plate frames, and rotation are those that hinder the segmentation work. A segmentation method is also known as a boundary box analysis. By this method, characters are assigned to connected components, and these are extracted using the boundary box analysis. The segmentation process is completed upon reduction of noise in the image.

Character recognition: The method of character recognition is completed by extracting the features of characters and their different classification techniques. A machine learning algorithm is used for recognition of characters from the number plate (Figure 13.9).

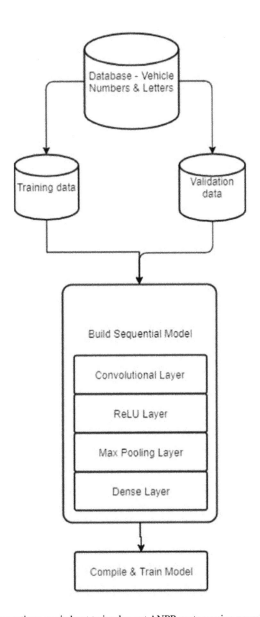

FIGURE 13.9 Sequence of operations carried out to implement ANPR system using neural networks.

13.5 Results

The system was implemented using simple image-processing techniques due to the lack of large and relevant dataset for training deep learning models. This had a low accuracy, as the solution is very sensitive to external conditions. For instance, the Tesseract OCR engine works best only when input imaged are clean, preprocessed with a good resolution. In real-world implementations, images may be grainy or low quality, or the driver of a given vehicle may have a special cover on their license plate to obfuscate the view of it, making ANPR *even more* challenging. To construct a deep learning model, a tailor-made dataset was generated by capturing images and videos of vehicles on the college campus to customize the model for the college gate monitoring system. However, it was neither large enough nor did it cover all possible scenarios of use case, thus lacking accuracy during implementation. The system achieved an accuracy in the range of 80% considering the limited test set it was evaluated against. The neural network was optimized by updating the weight values through back propagation. This doesn't ensure high accuracy, as deep learning methods heavily rely on high-quality dataset. Thus, the proposed system requires an efficient dataset to achieve high levels of accuracy.

13.6 Conclusion

The above system can be considered more of a *starting point* in building ANPR systems. The techniques used rely on basic computer vision and image-processing techniques to localize a license plate in an image, including morphological operations, image gradients, thresholding, bitwise operations, and contours. The methods used will work well only in controlled conditions and predictable environments, like when lighting conditions are uniform across input images and license plates are standardized, such as dark characters on a light license plate background.

But to build a system that works in *uncontrolled* environments, it is necessary to replace components (namely license plate localization, character segmentation, and character OCR) with more advanced machine learning and deep learning models. There are many challenges in running a 24/7 ANPR operation with hundreds of thousands of cars passing by the station every single day.

HOG and linear SVM are found to be effective for plate localization mainly when input license plates have a viewing angle that doesn't change more than a few degrees. For the system to work in unconstrained environments with drastic changes in view angles, deep learning–based models such as Faster R-CNN, SSDs, and YOLO are likely to generate better accuracy. A custom-designed OCR model may also significantly improve the accuracy of vehicle number recognition.

REFERENCES

Agarwal, T. (2014), "The evolution and future scope of augmented reality", *International Journal of Computer Science Issues*, *11*(6).

Andrew, N. G., Ngiam, J., Yu Foo, C., Mai, Y., Suen, C., Coates, A., et al., (2020), Deep Learning Tutorial published by Stanford, February 2017.

ANPR systems and its solutions, ANPR Systems | Automatic License Plate Recognition Systems (titanhz.com).

Bousquet, O., & Pez-Cruz, F. (2003), *"Kernel methods and their applications to signal processing,"* 2003 IEEE International Conference on Acoustics, Speech, and Signal Processing, 2003. Proceedings. (ICASSP'03), pp. 4–860. doi:10.1109/ICASSP.2003.1202779.

Bouwmans, T., Javed, S., Sultana, M., & Jung, S. K. (2019), "Deep neural network concepts for background subtraction: A systematic review and comparative evaluation", *Neural Networks*, *117*, 8–66.

Bushkovskyi, O. (2020), "What is pattern recognition and why it matters?"

Chand, D., Gupta, S., & Kavati, I. (2020), *Computer Vision based Accident Detection for Autonomous Vehicles*. India: Department of Computer Science and Engineering National Institute of Technology Warangal.

Dhote, S. (2020), "What is ANPR and How can we use it?" What is ANPR and How Can we Use it? – Proche (theproche.com).

Fogel, L. J., Owens, M. J., & Walsh, M. J. (1965), "Artificial intelligence through a simulation of evolution", in *Biophysics and Cybernetic Systems Maxfield*, Washington, DC: Spartan Books.

Gao, X., Zhang, J., & Wei, Z. (2018), *"Deep learning for sequence pattern recognition"*, *IEEE 15th International Conference on Networking, Sensing and Control (ICNSC)*, Zhuhai, doi:10.1109/ICNSC.2018.8361281.

Ghadage, S. S., & Khedkar, S. K. (2019), "A review paper on automatic number plate recognition system using machine learning algorithms", *International Journal of Engineering Research & Technology (IJERT)*, *8*(12).

Giraldo, J. H., & Bouwmans, T. (2020), "GraphBGS: Background Subtraction via Recovery of Graph Signals". arXiv preprint arXiv:2001.06404.

Golden, R. M. (2002), *Statistical Pattern Recognition*. University of Texas at Dallas Richardson.

Gowshalya Shri, A. M., & Arulprakash, M. (2014), "A Scheme for Detection of License Number Plate by the Application of Genetic Algorithms", *International Journal of Innovative Research in Computer and Communication Engineering*, *2*(1).

He, K., Zhang, X., Ren, S., & Sun, L. (2016), *"Deep Residual Learning for Image Recognition"*, *IEEE Conference on Computer Vision and Pattern Recognition (CVPR)*, Las Vegas, NV, doi:10.1109/CVPR.2016.90.

Jiang, S., Qin, H., Zhang, B., & Zheng, J. (2011), *Optimized Loss Functions for Object detection: A Case Study on Nighttime Vehicle Detection*. Hefei, China: School of Automotive and Transportation Engineering, Hefei University of Technology.

Makkar, S., & Sharma, L. (2019), *"A Face Detection using Support Vector Machine: Challenging Issues, Recent trend, solutions and proposed framework"*, in *Third International Conference on Advances in Computing and Data Sciences*, Inderprastha Engineering College, Ghaziabad.

Patel, C., Shah, D., & Patel, A. (2013), Automatic Number Plate Recognition System (ANPR): A survey. *International Journal of Computer Applications (IJCA)*, *69*, 21–33.

Prasad TVSNVC (2020), "Pattern Recognition: The basis of Human and Machine Learning", Pattern Recognition I Importance of Pattern Recognition (analyticsvidhya.com).

Rao, M. S., & Reddy, B. E. (2011), *"Comparative analysis of pattern recognition methods: An overview"*, *Indian Journal of Computer Science and Engineering (IJCSE)*.

Roy, S. (2007), *Vehicle License Plate Extraction and Recognition*. Kharagpur: Department of Computer Science and Engineering, Indian Institute of Technology.

Safari, P., & Kleinsteuber, M. (2013), "Deep Learning for Sequential Pattern Recognition", Master Thesis submitted at Technische Universitat Munchen.

Samra, G. A. (2016), "Genetic algorithms based orientation and scale invariant localization of vehicle plate number", *International Journal of Scientific & Engineering Research*, *7*(4).

Sawon (2020), *An Overview of Neural Approach on Pattern Recognition*.

Sharma, L. (2020), "Human detection and tracking using background subtraction in visual surveillance", *Towards Smart World*. New York: Chapman and Hall/CRC, https://doi.org/10.1201/9781003056751. pp. 317–328.

Sharma, L. (Ed.) (2021), *Towards Smart World*. New York: Chapman and Hall/CRC, https://doi.org/10.1201/9781003056751.

Sharma, L., & Garg, P. (Eds.) (2020a), *From Visual Surveillance to Internet of Things*. New York: Chapman and Hall/CRC, https://doi.org/10.1201/9780429297922.

Sharma, L., & Garg, P. K. (2020b), "Block based adaptive learning rate for moving person detection in video surveillance", *From Visual Surveillance to Internet of Things*, Taylor & Francis Group, CRC Press, Vol. *1*, pp. 201.

Sharma, L. Singh, A., & Yadav, D. K. (2016), "Fisher's linear discriminant ratio based threshold for moving human detection in thermal video", *Infrared Physics and Technology*.

14

Design of a Chess Agent Using Reinforcement Learning with SARSA Network

Vallidevi Krishnamurthy, V. Sanjay Thiruvengadam, Shankar Narayanan, R. C. Vignesh, and K. Sreeram
Sri Sivasubramaniya Nadar College of Engineering, Chennai, India

CONTENTS

14.1 Introduction

Reinforcement learning (RL) [1] is a fledgling method to create a game-playing artificial intelligence (AI) software. In RL, the agent (a game player) is allowed to explore all the possible ways it could move in the environment (Figure 14.1). For each action done by the agent, a transition happens from the current state to another state. According to the action, a reward is given. When the agent makes a good move, it is given a high reward, whereas when it makes a bad move, a low reward is given. So the agent tries to get a high reward by making good moves. This method can be used for developing game-playing software.

14.2 Literature Survey

The use of reinforcement learning in playing chess was a remarkable improvement in the field of AI. Neumann, Turing, and Shannon initially questioned whether a machine could play chess. Alan Turing worked on making a machine play chess with the intention that this will act as a wedge in attacking other problems [2].

Turochamp [3] is a chess program developed by Turing and Champernowne in 1948. It was created as part of the pair's research in computer science and machine learning. It is capable of playing an entire chess game against a human player at a low level of play by calculating all potential moves and all potential player moves in response, assigning point values to each game state and selecting the move with the highest average point value possible.

A good-known success story of reinforcement learning is TD-gammon, a backgammon-playing program that learned entirely by reinforcement learning and self-plays, and achieved a superhuman level of play [4].

The SARSA, as one kind of on-policy reinforcement learning methods, is integrated with deep learning to solve the video games control problems [5].

DOI: 10.1201/9781003140351-14

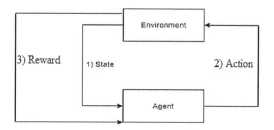

FIGURE 14.1 Typical reinforcement learning (RL) scenario.

DeepMind presented the first deep learning model to successfully learn control policies directly from high-dimensional sensory input using reinforcement learning. The model is a convolutional neural network, trained with a variant of Q-learning, whose input is raw pixels and whose output is a value function estimating future rewards. DeepMind applied the method to seven Atari 2600 games from the Arcade learning environment [6].

In 2016, DeepMind created an AI agent, Alpha Go [7], that was given a chance to play against the world's reigning human champion. AlphaGo won the match, 4–1, a triumph that sparked a wave of excitement regarding RL. It used a value and policy network and performed simulations using a model-free approach, like a Monte Carlo tree search. This helps in developing a robust agent that plays in any situation. This concept was extrapolated and used for building an agent Alpha Zero that was used to play chess. This agent defeated Stockfish, the reigning computer chess champion. Alpha Zero eliminated the need for using games played by human experts in developing its neural network model. It goes from zero to hero, i.e., it is only fed with the rules of the game and it develops its policy with self-play using reinforcement learning. It also uses only one neural network that combines the role of value and policy networks. In this chapter we attempt to explore the methods used in Alpha Zero to develop an agent capable of challenging amateur chess players [8].

In 2019, Arjan Groen created a library Reinforcement Learning Chess, which works in three chess environments, namely Move Chess, Capture Chess, and Real Chess. The agent primarily uses RL to learn to play in its environment [9].

14.3 Applications of Reinforcement Learning

- **Resources management in computer clusters**: Designing algorithms to allocate limited resources to different tasks is challenging and requires human-generated heuristics. The RL can be used to automatically learn to allocate and schedule computer resources to waiting jobs, with the objective of minimizing the average job slowdown [10].
- **Traffic light control**: Researchers tried to design a traffic light controller to solve the congestion problem.
- **Games:** RL is so well known these days because it is the mainstream algorithm used to solve different games and sometimes achieve superhuman performance. The most famous one must be Alpha Go and Alpha Go Zero. The Alpha Go, trained with countless human games, already achieved superhuman performance by using the value network and Monte Carlo tree search.
- **Personalized recommendations**: Guanjie et al. have applied RL in a news recommendation system in a paper titled "DRN: A Deep Reinforcement Learning Framework for News Recommendation" [11].

14.4 Architecture Design of the Proposed Chess Agent Using SARSA Network

The system chosen for the creation and testing of a chess agent is a SARSA network [7] that implements the concept of a Monte Carlo tree search (MCTS). Here the SARSA network implements the concepts of reinforcement learning as an on-policy technique. On-policy technique refers to an RL model that learns

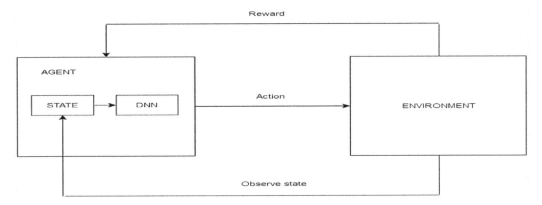

FIGURE 14.2 Architecture diagram of the proposed chess agent.

its policy as it interacts with the environment, so the interaction and learning of the agent with the environment happen simultaneously. The architecture of the proposed chess agent is depicted in Figure 14.2.

14.4.1 Components of the SARSA Network

States: Set of all states; here it is the set of all chess piece positions on the board.
Policy: It is the state-action table with the values corresponding to each pair.
Value function: The value function used in this study is the traditional Bellman's equation.

$$Q\left(S_t, A_t\right) \leftarrow Q\left(S_t, A_t\right) + \alpha\left[R_{t+1} + \gamma Q\left(S_{t+1}, A_{t+1}\right) - Q(S_t, A_t)\right] \tag{14.1}$$

Where $Q(S_t, A_t)$ – probability of winning from state S_t by making action A_t, R_t – Reward at the S_t, α – learning rate, and γ – discount factor (importance of future reward)

Actions: It contains all possible moves a certain piece can do from a position given all its environmental constraints.
Agent: The agent is the decision-making body of our system. It represents the AI bot that plays the chess game. It has the ability to perceive the environment and take actions that affect the environment
Environment: Here the environment is the chessboard. It contains all the pieces along with the rules that govern the movement of those pieces. It contains the idea of rewards, the value for each piece. It responds to the agent by giving the agent back a reward for each of its actions.
MCTS: This technique is used by the agent to look forward into the game by applying simulation of pieces. The looking forward is done using this MCTS. This tree contains:
Nodes: Chess board state.
Edges: Actions and the corresponding values connecting state S to S'.
This tree is built as the agent plays against a human opponent. The tree initially gets initialized with random values, but as more games are played, the tree gets updated with the moves along with its confidence score. It is this confidence score that is later used to apply the simulation again.
Expert policy: In MCTS, in order to make those simulations, there must be a policy that is used to choose successive children states required for the simulation. The expert policy used here is a densely layered neural network. This DNN is simultaneously trained along with the agent when the agent plays a chess game. The states stored in the MCTS are used as training for this model.
Working:
1.   When the chess game starts, the agent acts as the decision-making body that interacts with the chess board and "plays."
2.   The agent observes the state of the board and then starts an MCTS search process with the current state as the root node.

3. The MCTS tree keeps moving down the tree for 1 second. After 1 second, irrespective of whether the tree reaches the terminal state, the search process stops and then a back-propagation happens.

4. The return generated by the simulated episode is backed up to update, or to initialize, the action values attached to the edges of the tree traversed by the tree policy in this iteration of MCTS. No values are saved for the states and actions, visited by the rollout policy beyond the tree.

5. Now the agent, based on the simulation, decides the next move.

6. The agent updates its expert-policy DNN, after every ten-turn count.

7. After the game is over, the agent stops learning and then saves the expert-policy model.

14.5 Implementation of the Chess Agent Using SARSA Network-Module Description

Libraries used:
Keras: To load and store the expert-policy DNN model.
Python-chess: The chess simulation library for Python.

Components:
RLC_model.h5: The expert-policy Keras model.
Agent: contains all the required functions including the update strategy for the policy.
Environment: contains functions related to characteristics of a chess environment.
TD_search_m: contains the game simulation code, and the MCTS search code.

Environment:
Goal: To play chess effectively against a human opponent.
The environment is built using the library, called Python-chess. Figure 14.3 shows the chess board created by Python-chess. The library provides the following:

1. Board representation
2. Pieces representation
3. Move generation
4. Move validation

```
import chess
board = chess.Board()
board.push_san("e4")
board.push_uci("d7d5")
board.push_san("c3")
```

FIGURE 14.3   Chess environment built using python-chess library.

5. Support for UCI (universal chess interface), FEN (Forsyth-Edwards notation), SAN (standard algebraic notation) notations.

6. Render the chess board with SVG (Scalable Vector Graphics)

Code Snippet 14.1: Sample code for using the Python-chess library to build the environment.
Code Snippet 14.2: Pseudocode to calculate the reward given an action at a particular state.

```
getReward(action)
    currentPositionMaterialValue=board.getMaterialValue()
    board.push(action)
    nextPositionMaterialValue=board.getMaterialValue()
    reward=0
    extraReward=(nextPositionMaterialValue-currentPositionMaterialValue)*rewardFactor
    result=board.getResult()
    if(result=="White wins"):
        reward=100
    else if(result=="Black wins"):
        reward=-100
    reward+=extraReward
    return reward
```

Computation of Reward:

1. Assume the board is at some intermediate state of the game.
2. The board state is modeled as a node in a MCTS tree.
3. The system is iterated over all children of the node.
4. For each child, this same process is done again until a certain depth of the tree.
5. After reaching the leaf node
   a. If the reward = 1, if the system wins the game.
   b. If the reward = -1, if the system loses the game.
   c. The reward is 0.01 * difference in material value of initial and final state for all other situations.
6. Based on the end reward obtained, we update the state action value of the parent of this child with SARSA network formulae.
   a. Val(ParentState,Action) = Val(ParentState,Action) + LearningRate *(Reward of Child State + Discount_Factor *Val(ChildState,Action) − Val(ParentState,Action))
   b. Using the Val of Child State, we can update the value of parent state with learning rate and discount factor.
7. The StateAction pair for the current node with all the children is found, and the move that gives the best reward is chosen as the next move the agent plays.

Agent's Decision-Making Strategies – Policy:
The agent uses two different types of decision-making strategies to make a move. They are:

1. *Exploration*: It picks a random move out of the list of legal moves and explores the different states of the game.
2. *Exploitation*: It picks the move among the legal moves that has the highest probability to win.

At every state, the agent spends 1 second to expand the MCTS. The steps performed by the agent for the 1 second are shown in Figure 14.4. They are as follows:

1. *Selection*: Pick a move from the list of legal moves.
2. *Expansion*: Create a new child node if the node does not exist or move to the existing child node.
3. *Simulation*: Perform step 1 (Selection) and step 2 (Expansion) repeatedly until the time expires.
4. *Back propagation*: Once the time expires, the reward will be given. The parent nodes can be updated using the SARSA value function.

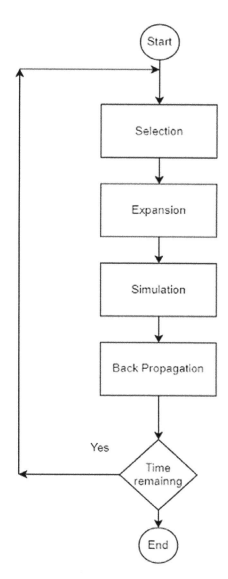

FIGURE 14.4   Steps performed by the agent.

14.6 Results

The graph in Figure 14.5 shows the reward that each move produced during the game. The rising pattern indicates that the agent has learned to make much better moves as it plays more games.

Figure 14.6 shows that as the games progressed, the material balance improved after each game. This indicates that the number of pieces remaining after each match is improving. This implies an improvement in the model's performance. Figure 14.7 shows a sample user interface of the chess game along with the results.

In [7]:

```
reward_smooth = pd.DataFrame(learner.reward_trace)
reward_smooth.rolling(window=1000,min_periods=1000).mean().plot(figsize=
(16,9),title='average reward over the last 1000 steps')
plt.show()
```

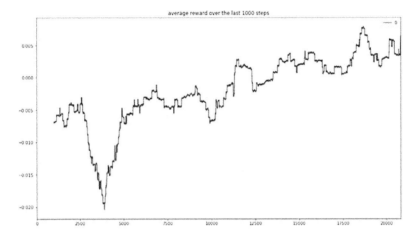

FIGURE 14.5   Average reward obtained by the agent over the last 1,000 steps.

In [8]:

```
reward_smooth = pd.DataFrame(learner.piece_balance_trace)
reward_smooth.rolling(window=50,min_periods=50).mean().plot(figsize=(16,
9),title='average piece balance over the last 50 episodes')
plt.show()
```

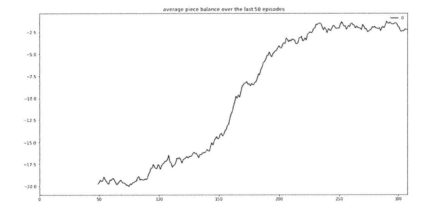

FIGURE 14.6   Average piece balance over the last 50 episodes.

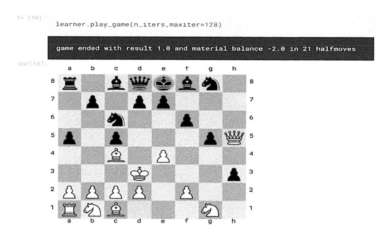

FIGURE 14.7   User interface of the chess game with the results.

14.7 Conclusion

In the attempt to create a chess-playing agent, the usefulness of a particular type of RL technique, namely the SARSA network, has been demonstrated. The SARSA network along with the Monte Carlo tree search technique does wonders because of its ability to play the game of chess by planning ahead and not just use policy that decides everything. The results obtained are positive, which indicates the success of this concept.

REFERENCES

1. Sutton, R. S., & Barto, A. G. (2015). *Reinforcement Learning: An Introduction*, MIT Press, Cambridge, MA.
2. Budd, C. (2018). Robots to play games, https://plus.maths.org/content/robots-play-games
3. Turing, A., & Champernowne, D. G. Turochamp, https://www.chessprogramming.org/Turochamp
4. Tesauro, G. (1995). Temporal difference learning and TD-Gammon. *Journal of the International Computer Games*, *18*, 88.
5. Zhao, D., Wang, H., Shao, K., & Zhu, Y. (2016). Deep reinforcement learning with experience replay based on SARSA. 1–6. doi:10.1109/SSCI.2016.7849837.
6. Mnih, V., Kavukcuoglu, K., Silver, D., Graves, A., Antonoglou, I., Wierstra, D., & Riedmiller, M. (2013). Playing Atari with deep reinforcement learning. arXiv preprint arXiv:1312.5602.
7. Silver, D., Schrittwieser, J., Simonyan, K., Antonoglou, I., Huang, A., Guez, A., Hubert, T., Baker, L., Lai, M., Bolton, A., Chen, Y., Lillicrap, T., Hui, F., Sifre, L., Driessche, G., Graepel, T., & Hassabis, D. (2017). Mastering the game of Go without human knowledge. *Nature*, *550*, 354–359. doi:10.1038/nature24270.
8. Silver, D., Hubert, T., Schrittwieser, J., Antonoglou, I., Lai, M., Guez, A., & Hassabis, D. (2017). Mastering chess and shogi by self-play with a general reinforcement learning algorithm. arXiv preprint arXiv:1712.01815.
9. Groen, A. Reinforcement Learning Chess, https://github.com/arjangroen/RLC
10. Li, Y. (2019). Reinforcement learning applications. arXiv preprint arXiv:1908.06973.
11. Zheng, G., Zhang, F., Zheng, Z., Xiang, Y., Yuan, N. J., Xie, X., & Li, Z. (2018). *DRN: A deep reinforcement learning framework for news recommendation.* In *Proceedings of the 2018 World Wide Web Conference (WWW '18).* International World Wide Web Conferences Steering Committee, Republic and Canton of Geneva, CHE, 167–176. https://doi.org/10.1145/3178876.3185994

15

Moving Objects Detection in Video Processing: A Graph Signal Processing Approach for Background Subtraction

Jhony H. Giraldo and Thierry Bouwmans
Laboratoire MIA, Université La Rochelle, France

CONTENTS

15.1 Introduction

Background subtraction is a crucial topic in video analysis, with multiple applications in video surveillance, intelligent transportation, and industrial vision, among others [1]. Background subtraction allows to distinguish the foreground (i.e., moving objects) from the background in video sequences [2]. Unsupervised, semi-supervised, and supervised methods have been proposed trying to solve this problem [3–5]; however, there is no method that is able to effectively handle all real-time background subtraction challenges such as camouflage, dynamic background, illumination variations, and camera jitter, to name a few [1–3, 5, 6, 31].

Several deep learning methods have been proposed for background/foreground separation in the literature [4]. These methods are based on different architectures such as Restricted Boltzman Machine [7], Deep Encoder-Decoder [8–10], Convolutional Neural Networks [11–16], and Generative Adversarial Networks [17–22]. Even though these methods have shown good results [4], they still require a lot of labeled data. As a matter of fact, there are no general answers in the literature about the sample complexity needed in deep learning regimen [23]. Moreover, most of the deep learning techniques for background subtraction have not evaluated their performance on unseen videos, leading to a limited analysis of the impact of these algorithms in real scenarios [6]. Recently, Giraldo and Bouwmans [5, 39, 40] proposed a graph based semi-supervised method for background subtraction (GraphBGS), which is based on the notions of graph signal processing (GSP). GraphBGS has the advantage of requiring less data than deep learning algorithms while adapting to complex scenarios unlike unsupervised methods. In this chapter, we propose a new active semi-supervised learning algorithm for background subtraction on unseen videos (ActiveBGS), using concepts of the theory of sampling in GSP. ActiveBGS can choose what are the best instances from the videos that should be labeled to get better performance. In other words, this chapter stands in a scenario in which we have a raw database of videos, and we want to get the best possible performance in background subtraction with minimum intervention, and by labeling just some of the instances.

This work lies in the semi-supervised learning domain by treating background subtraction as a problem of sampling and recovery of graph signals. ActiveBGS models the instances in videos for background subtraction on unseen videos as nodes embedded in a graph, with the background or foreground related to a graph signal. ActiveBGS is composed of instance segmentation, background initialization, construction of a graph, as well as sampling and reconstruction of graph signals. More specifically, our method builds on the approach of blue-noise sampling on graphs [24]. Blue-noise tries to sample nodes as far as possible between each other. ActiveBGS outperforms random sampling–based algorithms for some challenges of the Change Detection 2014 (CDNet2014) [25] database. The main contributions of this chapter can be summarized as follows: (1) to the best of our knowledge, the theory of active semi-supervised learning is employed for the first time in background subtraction; and (2) blue-noise sampling on graphs is extended to unseen sampling schemes in background subtraction.

The rest of the chapter is organized as follows. Section 15.2 explains the basic concepts and the proposed algorithm. Section 15.3 introduces the experimental framework. Finally, Sections 15.4 and 15.5 present the results and conclusions, respectively.

15.2 Active Background Subtraction

This section introduces the notation, basic concepts, and the proposed ActiveBGS. Figure 15.1 shows the pipeline of ActiveBGS.

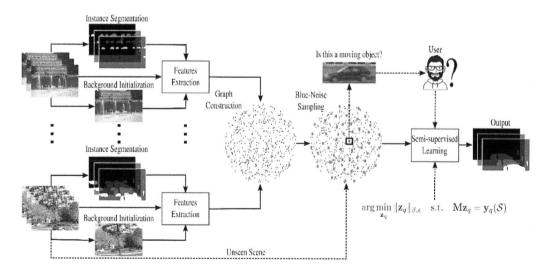

FIGURE 15.1 Pipeline of ActiveBGS. This algorithm uses temporal median filter and mask region convolutional neural network [26]. Each mask represents a node in the graph, and the representation is obtained with optical flow, intensity, and texture features. At the end, some nodes are sampled with blue noise, and the semi-supervised algorithm provides the labels in the graph.

15.2.1 Notation

Uppercase and lowercase boldface letters such as $\mathbf{W}$ and $\mathbf{y}$ represent matrices and vectors, respectively. Calligraphic letters such as $\mathcal{E}$ denote sets, and $|\mathcal{V}|$ corresponds to its cardinality. $(\cdot)^\mathsf{T}$ denotes transposition. $\sigma_{\min}(\mathbf{A})$ and $\sigma_{\max}(\mathbf{A})$ correspond to the minimum and maximum singular values of the matrix $\mathbf{A}$, respectively. Finally, $\operatorname{diag}(\mathbf{x})$ is a diagonal matrix with entries $x_1, \ldots, x_n$.

15.2.2 Background

Let $G = (\mathcal{V}, \mathcal{E})$ be an undirected, weighted, and connected graph. $\mathcal{V} = \{1, \ldots, N\}$ is the set of nodes with $|\mathcal{V}| = N$. $\mathcal{E} = \{(i,j)\}$ is the set of edges, where (i,j) is the edge connecting nodes i and j. A graph signal is defined as a function $y : \mathcal{V} \to \mathbb{R}$ on the nodes of G; this signal can be represented as $\mathbf{y} \in \mathbb{R}^N$ where $\mathbf{y}(i)$ is the function evaluated on the i - th node. The adjacency matrix of G is $\mathbf{W} \in \mathbb{R}^{N \times N}$ such that $\mathbf{W}(i,j) = w_{ij} \in \mathbb{R}^+$ is the weight between the nodes i and j. Similarly, $\mathbf{D} \in \mathbb{R}^{N \times N}$ is the degree matrix of the graph such that:

$$\mathbf{D}(i,j) = \sum_{j=1}^{N} \mathbf{W}(i,j) \ \forall \ i = 1, \ldots, N, \tag{15.1}$$

where $\mathbf{D}$ is a diagonal matrix. Furthermore, the combinatorial Laplacian operator is a positive semi-definite matrix defined as $\mathbf{L} = \mathbf{D} - \mathbf{W}$. Since $\mathbf{W}$ is symmetric for undirected graphs, $\mathbf{L}$ is a positive semi-definite matrix with a set of real non-negative eigenvalues $0 = \lambda_1 \leq \cdots \leq \lambda_N$, and a full set of orthogonal eigenvectors $\{\mathbf{u}_1, \ldots, \mathbf{u}_N\}$. The graph Fourier basis of G is defined by the spectral decomposition of the Laplacian matrix such that $\mathbf{L} = \mathbf{U} \mathbf{\Lambda} \mathbf{U}^\mathsf{T}$, where $\mathbf{U} = [\mathbf{u}_1, \ldots, \mathbf{u}_N]$ and $\mathbf{\Lambda} = \operatorname{diag}(\lambda_1, \ldots, \lambda_N)$. Thus, the Graph Fourier Transform (GFT) of a signal $\mathbf{y}$ is defined as $\hat{\mathbf{y}} = \mathbf{U}^\top \mathbf{y}$, and the inverse GFT is such that $\mathbf{y} = \mathbf{U}\hat{\mathbf{y}}$.

DEFINITION 1

$\mathbf{y}$ *is called bandlimited if* $\exists \rho \in \{1, \ldots, N-1\}$ *given that its GFT satisfies* $\hat{\mathbf{y}}(i) = 0 \ \forall \ i > \rho$.

The Paley-Wiener space of all ω-*bandlimited* signals is defined as $PW\omega(G) = \operatorname{span}(\mathbf{U}\rho : \lambda\rho \leq \omega)$, where $\mathbf{U}\rho = [\mathbf{u}_1, \ldots, \mathbf{u}\rho]$. Using Definition 1 and the notions of Paley-Wiener spaces, a signal $\mathbf{y}$ has cutoff frequency ω and bandwidth ρ when $\mathbf{y} \in PW\omega(G)$. Given the notion of bandlimitedness in GSP, the next step is to find a minimum bound for the sampling rate that allows perfect recovery of $\mathbf{y} \in PW\omega(G)$. The sampling rate is determined with a subset of nodes $\mathcal{S} \subset \mathcal{V}$ in G such that $\mathcal{S} = \{s_1, \ldots, s_m\}$, where $m = |\mathcal{S}| \leq N$ is the number of samples. $\mathbf{y}(\mathcal{S}) = \mathbf{M}\mathbf{y}$ is the sampled graph signal, where $\mathbf{M}$ is a binary decimation matrix whose entries are given by $\mathbf{M} = [\boldsymbol{\delta}_{s1}, \ldots, \boldsymbol{\delta}_{sm}]^\mathsf{T}$ and $\boldsymbol{\delta}_v$ is the N–dimensional Kronecker column vector centered at v. Perfect recovery from $\mathbf{y}(\mathcal{S})$ is possible when $|\mathcal{S}| \geq \rho$ [27]. In this chapter, $\mathcal{S}$ is a set of minority nodes, and its complement $\mathcal{S}^c$ is the set of majority nodes if $|\mathcal{S}| / N < 0.5$. Finally, $\mathbf{s} : \mathcal{V} \to \{0, 1\}^N$ is the sampling pattern vector associated with $\mathcal{S}$ such that $\mathbf{s}(i) = 1 \ \forall \ i \in \mathcal{S}$ and $\mathbf{s}(j) = 0 \ \forall \ j \in \mathcal{S}^c$. Intuitively, a graph signal is in the Paley-Wiener space of the graph when $\mathbf{y}$ is smooth in the vertex domain of G.

There is no simple way to choose the best subset of sampling nodes $\mathcal{S}$, because graphs do not have a regular structure. Previous methods for sampling set selection on graphs have mainly used spectral information [27, 28]. In contrast, authors like Puy et al. [29] proposed random sampling strategies, albeit with lower performance. Recently, sampling nodes selected to be further apart from each other has shown better results than random techniques [24]. These methods have shown a behavior dominated by high frequencies, providing the name of blue noise.

Intuitively, since the graph signal is smooth when $\mathbf{y} \in PW\omega(G)$, selecting a sampling set with nodes that are close to each other gives redundant information. In contrast, choosing nodes further apart between each other can lead to more robust sampling sets. Formally, the problem of optimally sampling a graph-signal can be expressed as choosing $\mathcal{S}$ such that the available bandwidth of $\mathbf{y}(\mathcal{S})$ is maximized. To this purpose, Pesenson [30] defined the concept of Λ-removable sets: for a given $\Lambda > 0$, $\mathcal{S}$ is a Λ-removable set if any $\mathbf{y} \in L_2(\mathcal{S})$ presents a Poincaré-type inequality with the constant Λ^{-1}, such that:

$$\|\mathbf{y}\|_2 \leq \left(\Lambda^{-1}\right)\|\mathbf{L}\mathbf{y}\|_2 \ \forall \ \mathbf{y} \in L_2(\mathcal{S}),$$ (15.2)

where $L_2(\mathcal{S})$ is the set of all signals $\mathbf{y}$ with support in $\mathcal{S}$ and finite ℓ_2-norm. The best constant Λ in this inequality is denoted by $\Lambda_\mathcal{S}$, and determines the importance of a given sampling set $\mathcal{S}$ in the sampling of a graph signal with a specific bandwidth. The relationship between the sampling problem and the removable sets was given in the following theorem by Pesenson [30]:

THEOREM 1

If for a set $\mathcal{S} \subset \mathcal{V}$ its complement $\mathcal{S}^c$ is a $\Lambda_{\mathcal{S}^c}$-removable set for any space $PW\omega(G)$ with $0 < \omega < \Lambda_{\mathcal{S}^c}$, In particular, every $\mathbf{y} \in PW\omega(G)$ is completely determined by its values on $\mathcal{S}$.

Theorem 1 states that is possible to obtain a good sampling set $\mathcal{S}$ as the one which minimizes the constant $\Lambda_{\mathcal{S}^c}$. Let $R_\mathbf{s}$ be the redness of a sampling pattern $\mathbf{s}$ such that:

$$R_\mathbf{s} = \frac{1}{m} \sum_{i=2}^{N} \frac{\hat{\mathbf{s}}^2(i)}{\lambda_i},$$ (15.3)

where $\hat{\mathbf{s}}$ is the GFT of $\mathbf{s}$ and $R_\mathbf{s}$ is assumed to take a minimum value for blue-noise sampling patterns. Using the concepts of Λ-removable and redness in graphs, Parada-Mayorga et al. [24] showed that blue-noise sampling patterns encourage high values of $\Lambda_{\mathcal{S}^c}$ from the following theorem:

THEOREM 2

Let $\mathbf{s} : \mathcal{V} \to \{0, 1\}^N$ be a sampling pattern and $|\mathcal{S}| = m$, then the $\Lambda_{\mathcal{S}^c}$-constant of the set $\mathcal{S}^c$ satisfies:

$$\Lambda_{\mathcal{S}^c} > C_\delta \left(\frac{R_\mathbf{s}}{\operatorname{tr}(\mathbf{D})R_\mathbf{s} - m^2\left(1 - \frac{m}{N}\right)^2} \right)^{\frac{2}{\delta}},$$ (15.4)

where $R_\mathbf{s}$ is the redness of $\mathbf{s}$ from Equation (15.3), $\operatorname{tr}(\mathbf{D})$ is the trace of $\mathbf{D}$, δ is the isoperimetric dimension of G [15], and $C\delta$ is a constant that depends only on δ. Proof: see [8].

Theorem 2 indicates that the best sampling sets $\mathcal{S}$ maximize $\Lambda_{\mathcal{S}^c}$. Therefore, by minimizing R_s we are promoting *good* sampling sets. In this chapter, we design an extension of blue-noise sampling on graphs for the specific case of background subtraction on unseen videos.

15.2.3 Instance Segmentation

ActiveBGS uses a Mask Region Convolutional Neural Network (Mask R-CNN) [26] in the instance segmentation step. The Mask R-CNN in ActiveBGS is trained in the Common Objects in Context (COCO) 2017 dataset [32], and the following objects are discarded to reduce computational complexity: traffic light, fire hydrant, stop sign, parking meter, bench, chair, couch, potted plant, bed, dining table, toilet, TV, microwave, oven, toaster, sink, refrigerator, clock, vase. Each output-mask from the videos is a node in ActiveBGS.

The definition of what is a node in the graph is very important in ActiveBGS. Other algorithms, such as super-pixel, can be used to represent the nodes in the problem of background subtraction. However, super-pixel was not used in this chapter because the number of nodes resulting from such an approach would lead to an extremely huge graph. Currently, there are no techniques in the state-of-the-art of GSP to handle such big graphs.

15.2.4 Background Estimation and Nodes Representation

The background initialization algorithm in ActiveBGS is the temporal median filter, and the videos are processed in grayscale. The representation of the nodes is then obtained from intensity, texture features, and optical flow. Let $\mathbf{I}_v^t$ and $\mathbf{I}_v^{t-1}$ be the grayscale Region of Interest (ROI) images corresponding to the node $v \in \mathcal{V}$ in the current (t) and previous ($t-1$) frames, respectively. Let $\mathbf{B}_v$ and $\mathcal{P}_v$ be the ROI image of the background, and the set of pixel indices corresponding to the node v, respectively. Furthermore, let $\mathbf{v}_x^t(\mathcal{P}_v)$ and $\mathbf{v}_y^t(\mathcal{P}_v)$ be the optical flow vectors of the current image with support on $\mathcal{P}_v$ for the horizontal and vertical direction, respectively, where the optical flow is obtained from the Lucas-Kanade method [33]. $\mathbf{v}_x^t(\mathcal{P}_v)$ and $\mathbf{v}_y^t(\mathcal{P}_v)$ are used to compute histograms and some descriptive statistics (the mean, minimum, maximum, range, standard deviation, mean absolute deviation). Furthermore, intensity histograms are computed in $\mathbf{I}_v^t(\mathcal{P}_v)$, $\mathbf{I}_v^{t-1}(\mathcal{P}_v)$, $\mathbf{B}_v(\mathcal{P}_v)$, and $\left|\mathbf{I}_v^t(\mathcal{P}_v) - \mathbf{B}_v(\mathcal{P}_v)\right|$. Finally, the texture representation is computed in $\mathbf{I}_v^t$, $\mathbf{I}_v^{t-1}$, $\mathbf{B}_v$, and $\left|\mathbf{I}_v^t - \mathbf{B}_v\right|$, using local binary patterns [34]. The representation of the node is obtained from the concatenation of all the previous features, i.e., optical flow, intensity, and texture features. Each instance v is given by a 853-dimensional vector $\mathbf{x}_v$.

15.2.5 Graph Construction

Let $\mathbf{X} \in \mathbb{R}^{N \times M}$ be the matrix of N instances, where $\mathbf{X} = [\mathbf{x}_1, \mathbf{x}_2, \ldots, \mathbf{x}_N]^\mathsf{T}$. First, a k-nearest neighbors algorithm with k = 30 is employed to connect the nodes in the graph. Thus, vertices are connected to obtain an undirected graph. The weight between two connected nodes i, j is expressed as follows:

$$w_{ij} = \exp-\frac{d(i,j)^2}{\sigma^2}$$

where $d(i,j) = \|\mathbf{x}_i - \mathbf{x}_j\|_2$ and σ^2 is the standard deviation of the Gaussian function computed as:

$$\sigma = \frac{1}{|\mathcal{E}| + N} \sum_{(i,j) \in \mathcal{E}} d(i,j) \tag{15.5}$$

15.2.6 Blue-Noise Sampling for Unseen Videos

An algorithm to compute blue-noise sampling on graphs has been introduced by Parada-Mayorga et al. [24]. This method employed void-and-cluster, which was originally proposed in [35] for digital halftoning. Void-and-cluster generates homogeneous sampling patterns on graphs by iteratively measuring the concentration of minority nodes in a sampling pattern. Afterwards, void-and-cluster swaps the minority node in the zone of highest concentration with the non-minority node in the zone of least concentration. In this chapter, we extend the blue-noise algorithm on graphs for background subtraction of unseen videos by modifying the set of majority nodes $\mathcal{S}$ when $|\mathcal{S}|/N < 0.5$ in each iteration of the void-and-cluster algorithm.

Let $\mathbf{W_d} \in \mathbb{R}^{N \times N}$ be the matrix of distances of G such that $\mathbf{W_d}(i,j) = \|\mathbf{x}_i - \mathbf{x}_j\|_2 \Leftrightarrow (i,j) \in \mathcal{E}$, and let $\mathbf{\Gamma} \in \mathbb{R}^{N \times N}$ be the matrix of shortest-path distances of all node pairs of $\mathbf{W_d}$ (geodesic distance matrix). In order to exploit Theorem 2 for generating blue-noise sampling patterns, a Gaussian kernel is evaluated on $\mathbf{\Gamma}$ such that:

$$\mathbf{K}(i,j) = \exp\left(-\frac{\mathbf{\Gamma}(i,j)^2}{\sigma^2}\right) \forall \ 1 \le i, \ j \le N$$

where σ is the standard deviation of the Gaussian filter defined in Equation (15.4). Let $\mathbf{v} \in \mathbb{R}^N$ be the vector representing the concentrations of sampled nodes such that:

$$\mathbf{v}(i) = \sum_{j=1}^{N} \mathbf{s}(j)\mathbf{K}(i,j) \ \forall \ 1 \leq i \leq N. \tag{15.6}$$

Equation (15.6) is adding all the kernelized geodesic distances between each node i and the other vertices $j \in \mathcal{S}$ for all $1 \leq i \leq N$, i.e., $\mathbf{v}(i)$ is a measure of the concentration of sampled nodes around i. In the original void-and-cluster algorithm [8], for $|\mathcal{S}|/N \leq 0.5$ the location of the tightest cluster (the sampled node in the highest concentration) can be found as $\mathrm{argmax}_i\{\mathbf{v}(i)\}$ for $i \in \mathcal{S}$, and the position of the largest void (the non-sampled node in the lowest concentration) as $\mathrm{argmin}_j\{\mathbf{v}(j)\}$ for $j \in \mathcal{S}^c$. Taking into account the unseen videos in background subtraction, let $\mathcal{U}_k$ be the set of nodes corresponding to a specific video sequence k, then the position of the largest void is found as $\mathrm{argmin}_j\{\mathbf{v}(j)\}$ for $j \in \mathcal{S}^c \setminus \mathcal{U}_k$. Algorithm 1 shows the pseudo-code for the void-and-cluster method for sampling of graph signals for background subtraction on unseen videos, where $\mathbf{s_r}$ is a random sampling pattern with $\mathbf{s_r}(\mathcal{S}) = 1$ such that $\mathcal{S}$ does not contain any element of $\mathcal{U}_k$.

15.2.7 Semi-supervised Learning Algorithm

The semi-supervised learning algorithm of this chapter is based on the minimization of the Sobolev norm of $\mathbf{y}$ [36].

DEFINITION 2

The Sobolev norm is defined as follows:

$$\left\| \mathbf{y}_{\beta,\epsilon} \right\| = \left\| \left(\mathbf{L} + \epsilon\mathbf{I}\right)^{\beta/2} \mathbf{y} \right\|_2, \epsilon \geq 0, \ \beta \in \mathbb{R}. \tag{15.7}$$

The variational problem can be stated a follows: find a vector $\mathbf{z}_q$ from $L_2(G)$ such that: $\mathbf{z}_q(\mathcal{S}) = \mathbf{M}\mathbf{z}_q = \mathbf{y}_q(\mathcal{S})$, and $\mathbf{z}_q$ minimizes the Sobolev norm $\mathbf{z}_q \rightarrow \|(\mathbf{L} + \epsilon\mathbf{I})\beta/2\mathbf{z}_q\|_2$, where $\mathbf{y}_q$ is the graph signal associated with the q - th class. Put differently, the variational problem tries to solve:

Algorithm 1 Void-and-cluster in Graphs

Input: $\mathbf{K}$, $m = |\mathcal{S}|$, max_iter
Output: s: sampling pattern
Initialize: id_cl$^-$ = 1, id_vd$^-$ = 1, id_cl = 0,
id_vd = 0, cont = 1, s = $\mathbf{s_r}$
1: **while**   ((id_cl $\neq$ id_vd$^-$)**or**(id_vd $\neq$ id_cl$^-$))
    and (cont < max_iter) **do**
2:      id_vd$^-$ = id_vd; id_cl$^-$ = id_cl
3:      $\mathbf{v}(l) = \sum_{j=1}^{N} \mathbf{s}(j)\mathbf{K}(l,j) \ \forall \ 1 \leq l \leq N$
4:      id_cl = $\arg\max_i\{\mathbf{v}(i)\} \ \forall \ i \in \mathcal{S}$
5:      id_vd = $\arg\min_j\{\mathbf{v}(j)\} \ \forall \ j \in \mathcal{S}^c \setminus \mathcal{U}_k$
6:      s(id_cl) = 0; s(id_vd) = 1
7:      cont++
8: **end while**
9: **return** s

$$\mathrm{arg\,min}_{\mathbf{z}_q} \mathbf{z}_q^\top \left(\mathbf{L} + \epsilon\mathbf{I}\right)^\beta \mathbf{z}_q \quad \text{s.t.} \quad \mathbf{M}\mathbf{z}_q - \mathbf{y}_q(\mathcal{S}) = 0. \tag{15.8}$$

Equation (15.8) is solved for $q = 1, 2$ in this chapter. The term $(\mathbf{L} + \epsilon\mathbf{I})$ in Equation (15.7) is always invertible for $\epsilon > 0$ [5].

THEOREM 3

Let $\mathbf{\Psi} \in \mathbb{R}^{N \times N}$ *be a perturbation matrix. Given a combinatorial Laplacian matrix* $\mathbf{L}$, *the term* $\mathbf{L} + \mathbf{\Psi}$ *presents a lower and an upper bound in the condition number as follows:*

$$\frac{\sigma_{\max}(\mathbf{L}+\mathbf{\Psi})}{\sigma_{\max}(\mathbf{\Psi})} \leq k(\mathbf{L}+\mathbf{\Psi}) \leq \frac{\sigma_{\max}(\mathbf{L})+\sigma_{\max}(\mathbf{\Psi})}{\sigma_{\min}(\mathbf{L}+\mathbf{\Psi})} \tag{15.9}$$

where $k(\mathbf{L} + \mathbf{\Psi})$ *is the condition number of* $\mathbf{L} + \mathbf{\Psi}$. *Proof: Please see [5].*

Theorem 3 gives a lower and an upper bound in the condition number of $\mathbf{L} + \mathbf{\Psi}$. In the same way, Weyl's inequality [37] shows the way the eigenvalues of a perturbed matrix change and is presented as follows:

THEOREM 4

Let $\mathbf{L}$ *and* $\mathbf{\Psi}$ *be Hermitian matrices with a set of eigenvalues* $\{\lambda_1, ..., \lambda_N\}$ *and* $\{\psi_1, ..., \psi_N\}$, *respectively. The matrix* $\mathbf{L} + \mathbf{\Psi}$ *has a set of eigenvalues* $\{v_1, ..., v_N\}$ *where the following inequalities hold for* $i = 1, ..., N$:

$$\lambda_i + \psi_1 \leq v_i \leq \lambda_i + \psi_N \tag{15.10}$$

Proof: Please see [37].

According to Theorem 4, if $\mathbf{\Psi} \succ 0 \rightarrow v_i > \lambda_i \ \forall \ i \in \mathcal{V}$. If we want to get $\det(\mathbf{L} + \mathbf{\Psi}) \neq 0$, $\mathbf{\Psi}$ should be positive definite. If $\mathbf{\Psi} = \epsilon\mathbf{I}$, where $\epsilon \in \mathbb{R}^+$ and $\mathbf{I}$ is the identity matrix, then $\sigma_{\max}(\epsilon\mathbf{I}) = \epsilon$. Moreover, $\sigma_{\min}(\mathbf{L} + \epsilon\mathbf{I}) > 0$ according to Theorem 4. Consequently, the upper bound in Equation (15.8) for $\mathbf{\Psi} = \epsilon\mathbf{I}$ is

$$k(\mathbf{L}+\epsilon\mathbf{I}) \leq \frac{\sigma_{\max}(\mathbf{L})+\epsilon}{\sigma_{\min}(\mathbf{L}+\epsilon\mathbf{I})} < \infty$$

Since the Sobolev norm is invertible for $\epsilon > 0$ (based on Theorems 3 and 4), the optimization problem in Equation (15.10) has a closed-form solution given by:

$$\tilde{\mathbf{Z}} = \left((\mathbf{L}+\epsilon\mathbf{I})^{-1}\right)^{\beta} \mathbf{M}^{\mathrm{T}} \left(\mathbf{M}\left((\mathbf{L}+\epsilon\mathbf{I})^{-1}\right)^{\beta} \mathbf{M}^{\mathrm{T}}\right)^{-1} \mathbf{Y}(\mathcal{S}) \tag{15.11}$$

where $\tilde{\mathbf{Z}} = [\mathbf{z}_1, \mathbf{z}_2]$, and $\mathbf{Y}(\mathcal{S})$ is the sub-matrix of $\mathbf{Y}$ with rows indexed by $\mathcal{S}$. The final classification of a node i is calculated as the class corresponding to the maximum value in the i - th row of $\tilde{\mathbf{Z}}$. In this chapter, the semi-supervised learning problem is solved using the GSP toolbox [38].

15.3 Experimental Framework

This section presents the dataset, the evaluation measures, and the experimental results of ActiveBGS.

15.3.1 Dataset and Evaluation Measures

ActiveBGS is evaluated in CDNet2014 using the precision, recall, and F-measure metrics. These metrics are computed as follows:

$$\text{Recall} = \frac{TP}{TP + FN}, \quad \text{Precision} = \frac{TP}{TP + FP},$$
$$\text{F measure} = 2 \frac{\text{Precision} \times \text{Recall}}{\text{Precision} + \text{Recall}}, \tag{15.12}$$

where FN, FP, and TP are the number of False Negatives, False Positives, and True Positives pixels, respectively.

15.3.2 Experiments

A graph is constructed for each challenge of the CDNet2014. In addition, for each sequence a percentage of the number of nodes (from 0.1% to 10%) in the graph is sampled using random and blue-noise sampling with an unseen scheme. Eigendecomposition-based algorithms [27, 28] for sampling of graph signals are discarded because the number of nodes of each challenge is huge (from 4,336 up to 33,291 nodes). Finally, the semi-supervised learning classifies the non-labeled nodes. Since blue-noise is a stochastic sampling technique [24], the experiments are repeated ten times for each sequence and each sampling density. The metrics are computed on the unseen videos.

15.4 Results and Discussions

Figure 15.2 provides the results comparing random and blue-noise sampling on graphs for background subtraction. Blue-noise sampling on graphs has better performance than random in the challenges shadow and low frame rate; in the low sampling densities of dynamic background, bad weather, and intermittent object motion; and finally, in the high sampling densities of thermal and camera jitter. On the other hand, random is better than blue-noise sampling in the high sampling densities of intermittent object motion, dynamic background, and bad weather, and in the low densities of thermal, and camera jitter. Finally, blue-noise and random sampling have approximately the same performance in the challenge baseline.

GraphBGS [5], for example, constructed a graph for the whole CDNet2014 database. In this chapter, a graph is built for each critical challenge because the approach of GraphBGS leads in a big graph, with a huge memory requirement for the matrix $\mathbf{K}$ in Algorithm 1 (more than 500 gigabytes of memory with variables type double). A direct comparison between ActiveBGS and GraphBGS can be done once we have a blue-noise sampling technique able to handle such big graphs. To this end, Theorem 2 should be exploited trying to avoid the computation of $\mathbf{K}$ and the vector of concentrations $\mathbf{v}$, leading to a fast blue-noise sampling technique for background subtraction, and other graph signal sampling applications.

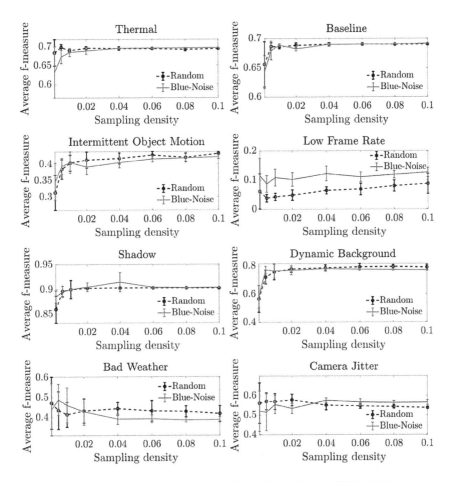

FIGURE 15.2 Average f-measure compared to sampling density in eight challenges of CDNet2014 using random and blue-noise sampling on graphs. Each point in the plots is a Monte Carlo cross-validation experiment with ten repetitions.

15.5 Conclusion

This chapter introduced a new active semi-supervised learning algorithm for background subtraction (ActiveBGS). This algorithm involves instance segmentation, background initialization, feature extraction, construction of the graph, blue-noise sampling on graphs, and finally a Sobolev minimization semi-supervised learning approach. ActiveBGS outperforms random-based techniques in some challenges of CDNet2014.

REFERENCES

1. B. Garcia-Garcia, T. Bouwmans, A. J. R. Silva (2020). Background subtraction in real applications: Challenges, current models and future directions. *Computer Science Review*, *35*, 100204.
2. T. Bouwmans, A. Sobral, S. Javed, S. K. Jung, E. H. Zahzah (2017). Decomposition into low-rank plus additive matrices for background/foreground separation: A review for a comparative evaluation with a large-scale dataset. *Computer Science Review*, *23*, 1–71.
3. T. Bouwmans (2014). Traditional and recent approaches in background modeling for foreground detection: An overview. *Computer Science Review*, *11*, 31–66.

4. T. Bouwmans, S. Javed, M. Sultana, S. K. Jung (2019). Deep neural network concepts for background subtraction: A systematic review and comparative evaluation. *Neural Networks*, *117*, 8–66.

5. J. H. Giraldo, T. Bouwmans (2020). GraphBGS: Background Subtraction via Recovery of Graph Signals. arXiv preprint arXiv:2001.06404.

6. O. Tezcan, P. Ishwar, J. Konrad (2020). *BSUV-Net: A fully-convolutional neural network for background subtraction of unseen videos*. In *The IEEE Winter Conference on Applications of Computer Vision* (pp. 2774–2783).

7. J. Gracewell, M. John (2020). Dynamic background modeling using deep learning autoencoder network, *Multimedia Tools and Applications*, *79*, 4639–4659.

8. S. Choo, W. Seo, D. Jeong, N. Cho (2018). *Multi-scale recurrent encoder-decoder network for dense temporal classification, IAPR International Conference on Pattern Recognition, ICPR 2018*, Beijing, China.

9. A. Farnoosh, B. Rezaei, S. Ostadabbas (2021). *DEEPBM: Deep probabilistic background model estimation from video sequences, International Conference on Pattern Recognition (ICPR 2020)*, Milan, Italy.

10. S. Ammar, T. Bouwmans, N. Zaghden, M. Neji (2019). *Moving objects segmentation based on DeepSphere in video surveillance, International Symposium on Visual Computing, ISVC 2019*, Tahoe City, USA.

11. M. Braham, M. Van Droogenbroeck (2016), *Deep background subtraction with scene-specific convolutional neural networks, IWSSIP 2016*, Bratislava, Slovakia.

12. T. Minematsu, A. Shimada, H. Uchiyama, R. Taniguchi (2018). Analytics of Deep Neural Network-based Background Subtraction, *MDPI Journal of Imaging*.

13. T. Minematsu, A. Shimada, R. Taniguchi (2020). *Rethinking Background and Foreground in Deep Neural Network-Based Background Subtraction, IEEE International Conference on Image Processing, ICIP 2020*, Abu Dhabi, UAE.

14. P. Patil, S. Murala (2018). MSFgNet: A novel compact end-to-end deep network for moving object detection, *IEEE Transactions on Intelligent Transportation Systems*.

15. I. Osman, M. Shehata (2020). *MODSiam: Moving object detection using siamese networks, IEEE Canadian Conference on Electrical and Computer Engineering, CCECE 2020* (pp. 1–6).

16. M. Mandal, V. Dhar, A. Mishra, S. Vipparthi, M. Abdel-Mottaleb (2021). 3DCD: Scene independent end-to-end spatiotemporal feature learning framework for change detection in unseen videos, *IEEE Transactions on Image Processing*, *30*, 546–558.

17. M. Sultana, A. Mahmood, S. Javed, S. Jung (2019). Unsupervised deep context prediction for background estimation and foreground segmentation, *Machine Vision and Applications, 30*, 375–395.

18. M. Sultana, A. Mahmood, T. Bouwmans, S. Jun (2019). *Complete moving object detection in the context of robust subspace learning*, Workshop on Robust Subspace Learning and Computer Vision, ICCV 2019, Seoul, South Korea.

19. M. Sultana, A. Mahmood, T. Bouwmans, S. Jung (2020). *Unsupervised adversarial learning for dynamic background modeling, International Workshop on Frontiers of Computer Vision, IW-FCV 2020*, Ibusuki, Japan.

20. M. Sultana, A. Mahmood, T. Bouwmans, S. Jung (2020). *Dynamic background subtraction using least square adversarial learning, ICIP 2020*, Abu Dhabi, UAE.

21. H. Didwania, S. Ghatak, S. Rup (2019). *Multi-frame and Multi-scale Conditional Generative Adversarial Networks for Efficient Foreground Extraction, International Conference on Computer Vision and Image Process, CVIP 2019* (pp. 211–222).

22. W. Yu, J. Bai, L. Jiao (2020). Background Subtraction Based on GAN and Domain Adaptation for VHR Optical Remote Sensing Videos, *IEEE Access*.

23. S. S. Du, Y. Wang, X. Zhai, S. Balakrishnan, R. R. Salakhutdinov, A. Singh (2018). How many samples are needed to estimate a convolutional neural network? In *Advances in Neural Information Processing Systems* (pp. 373–383).

24. A. Parada-Mayorga, D. L. Lau, J. H. Giraldo, G. R. Arce (2019). Blue-noise sampling on graphs. *IEEE Transactions on Signal and Information Processing over Networks*, *5*(3), 554–569.

25. N. Goyette, P. M. Jodoin, F. Porikli, J. Konrad, P. Ishwar (2012, June). *Changedetection. net: A new change detection benchmark dataset*. In *2012 IEEE Computer Society Conference on Computer Vision and Pattern Recognition Workshops* (pp. 1–8).

26. K. He, G. Gkioxari, P. Dollár, R. Girshick (2017). *Mask R-CNN*. In *Proceedings of the IEEE International Conference on Computer Vision* (pp. 2961–2969).

27. S. Chen, R. Varma, A. Sandryhaila, J. Kovačević (2015). Discrete Signal Processing on Graphs: Sampling Theory. *IEEE Transactions on Signal Processing, 63*(24), 6510–6523.

28. A. Anis, A. Gadde, A. Ortega (2016). Efficient sampling set selection for bandlimited graph signals using graph spectral proxies. *IEEE Transactions on Signal Processing, 64*(14), 3775–3789.

29. G. Puy, N. Tremblay, R. Gribonval, P. Vandergheynst (2018). Random sampling of bandlimited signals on graphs. *Applied and Computational Harmonic Analysis, 44*(2), 446–475.

30. I. Z. Pesenson (2015, May). *Sampling solutions of Schrödinger equations on combinatorial graphs.* In *2015 International Conference on Sampling Theory and Applications (SampTA)* (pp. 82–85). IEEE.

31. F. R. Chung, F. C. Graham (1997). Spectral graph theory (No. 92). American Mathematical Soc.

32. T. Y. Lin, M. Maire, S. Belongie, J. Hays, P. Perona, D. Ramanan, C. L. Zitnick (2014, September). *Microsoft coco: Common objects in context.* In *European Conference on Computer Vision* (pp. 740–755). Springer, Cham.

33. Lucas, B. D., Kanade, T. (1981). An iterative image registration technique with an application to stereo vision.

34. T. Ojala, M. Pietikainen, T. Maenpaa (2002). Multiresolution gray-scale and rotation invariant texture classification with local binary patterns. *IEEE Transactions on Pattern Analysis and Machine Intelligence, 24*(7), 971–987.

35. R. A. Ulichney (1993, September). Void-and-cluster method for dither array generation. In *Human Vision, Visual Processing, and Digital Display IV* (Vol. *1913*, pp. 332–343). International Society for Optics and Photonics.

36. I. Pesenson (2009). Variational splines and Paley–Wiener spaces on combinatorial graphs. *Constructive Approximation, 29*(1), 1–21.

37. R. A. Horn, C. R. Johnson (2012). *Matrix analysis.* Cambridge University Press.

38. N. Perraudin, J. Paratte, D. Shuman, L. Martin, V. Kalofolias, P. Vandergheynst, D. K. Hammond (2014). GSPBOX: A toolbox for signal processing on graphs. arXiv preprint arXiv:1408.5781.

39. J. H. Giraldo, T. Bouwmans (2020). *Semi-supervised Background Subtraction of Unseen Videos: Minimization of the Total Variation of Graph Signals,* IEEE International Conference on Image Processing.

40. J. H. Giraldo, S. Javed, T. Bouwmans (2020). Graph moving object segmentation, *IEEE Transactions on Pattern Analysis and Machine Intelligence,* October 25–28.

16

Application of Artificial Intelligence in Disaster Response

Alok Bhardwaj

Indian Institute of Technology, Roorkee, India

CONTENTS

16.1 Introduction

Centre for Research on the Epidemiology of Disasters (CRED) defines a disaster as "a situation or event that overwhelms local capacity, necessitating a request at the national or international level for external assistance; an unforeseen and often sudden event that causes great damage, destruction and human suffering." Disaster management is a prerequisite to manage and prevent disasters in the future, as disasters in today's age cause both monetary damages in billions of dollars as well as casualties in large numbers; for example, $125 billion damages and 88 casualties during Hurricane Harvey in 2017 (Stephens et al., 2020). Disaster management comprises four phases, namely disaster mitigation, preparedness, response, and recovery (Sun et al., 2020a). Among all the four phases, AI is particularly used in the response phase (Sun et al., 2020b). Specifically, response phase refers to situations during a disaster that includes searching for affected people and subsequently planning rescue operations; planning first-aid kits based on the knowledge of affected area; event mapping; and resource allocation. In addition, damage assessment is an integral part of disaster response that is required to assess the level of damage done by hazards such as floods, earthquakes, volcanoes, or hurricanes. Damage assessment essentially includes estimating the number of damaged building, roads, bridges, and other infrastructure.

AI is an overarching blanket that includes what is commonly referred to as machine learning and deep learning. Machine learning differs from deep learning in the sense that features are needed to be defined for operating machine learning methods whereas deep learning methods do not have a prerequisite for defining data features. Nowadays, large volumes of data related to a certain disaster are collected from various sources such as remote sensing sensors, social media such as Twitter, Internet of Things (IoT) devices, and news reports (Yu et al., 2018). Analysis of data originating from such sources to extract meaningful information is a tremendous task. AI methods of machine and/or deep learning are designed to extract useful patterns of information from datasets originating from many sources. Such datasets are also complex in nature, i.e., such datasets represent phenomena whose behavior and/or characteristics are hard to define. A disaster is a phenomenon that is a result of an interplay of many factors that makes analyzing a disaster a complex problem. For example, generally a flash flood in a mountainous terrain occurs after heavy rainfall, but factors such as geometry of a valley, formation and breaking up of landslide dams that might cause landslide lake outburst floods, size of boulders that may temporarily block the flow of water in a valley, contribution of groundwater to the flood, and many other such factors make a flood-related disaster complex in nature and hard to analyze. Therefore, application of AI in analyzing data from different sources to arrive at some meaningful information to respond to disasters is critical and highly crucial.

Referring back to the flood disaster example, AI can be used as a disaster response tool to identify damaged infrastructure (damaged building, roads, monuments, etc.) and flood extent (Tay et al., 2020). In today's world, a firsthand rough estimate of damage in hazard-affected areas can be gauged rather quickly from remote sensing images by comparing pre- and either during-event or post-event images to highlight the damaged sections. In addition, change detection methods are also used to assess damage using pre- and post-event images. In a survey conducted by Izumi et al. (2019), it was found that many experts accepted the potential of AI in reducing risks due to disasters by providing meaningful information on early warning and evacuation. Therefore, it is with high possibility that the use of AI might make changes in the current state of disaster risk reduction methods. For example, Sublime et al. (2019) applied unsupervised AI methods to determine damage caused by the 2011 Tohoku earthquake and tsunami.

Before delving into details of the application of AI to respond to different types of disasters such as earthquakes, floods, volcanoes, landslides, or fire-related disasters, it is critical to have an understanding of a generic subdivision and categories of AI methods. AI methods can be subdivided into six categories based on the availability of output datasets, model architecture, and mode of interpretation of input data, namely supervised, unsupervised, reinforcement learning, deep learning, deep reinforcement learning, and optimization learning (Sun et al., 2020a, 2020b). As it is not possible to delve into detail of each category, a short description is provided here for a general understanding of each category. Supervised learning can be used with datasets with known input and output data, where output data is also known as labeled data. If a priori information is already available, then certain algorithms can be trained with input and labeled data to generate trained models. Such trained models can be applied on unseen and/or test datasets related to similar disasters to determine outputs such as number of damaged buildings, flood extent, wildfire extent, etc.

On the other hand, when labeled data is not available to train an AI model, statistical algorithms are used to extract meaningful information or hidden structure from input datasets. This category of AI learning that is applied on unlabeled datasets, when a priori information is not available, is known as unsupervised learning. Clustering is an example of unsupervised learning that is used to identify different clusters in a given dataset where each cluster shares a common characteristic. For example, clustering can be used to identify damaged buildings based on material such as RCC, masonry buildings, etc., or purpose of a building such as school, office, or residence buildings.

Next, deep learning is a subcategory of AI whose architecture is based on building blocks inspired by human cognitive learning. For example, deep learning method of artificial neural networks (ANNs) consist of multiple layers, and each layer has cells known as neurons. In an ANN network, each neuron is connected to the previous layer's neurons and the next layer's neurons. Neurons contain activation functions that transforms an input to a neuron into an output in a specified range. Next, during the training of an ANN, weights on the connection between each set of neurons are updated. Training is stopped when a certain criterion is met as set by the architect of the ANN architecture. Some applicatons of ANN and other deep learning methods include facial recognition, natural language processing, disaster response, self-driving automation in vehicles, and many other complex tasks.

The next subcategory of AI is reinforcement learning, where an agent, which is a software, takes actions based on observations made within an environment (Géron, 2017). The objective of the agent is to learn a way to maximize positive rewards and minimize negative rewards. The algorithm used by the agent is known as its policy. Policy could be an AI method, either a machine learning method or a deep learning method (hence the name "deep reinforcement learning").

The next subcategory of AI methods is optimization, which refers to the selection of the best AI model based on some selection criterion such as an objective function. Collectively, based on the comprehensive review conducted by W. Sun et al. (2020b), some of the AI methods in each of the six subcategories are as follows: (a) supervised learning, namely logistic regression (LR), support vector machine (SVM), Naïves Bayes, decision trees, k-nearest neighbors, random forest, and linear and nonlinear regression; (b) unsupervised learning, namely clustering methods including hierarchical, k-means, fuzzy, principle component analysis, and hidden Markov modeling; (c) deep learning, namely convolutional neural networks and recurrent neural networks; (d) reinforcement learning such as Q-learning and policy gradient; (e) Deep Q networks for deep reinforcement learning; and (f) optimization, namely genetic algorithm, particle swarm optimization, and simulated annealing.

In the following subsections, different disasters are described, along with the application of AI to analyze those disasters. Section 16.2.1 describes the application of AI to understand earthquakes, Section 16.2.2 application of AI to understand floods, Section 16.2.3 application of AI to understand volcanoes, Section 16.2.4 application of AI to understand landslides, and Section 16.2.5 application of AI to understand wildfires. Section 16.3 provides recommendations as well as limitations on use of AI techniques in disaster response. Section 16.4 concludes the chapter with the major learnings from the application of AI in disaster response.

16.2 Application of AI to Understand Natural Disasters

16.2.1 Applicaton of AI to Understand Earthquakes

Earthquake prediction is often considered as the Holy Grail of seismology, which in a general sense refers to the prediction of earthquakes considered as point-sources (also known as seismicity) (Mignan & Broccardo, 2019a). Shallow ANN models such as support vector machines (SVM), decision tree, K-means and hierarchical clustering and others, dated back to 1994, were used to predict mainshock event magnitudes, location of mainshock events by determining latitude and longitude of event locations, and mainshock occurrence time (Al Banna et al., 2020; Mignan & Broccardo, 2019a). Since 1994, complexity of ANN models has increased tremendously with recent applications of complex models of recurrent neural networks (RNN), long short term memory (LSTM), and convolutional neural networks (CNNs) to earthquake prediction (Mignan & Broccardo, 2019a). For example, Perol et al. (2017) developed a highly scalable CNN, known as ConvNetQuake, for low-magnitude earthquake detection using a single waveform from seismograms, and demonstrated that ConvNetQuake was superior to other traditional methods used to detect earthquakes. Mignan & Broccardo (2019a) suggested that deep learning can be used for improved event detection using unstructured seismic-waveform analysis. Seydoux et al. (2020) applied an unsupervised deep learning approach to detect and cluster seismic signals from continuous seismic records that can be used for forecasting earthquakes or seismic activity in earthquake-prone areas. DeVries et al. (2018) showed that a deep neural network with six hidden layers, each layer with 50 neurons, and each neuron containing a hyperbolic tangent activation function, after training on 131,000 mainshock-aftershock pairs, was able to predict spatial locations of aftershocks with good accuracy (area under the curve value of 0.849). However, the results of DeVries et al. (2018) were contested by Mignan and Broccardo (2019b), as the latter showed better performance in predicting spatial locations of aftershocks using a two-parameter LR model.

Al Banna et al. (2020) conducted a comprehensive review on the use of AI techniques such as rule-based fuzzy logic, shallow and deep learning methods such as ANN, CNN, SVM, etc. in earthquake prediction. Some salient results from Al Banna et al. (2020) are as follows: (a) separately training AI models on rare earthquake events with magnitude greater than 6 from smaller magnitude events to reduce the error in prediction of magnitude; (b) use of attention-based AI architecture for prediction of time of occurrence of earthquakes by considering earthquake sequences as a time-series sequence; (c) creation of benchmark datasets for proper training of deep learning models and compare its performance with other methods and/or models applied on the same benchmark datasets; (d) use of similar parameters and specific scale to measure earthquake magnitudes is required for proper training of

AI models so that such models can perform similarly on different datasets. It is believed that earthquake prediction could be achieved using deep learning methods, but simpler, non-AI models such as an exponential law for magnitude prediction and a power law for aftershock prediction in space are also able to provide similar predictions to those obtained from ANNs (Mignan & Broccardo, 2019a). Therefore, the application of AI in earthquake prediction is still nascent and needs further development in the near future.

16.2.2 Application of AI to Understand Floods

Urban flooding is globally affecting many cities causing huge economic and personal losses (R. Q. Wang et al., 2018). For example, economic losses in 1998 and 2010, considered exceptional years with huge flooding events, exceeded $40 billion each (Jha et al., 2012). AI is a potential tool that can be used to learn patterns and extract information useful to understand flood behaviors from past and present flood events for improving resilience and prevention of ensuing damage (Saravi et al., 2019). Floods are highly complex phenomena, and AI techniques in particular can prove to be useful for flood detection to understand and model flood nonlinearity using historical information without needing any information of the underlying physical processes (Mosavi et al., 2018). Some of the widely used AI techniques used for flood modeling include ANNs such as feed-forward neural networks, RNN, extreme learning machine, neuro-fuzzy, adaptive neuro-fuzzy inference systems (ANFIS), SVM, wavelet neural networks, and decision trees (Mosavi et al., 2018).

During floods, social media can be used by people in flood-affected and non-affected areas to post information on media platforms such as Facebook and Twitter. Social media data provides *hyper-resolution data* that depicts flood flows at the street or parcel level required for proper understanding of urban flood flow problems. A big advantage of using social media is to use the firsthand data of text and photos to train AI models for real-life flooding situations (R. Q. Wang et al., 2018). AI can leverage the power of social media by collating this information to extract meaningful patterns such as flood hotspots and number of people stuck in flood affected areas. AI-enabled social media analysis can help in capturing, geolocating, and classifying text and images to respond during emergency situations (Wang et al., 2018). However, data reliability is a major issue when data is obtained from social media, and one needs to be aware of this limitation when using social media for flood response. In addition, disaster-responding agencies can utilize mobile data for sentiment analysis to identify distressed calls from flood-affected areas and respond accordingly (Sun et al., 2020b).

In addition to social media data, weather forecast information at flood-affected locations can be used for flood response. For example, Saravi et al. (2019) used a random forest (RF) technique for classification of different kind of floods such as flash flood, lakeshore flood, and coastal flood using weather forecast information. Next, high-resolution aerial imageries collected from unmanned aerial vehicles (UAVs) can also be used for response. Ofli et al. (2016) recommends application of supervised machine learning methods on crowdsourcing data collected on large volumes of UAV images such as damaged areas, blocked roads, etc. for an effective disaster response.

In flood response, it is important to consider the duration of lead-time at which the prediction is made. Mosavi et al. (2018) conducted a comprehensive review and suggested that AI techniques, such as ANN, ELM, SVM, etc., and hybrid AI models such as wavelet-neuro-fuzzy are more accurate than traditional hydrological models and statistical methods in prediction of floods with short as well as long (ranging from week to months and annual) lead-times. Mosavi et al. (2018) suggested that ANNs are the most widely used AI method for flood prediction, but it struggles with generalization of the trained ANN networks, unlike SVM that showed better performance than ANN in terms of generalization of flood prediction. A recent example of the use of a deep learning method for flood response is by Hu et al. (2019), who applied LSTM, together with reduced-order model and proper orthogonal decomposition, for spatial and temporal prediction of flood inundations. Moreover, hybrid models can be utilized for flood prediction and can be prepared by combining two or more AI methods, or AI methods with conventional methods. Hosseini et al. (2020) recently showed such hybridization of AI models by using state-of-the-art ensemble machine learning models of generalized linear model (GLMBoost), random forest, and Bayesian generalized linear model (BayesGLM) for modeling spatial location of flash floods using a set of predictor variables selected using simulated annealing. In summary, it has been shown by researchers that AI proves to be a better tool for flood prediction than conventional methods are.

16.2.3 Application of AI to Understand Volcanoes

According to Furtney et al. (2018), globally less than 10% of the volcanoes are monitored on a regular basis. Satellite systems can be used for global observation of active and dormant volcanoes at regular intervals, including deformations using synthetic aperture radar (SAR), thermal features from thermal infrared (TIR), and SO_2 emissions from ultraviolet spectroscopy (UV). From a disaster response point of view, measurement of ground deformation after a volcanic eruption is critical. Interferometric synthetic aperture radar (InSAR) remote sensing technique utilizes differential phase information between two SAR scenes to measure deformation in the line of sight of the signal received from affected areas.

AI methods are used by researchers with aforementioned techniques to respond to volcanic disasters. In particular, CNN is used with interferometric SAR datasets to detect rapid deformation signals of certain volcanoes (Anantrasirichai et al., 2018). To determine surface deformations following a volcanic eruption, Anantrasirichai et al. (2018) applied CNN to interferograms (interferogram is an image of differential phase) developed using InSAR data at over 900 volcanoes around the world and found that a pre-trained CNN known as AlexNet was capable to detect large, rapid deformations. The input interferogram to AlexNet had dimensions of 224×224 followed by five two-dimensional (2D) convolutional layers with ReLU activation unit and max-pooling, followed by three fully connected layers and the output layer containing a binary value of either unrest of the volcano or no unrest. Anantrasirichai et al. (2018) found that AlexNet had the largest AUC of 0.995 and performed better than ResNet50 and InceptionV3 for detection of volcanic activity using InSAR interferograms. Bueno et al. (2020) used the probabilistic Bayesian neural networks (BNN) method to identify unrest at two active volcanoes in Helens, Washington, USA, and Bezymianny, Kamchatka, Russia, and found that BNNs can be used via transfer learning to identify unrest across different periods of eruption. Bernardinetti and Bruno (2019) used K-means clustering with Silhouette Index to identify uncertainty in K-mean clustering graphically by altering the color scheme, and self-organizing maps (SOM) to interpret geophysical data from the Solfatara volcano in Italy. Lara et al. (2021) developed an automated system to recognize microearthquakes from the Cotopaxi volcano using pre-trained models of AlexNet, SqueezeNet, GoogLeNet, and ResNet-50 on ImageNet to classify spectrograms into either event or no-event category, i.e., to determine the presence of microearthquakes. J. Sun et al. (2020a) applied CNN with modified U-net to remove atmospheric signals from unwrapped surface displacements from InSAR. Titos et al. (2020) used LeNet with transfer learning concepts to fine-tune certain layers of LeNet to classify volcano-seismic events with 94% accuracy. Collectively, researchers have successfully applied AI to identify deformations following a volcanic activity, which is required during response.

16.2.4 Application of AI to Understand Landslides

Landslides are downward and outward movement of slope materials, such as rocks debris, and soil, under the influence of gravity, that is developed in time via several stages (Hungr et al., 2014). Early warning systems are required to predict imminent landslides for at-risk communities, which can be achieved using deep learning as a tool to determine instabilities in hillslopes (Orland et al., 2020). Many studies have confirmed that deep learning–based methods outperform traditional methods for landslide detection. For example, Dou et al. (2020) applied an LR model and a deep neural network to calculate landslide susceptibility for 2018 Hokkaido earthquake, and found that the deep neural network is an appropriate method for calculating susceptibilities even when sufficient landslide inventory is not available. Within southeast Asia, Vietnam is particularly susceptible to landslides, and some of the most recent landslide susceptibility studies have found that AI outperforms traditional methods in detection of landslides. For example, Bui et al. (2020a) found that a deep neural network outperformed SVM, RF, MLP-Neural Network, and C4.5 for detection of landslide susceptibility in Kom Tum province of Vietnam; Dao et al. (2020) found higher performance of a deep neural network in detecting landslides in Muong Lay district; Nhu et al. (2020) found that a deep neural network with Adam optimization method and lost function of mean squared error was the best model to detect landslides in Ha Long, Vietnam. Next, Wang et al. (2021) found CNN with 11 layers as the best-performing AI model compared to LR, SVM, RF, discrete AdaBoost, LogitBoost, Gentle AdaBoost, and CNN with 6 layers in predicting landslides on Lantau Island, Hong Kong.

Orland et al. (2020) demonstrated that an LSTM-based prediction model was successful in accurately predicting pour water pressure timing and magnitudes of rainfall in 36-hour intervals, implying development of an early warning system based on AI. Meng et al. (2020) applied LSTM to predict water-induced landslide displacements without decomposing displacement time series. Ye et al. (2019) applied a deep belief network to obtain spectral as well as spatial features of landslides and used these features within a LR model to map landslides. The authors found that deep belief network showed better performance that spectral information divergence method, spectral angle match, and linear SVM to detect landslides. In another recent study, W. Wang et al. (2020) applied a deep belief network to detect landslides and found to have a better performance than that of BPNN and LR model. Bui et al. (2020b) proposed Hue-Bi-dimensional Empirical Mode Decomposition (H-BEMD) method, an image transformation method, to be used within a CNN framework to detect landslide sizes using optical satellite images. The authors found that the CNN enhances the localization of landslides and increases the chances of landslide detection. Fang et al. (2021) applied four ensemble methods to integrate CNNs, RNNs, SVM, and LR for landslide susceptibility mapping, and found that ensemble classifiers greatly improves prediction ability of base deep learning and machine learning classifiers. Huang et al. (2020) applied fully connected sparse auto-encoder for landslide detection and found it to have a better performance than SVM and BPNN. Prakash et al. (2020) found that U-Net-based landslide prediction model outperformed traditional methods including ANN, LR, and RF. Ma et al. (2020b) presented a comprehensive survey of the application of machine learning for prevention of landslides, including methods of ANN, CNN, DBN, extreme learning machine, generative adversarial networks, graph neural networks, naïve Bayes for detection of landslides, and particularly focussing on determining landslide susceptibility and development of early warning systems.

16.2.5 Application of AI to Understand Wildfires

Increase in wildfires and/or bushfires in recent decades, including the 2009 and 2013 bushfires in Australia, is testimony of the required effective disaster response to wildfire events (Ma et al., 2020a). AI is one of the potential ways to extract meaningful information from varied sources for analyzing wildfires and associated damages. Ma et al. (2020a) has applied AI-based Hybrid Step XGBoost to develop a real-time alarm system for wildfires by studying the process of contact to ignition between upper story vegetation and powerlines. The researchers found that AI-based XGBoost was a better method than traditional methods for bushfire detection.

Wildfires/bushfires release a lot of smoke and dust in the environment. Therefore, it is equally important to measure the concentration of $PM_{2.5}$ particles in the environment to control air pollution caused by wildfire smoke. Li et al. (2020) applied autoencoder-based full residual deep learning method to predict $PM_{2.5}$ concentrations in California using input data of aerosol optical depth; meteorological variables such as temperature, wind speed, specific humidity, precipitation etc.; and wildfire smoke plume data over California. Reid et al. (2015) applied different machine learning methods including generalized boosting model, bagged trees, support vector machines, k-nearest neighbors, and others to detect $PM_{2.5}$ concentration following the 2008 northern California wildfires, and found that generalized boosting model was the best method to predict spatial and temporal behavior of the fine particulates following the wildfire. Bruni Zani et al. (2020) applied deep neural networks to detect concentration of PM_{10} particulates using satellite estimates of particulate matter in equatorial Asia.

Farasin et al. (2020) developed a supervised deep learning method that is a combination of a classification method and a regression method called as double U-Net where "double" refers to binary classification and regression steps. This double U-Net was used for estimation of wildfire-induced damage as well as damage severity. Govil et al. (2020) trained a machine learning model from scratch on thousands of HPWREN (High performance Wireless Research & Education Network) camera images of wildfire smoke to develop a wildfire smoke detection system. Pundir and Raman (2019) applied dual deep learning model including CNN and SVM for detection of smoke images. First, a pre-built AlexNet is trained on images containing smoke and smoke velocity extracted from frames of smoke videos using the optical flow method. In the next step, authors trained an SVM on the output from AlexNet for identifying smoke in images. Hence, AI proves to be a useful tool to not only estimate spatial extent of wildfires but also measure a component of pollution created by wildfires.

16.3 Caution in Using AI for Disaster Response

Training data is considered a major component of the AI pipeline that is required to successfully execute machine and deep learning methods (Alemohammad et al., 2020). Collection and documentation of training data, after following bias-free sampling, is therefore of utmost importance and is the main bottleneck in the application of AI-based techniques for disaster response. In addition to documentation of datasets, the right-sized training set is required to achieve good performance in identifying or "learning" patterns by AI models (Perol et al., 2017). For example, class imbalance is often encountered in training datasets, for instance in a flood-related dataset, flood class would be small in number as compared to non-flood class. Hence, class imbalance should be managed in a training set, and this training set should then be used for AI modeling. Otherwise, AI models mis-predict if the disaster-related section is imbalanced in the training set. Further, Alemohammad et al. (2020) recommends the development and application of those AI models that utilize smaller amount of training data, as well as use of simulated training datasets to run AI models. In addition, integration of training datasets with different resolution, scales, and dimensions is required for building a robust dataset that is also known as analysis-ready data (ARD). The ARD format is free from errors, reprojected and re-gridded to match the data from different instruments, and the unused data is usually masked out (Alemohammad et al., 2020). Lastly, it has become essential to develop and share benchmark datasets for comparison of different AI models.

A major limitation in the application of AI models to disaster response is over-fitting of training datasets by AI models, i.e., AI models fit a given training dataset fairly well but do not generalize well beyond the dataset on which they are trained. This may occur due to various reasons such as class imbalance, improper selection of AI model, noise in training data, and too many parameters included in the model that increase the variance and thus lead to over-fitting. On the other hand, too few parameters should also be avoided in the model, as this leads to high bias and thus under-fitting of data by AI models. Utmost care should be taken during the training phase to ensure that models are neither over-fitting nor under-fitting a given dataset by undertaking certain steps such as early stopping, choosing the correct value of learning parameter, choosing the number of epoch cycles to train the model, regularization, data augmentation, label smoothing, etc. Hyper-parameter tuning helps reduce both over- and under-fitting to a large extent by identifying optimized values of hyper-parameters. These hyper-parameters with their optimized values can then be used for training AI models.

To ensure reproducibility of AI models and reduce the task of rebuilding AI models for responding to different disasters, AI models should be documented in a structured manner and shared based on the principles of FAIR (i.e., findability, accessibility, interoperability, and reusability) (Wilkinson et al., 2016). For more details on the ethics to be followed for reproducibility of AI models, see Wilkinson et al. (2016).

Collectively, a flow diagram is prepared to show major steps of the application of AI for disaster response (Figure 16.1). According to the flow diagram, the first major step is to collate datasets from varied sources including satellite sensors, social media, field data, or even simulated datasets. After all the relevant datasets have been collected, they must also be visualized. One of the ways to visualize spatial datasets is to use a geographic information system (GIS) – for instance, ArcGIS software. Visualization of spatial datasets is required during the initial stage to appreciate the strength and weakness of collated datasets in terms of providing relevant information. Simultaneously, datasets should be cleaned and wrangled (process to transform data) to remove undesirable and spurious values. After cleaning, datasets are ready for the preparation of training, test, and validation datasets. These datasets can be prepared by subdividing the complete input dataset using some fixed proportion, i.e., 70% of the input dataset for training (and 10% out of that 70% for validation) and remaining 30% as a test dataset. Next, either a pre-trained AI model or a model built from scratch using hyper-parameter tuning is selected. If the AI model is built from scratch, comprehensive hyper-parameter tuning is required using either grid or

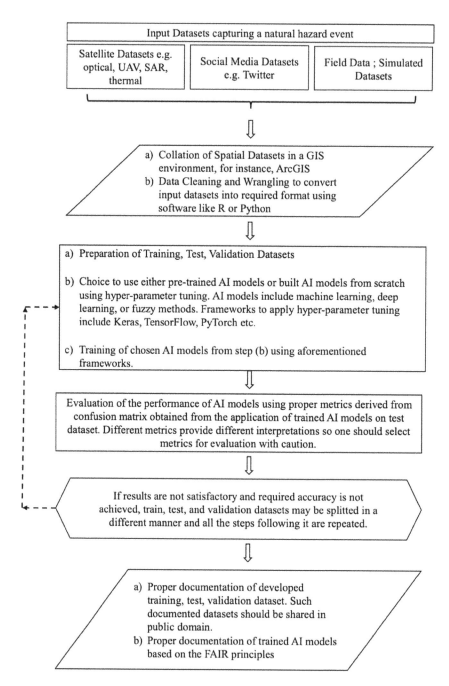

FIGURE 16.1 Flow diagram on the preparation of an AI model for disaster response. Major steps of the flow diagram include collation of different datasets; data cleaning; preparation of training, test, and validation datasets; selection of AI model; training of chosen AI model; evaluation; and documentation of datasets and trained AI model.

random search to select optimized values of hyper-parameters. Finally, the selected AI model is trained on the input training dataset, validated on the validation dataset, and then tested on the test dataset. After the AI model is executed on the test dataset, a test confusion matrix is generated that is generally used for evaluation. Evaluation is a critical step in the whole pipeline, as different metrics such as overall accuracy, commission error, omission error, kappa, receiver operating curves, and many other provide different perspectives on evaluation. Therefore, one should be cautioned that one metric representing good evaluation results in a specific problem might not be the right choice for any other problem. After the application of the correct metric, if the accuracy is not achieved as expected, the whole process from subdivision of the whole dataset into train, test, and validation datasets should be repeated until the desired accuracy is achieved. Finally, a proper documentation of all datasets as well as the trained AI model should be prepared following the FAIR principles. The trained AI model should be shared in the public domain so that other users may use it as a benchmark result for further research.

16.4 Conclusion

This chapter highlighted the salient points related to the application of AI to disaster response for different types of natural disasters including earthquakes, floods, landslides, volcanoes, and wildfires. Disaster response is a critical step required to reduce casualties and economic loss during a disaster event. With the advancements in computing technology, accessibility to training datasets collated from variegated datasets on digital cloud, and availability of robust AI models specialized in exploring both spatial and temporal characteristics of datasets, AI is proving to be a potential tool that is bringing a real change in disaster response in today's world.

REFERENCES

Al Banna, M. H., Taher, K. A., Kaiser, M. S., Mahmud, M., Rahman, M. S., Hosen, A. S. M. S., & Cho, G. H. (2020). Application of artificial intelligence in predicting earthquakes: State-of-the-art and future challenges. *IEEE Access*, *8*, 192880–192923. https://doi.org/10.1109/ACCESS.2020.3029859

Alemohammad, H., Maskey, M., Estes, L., Gentine, P., Lunga, D., & Yi, Z.-F. (Nana). (2020). *Advancing Application of Machine Learning Tools for NASA's Earth Observation Data*. https://cdn.earthdata.nasa.gov/conduit/upload/14287/NASA_ML_Workshop_Report.pdf

Anantrasirichai, N., Biggs, J., Albino, F., Hill, P., & Bull, D. (2018). Application of machine learning to classification of volcanic deformation in routinely generated InSAR data. *Journal of Geophysical Research: Solid Earth*, *123*(8), 6592–6606. https://doi.org/10.1029/2018JB015911

Bernardinetti, S., & Bruno, P. P. G. (2019). The hydrothermal system of Solfatara Crater (Campi Flegrei, Italy) inferred from machine learning algorithms. *Frontiers in Earth Science*, *7*(November), 1–18. https://doi.org/10.3389/feart.2019.00286

Bruni Zani, N., Lonati, G., Mead, M. I., Latif, M. T., & Crippa, P. (2020). Long-term satellite-based estimates of air quality and premature mortality in Equatorial Asia through deep neural networks. *Environmental Research Letters*, *15*(10). https://doi.org/10.1088/1748-9326/abb733

Bueno, A., Benitez, C., De Angelis, S., Diaz Moreno, A., & Ibanez, J. M. (2020). Volcano-Seismic transfer learning and uncertainty quantification with Bayesian neural networks. *IEEE Transactions on Geoscience and Remote Sensing*, *58*(2), 892–902. https://doi.org/10.1109/TGRS.2019.2941494

Bui, D. T., Tsangaratos, P., Nguyen, V. T., Liem, N. Van, & Trinh, P. T. (2020a). Comparing the prediction performance of a Deep Learning Neural Network model with conventional machine learning models in landslide susceptibility assessment. *Catena*, *188*(July), 104426. https://doi.org/10.1016/j.catena.2019.104426

Bui, T. A., Lee, P. J., Lum, K. Y., Loh, C., & Tan, K. (2020b). Deep Learning for Landslide Recognition in Satellite Architecture. *IEEE Access*, *8*, 143665–143678. https://doi.org/10.1109/ACCESS.2020.3014305

Dao, D. Van, Jaafari, A., Bayat, M., Mafi-Gholami, D., Qi, C., Moayedi, H., Phong, T. Van, Ly, H. B., Le, T. T., Trinh, P. T., Luu, C., Quoc, N. K., Thanh, B. N., & Pham, B. T. (2020). A spatially explicit deep learning neural network model for the prediction of landslide susceptibility. *Catena*, *188*(November), 104451. https://doi.org/10.1016/j.catena.2019.104451

DeVries, P. M. R., Viégas, F., Wattenberg, M., & Meade, B. J. (2018). Deep learning of aftershock patterns following large earthquakes. *Nature*, *560*(7720), 632–634. https://doi.org/10.1038/s41586-018-0438-y

Dou, J., Yunus, A. P., Merghadi, A., Shirzadi, A., Nguyen, H., Hussain, Y., Avtar, R., Chen, Y., Pham, B. T., & Yamagishi, H. (2020). Different sampling strategies for predicting landslide susceptibilities are deemed less consequential with deep learning. *Science of the Total Environment*, *720*(February), 137320. https://doi.org/10.1016/j.scitotenv.2020.137320

Fang, Z., Wang, Y., Peng, L., & Hong, H. (2021). A comparative study of heterogeneous ensemble-learning techniques for landslide susceptibility mapping. *International Journal of Geographical Information Science*, *35*(2), 321–347. https://doi.org/10.1080/13658816.2020.1808897

Farasin, A., Colomba, L., & Garza, P. (2020). Double-step U-Net: A deep learning-based approach for the estimation of wildfire damage severity through sentinel-2 satellite data. *Applied Sciences (Switzerland)*, *10*(12). https://doi.org/10.3390/app10124332

Furtney, M. A., Pritchard, M. E., Biggs, J., Carn, S. A., Ebmeier, S. K., Jay, J. A., McCormick Kilbride, B. T., & Reath, K. A. (2018). Synthesizing multi-sensor, multi-satellite, multi-decadal datasets for global volcano monitoring. *Journal of Volcanology and Geothermal Research*, *365*, 38–56. https://doi.org/10.1016/j.jvolgeores.2018.10.002

Géron, A. 2017. *Hands-on machine learning with Scikit-Learn and TensorFlow. Concepts, tools, and techniques to built intelligent systems*. O'Reilly Media, Inc.

Govil, K., Welch, M. L., Ball, J. T., & Pennypacker, C. R. (2020). Preliminary results from a wildfire detection system using deep learning on remote camera images. *Remote Sensing*, *12*(1). https://doi.org/10.3390/RS12010166

Hosseini, F. S., Choubin, B., Mosavi, A., Nabipour, N., Shamshirband, S., Darabi, H., & Haghighi, A. T. (2020). Flash-flood hazard assessment using ensembles and Bayesian-based machine learning models: Application of the simulated annealing feature selection method. *Science of the Total Environment*, *711*, 135161. https://doi.org/10.1016/j.scitotenv.2019.135161

Hu, R., Fang, F., Pain, C. C., & Navon, I. M. (2019). Rapid spatio-temporal flood prediction and uncertainty quantification using a deep learning method. *Journal of Hydrology*, *575*(March), 911–920. https://doi.org/10.1016/j.jhydrol.2019.05.087

Huang, F., Zhang, J., Zhou, C., Wang, Y., Huang, J., & Zhu, L. (2020). A deep learning algorithm using a fully connected sparse autoencoder neural network for landslide susceptibility prediction. *Landslides*, *17*(1), 217–229. https://doi.org/10.1007/s10346-019-01274-9

Hungr, O., Leroueil, S., & Picarelli, L. (2014). The Varnes classification of landslide types, an update. *Landslides*, *11*(2), 167–194. https://doi.org/10.1007/s10346-013-0436-y

Izumi, T., Shaw, R., Djalante, R., Ishiwatari, M., & Komino, T. (2019). Disaster risk reduction and innovations. *Progress in Disaster Science*, *2*, 100033. https://doi.org/10.1016/j.pdisas.2019.100033

Jha, A. K., Bloch, R., & Lamond, J. (2012). Cities and flooding: A guide to integrated urban flood risk management for the 21st century by Abhas Jha, Robin Bloch, Jessica Lamond, and other contributors. *Journal of Regional Science*, *52*(5). https://doi.org/10.1111/jors.12006_6

Lara, F., Lara-Cueva, R., Larco, J. C., Carrera, E. V., & León, R. (2021). A deep learning approach for automatic recognition of seismo-volcanic events at the Cotopaxi volcano. *Journal of Volcanology and Geothermal Research*, *409*. https://doi.org/10.1016/j.jvolgeores.2020.107142

Li, L., Girguis, M., Lurmann, F., Pavlovic, N., McClure, C., Franklin, M., Wu, J., Oman, L. D., Breton, C., Gilliland, F., & Habre, R. (2020). Ensemble-based deep learning for estimating PM2.5 over California with multisource big data including wildfire smoke. *Environment International*, *145*, 106143. https://doi.org/10.1016/j.envint.2020.106143

Ma, J., Cheng, J. C. P., Jiang, F., Gan, V. J. L., Wang, M., & Zhai, C. (2020a). Real-time detection of wildfire risk caused by powerline vegetation faults using advanced machine learning techniques. *Advanced Engineering Informatics*, *44*(March), 101070. https://doi.org/10.1016/j.aei.2020.101070

Ma, Z., Mei, G., & Piccialli, F. (2020b). Machine learning for landslides prevention: A survey. *Neural Computing and Applications*, *8*. https://doi.org/10.1007/s00521-020-05529-8

Meng, Q., Wang, H., He, M., Gu, J., Qi, J., & Yang, L. (2020). Displacement prediction of water-induced landslides using a recurrent deep learning model. *European Journal of Environmental and Civil Engineering*, *0*(0), 1–15. https://doi.org/10.1080/19648189.2020.1763847

Mignan, A., & Broccardo, M. (2019a). Neural network applications in earthquake prediction (1994–2019): Meta-analytic insight on their limitations. *ArXiv*, *91*(4). https://doi.org/10.1785/0220200021. Introduction

Mignan, A., & Broccardo, M. (2019b). One neuron versus deep learning in aftershock prediction. *Nature*, *574*(7776), E1–E3. https://doi.org/10.1038/s41586-019-1582-8

Mosavi, A., Ozturk, P., & Chau, K. W. (2018). Flood prediction using machine learning models: Literature review. *Water (Switzerland)*, *10*(11), 1–40. https://doi.org/10.3390/w10111536

Nhu, V. H., Hoang, N. D., Nguyen, H., Ngo, P. T. T., Thanh Bui, T., Hoa, P. V., Samui, P., & Tien Bui, D. (2020). Effectiveness assessment of Keras based deep learning with different robust optimization algorithms for shallow landslide susceptibility mapping at tropical area. *Catena, 188*(November), 104458. https://doi.org/10.1016/j.catena.2020.104458

Ofli, F., Meier, P., Imran, M., Castillo, C., Tuia, D., Rey, N., Briant, J., Millet, P., Reinhard, F., Parkan, M., & Joost, S. (2016). Combining human computing and machine learning to make sense of big (Aerial) data for disaster response. *Big Data, 4*(1), 47–59. https://doi.org/10.1089/big.2014.0064

Orland, E., Roering, J. J., Thomas, M. A., & Mirus, B. B. (2020). Deep learning as a tool to forecast hydrologic response for landslide-prone hillslopes. *Geophysical Research Letters, 47*(16). https://doi.org/10.1029/2020GL088731

Perol, T., Gharbi, M., & Denolle, M. A. (2017). Convolutional neural network for earthquake detection and location. *ArXiv, 2016*(March), 2–10.

Prakash, N., Manconi, A., & Loew, S. (2020). Mapping landslides on EO data: Performance of deep learning models vs. traditional machine learning models. *Remote Sensing, 12*(3). https://doi.org/10.3390/rs12030346

Pundir, A. S., & Raman, B. (2019). Dual deep learning model for image based smoke detection. *Fire Technology, 55*(6), 2419–2442. https://doi.org/10.1007/s10694-019-00872-2

Reid, C. E., Jerrett, M., Petersen, M. L., Pfister, G. G., Morefield, P. E., Tager, I. B., Raffuse, S. M., & Balmes, J. R. (2015). Spatiotemporal prediction of fine particulate matter during the 2008 Northern California wildfires using machine learning. *Environmental Science and Technology, 49*(6), 3887–3896. https://doi.org/10.1021/es505846r

Saravi, S., Kalawsky, R., Joannou, D., Casado, M. R., Fu, G., & Meng, F. (2019). Use of artificial intelligence to improve resilience and preparedness against adverse flood events. *Water (Switzerland), 11*(5). https://doi.org/10.3390/w11050973

Seydoux, L., Balestriero, R., Poli, P., Hoop, M. de, Campillo, M., & Baraniuk, R. (2020). Clustering earthquake signals and background noises in continuous seismic data with unsupervised deep learning. *Nature Communications, 11*(1). https://doi.org/10.1038/s41467-020-17841-x

Stephens, W., Wilt, G. E., Lehnert, E. A., Molinari, N. M., & LeBlanc, T. T. (2020). A spatial and temporal investigation of medical surge in Dallas-Fort Wort during Hurricane Harvey. *Disaster Medicine and Public Health Preparedness, 14*(1), 111–118. https://doi.org/10.1017/dmp.2019.143.A

Sun, J., Wauthier, C., Stephens, K., Gervais, M., Cervone, G., La Femina, P., & Higgins, M. (2020a). Automatic detection of volcanic surface deformation using deep learning. *Journal of Geophysical Research: Solid Earth, 125*(9), 1–17. https://doi.org/10.1029/2020JB019840

Sun, W., Bocchini, P., & Davison, B. D. (2020b). Applications of artificial intelligence for disaster management. In *Natural Hazards* (Vol. *103*, No. 3). Netherlands: Springer. https://doi.org/10.1007/s11069-020-04124-3

Tay, C. W. J., Yun, S. H., Chin, S. T., Bhardwaj, A., Jung, J., & Hill, E. M. (2020). Rapid flood and damage mapping using synthetic aperture radar in response to Typhoon Hagibis, Japan. *Scientific Data, 7*(1), 1–9. https://doi.org/10.1038/s41597-020-0443-5

Titos, M., Bueno, A., García, L., Benítez, C., & Segura, J. C. (2020). Classification of isolated volcano-seismic events based on inductive transfer learning. *IEEE Geoscience and Remote Sensing Letters, 17*(5), 869–873. https://doi.org/10.1109/LGRS.2019.2931063

Wang, H., Zhang, L., Yin, K., Luo, H., & Li, J. (2021). Landslide identification using machine learning. *Geoscience Frontiers, 12*(1), 351–364. https://doi.org/10.1016/j.gsf.2020.02.012

Wang, R. Q., Mao, H., Wang, Y., Rae, C., & Shaw, W. (2018). Hyper-resolution monitoring of urban flooding with social media and crowdsourcing data. *Computers and Geosciences, 111*(September), 139–147. https://doi.org/10.1016/j.cageo.2017.11.008

Wang, W., He, Z., Han, Z., Li, Y., Dou, J., & Huang, J. (2020). Mapping the susceptibility to landslides based on the deep belief network: A case study in Sichuan Province, China. *Natural Hazards, 103*(3), 3239–3261. https://doi.org/10.1007/s11069-020-04128-z

Wilkinson, M. D., Dumontier, M., Aalbersberg, I. J., Appleton, G., Axton, M., Baak, A., Blomberg, N., Boiten, J. W., da Silva Santos, L. B., Bourne, P. E., Bouwman, J., Brookes, A. J., Clark, T., Crosas, M., Dillo, I., Dumon, O., Edmunds, S., Evelo, C. T., Finkers, R., … Mons, B. (2016). Comment: The FAIR guiding principles for scientific data management and stewardship. *Scientific Data, 3*, 1–9. https://doi.org/10.1038/sdata.2016.18

Ye, C., Li, Y., Cui, P., Liang, L., Pirasteh, S., Marcato, J., Goncalves, W. N., & Li, J. (2019). Landslide detection of hyperspectral remote sensing data based on deep learning with constrains. *IEEE Journal of Selected Topics in Applied Earth Observations and Remote Sensing*, *12*(12), 5047–5060. https://doi.org/10.1109/JSTARS.2019.2951725

Yu, M., Yang, C., & Li, Y. (2018). Big data in natural disaster management: A review. *Geosciences (Switzerland)*, *8*(5). https://doi.org/10.3390/geosciences8050165

17

Use of Robotics in Surgery: Current Trends, Challenges, and the Future

Mukesh Carpenter
Alshifa Hospital New Delhi, India

CONTENTS

17.1 Introduction

The era of robotic surgeries began in the 1980s: in 1985, the first Robotic Surgical system was used for stereotactic surgery, known as PUMA 560. Later, in 1988, PROBOT was used for prostate surgery. Then, in 1992, ROBODOC was developed for preparation of a "femur cavity." In 1990, a revolution has been witnessed, known as laparoscopic revolution, where several open surgeries were adapted to minimal access surgery (MAS), which results in less duration of hospital stay, less amount of postoperative pain, minimal chances of wound infection and better results, which have made surgeries such as laparoscopic cholecystectomy the standard of care for cholelithiasis, as shown in Figure 17.1 [1–4]. Due to their favorable outcomes, many surgeons used minimal invasive techniques for their surgical operations. In this technique there are open challenging issues such as unstable position of the camera while performing operations and provide a 2D vision of the domain. The primary surgeon was bound to difficult positions to operate with straight laparoscopic instruments.

During the early 21st century, the development of new technique or technology leads to further advances in MAS. Both robotic and telepresence surgery efficiently addresses the drawbacks of laparoscopic and thoracoscopic surgeries. Robotic surgery is supposed to remain a growing part of surgery [6, 7, 39, 40].

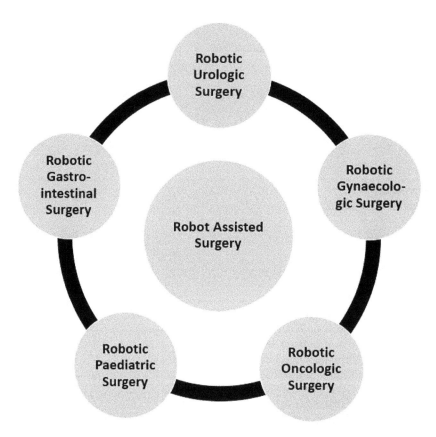

FIGURE 17.1   Different types of robot-assisted surgery.

It is also assumed that in future most of the surgery will be done by robots. Robotic surgery comprises of a robotic surgical system to perform surgery on patients. Like MAS for example, the Da Vinci system is the most used system and consists of three basic components: (1) the surgeon console, (2) the patient cart, and (3) the vision cart. A combination of all these components makes a surgeon understand the current scenario and then mimics the minutes to guide the instrument. As robotic surgery does not require specific training, it can definitely change the current surgical training model and remodel the learning paradigm of people by providing new possible solutions such as robotic telementoring and robotic surgical simulators [1, 8, 10–14].

These surgeries are found to be very effective for complex conditions that require extreme precision. In some conditions, the complex surgeries are not possible in the conventional manner. This technique is the only option left for surgeons because it provides the flexibility to deeply examine the part of the body being operated on. This also provides a clearer view that overall results in smoother procedures. Robotic surgeries also provide more control over what is being done, as shown in Figure 17.2 [15, 16, 18].

17.2 Literature Review

In literature, many surgical operations were done using the robotic surgery system [10–34]. Some of the surgical procedures done using robotics are presented in Table 17.1.

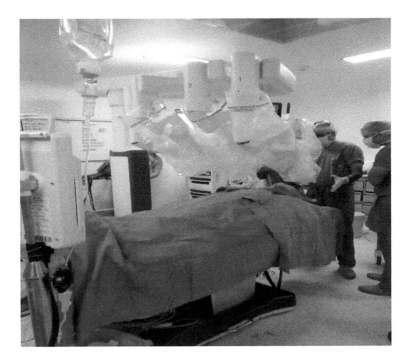

FIGURE 17.2 Surgical procedure using the Da Vinci system.

TABLE 17.1

Robotic Surgery Operations with their Domains

Domain	Robotic Surgery
Robotic gastrointestinal surgery [10–15]	1997: Himpens et al. [12] – first robotic cholecystectomy Antireflux operations, Heller's myotomy, gastric bypass, gastrojejunostomy, esophojectomy, gastric banding colectomy, splenectomy, adrenalectomy, and pancreatic resection reported to date
Robotic urologic surgery [16–19]	Radical robotic prostatectomy is the most common operation performed robotically and is gaining widespread recognition in the United States and Europe Nephrectomy and pelvic lymph node dissection also reported
Robotic gynaecologic surgery [20–22]	Robotic hysterectomy, salpingo-oophorectomy, and microsurgical fallopian tube reanastomosis
Robotic cardiothoracic surgery [23–29]	Surgical robots allow cardiothoracic surgeons to perform complex cardiothoracic procedures while avoiding the significant morbidity of sternotomy and thoracotomy Hundreds of robotic coronary bypasses have been performed to date Mitral valve repairs, atrial septal defect repair, pericardiectomy, lobectomies, and tumour enucleations
Robotic oncologic surgery [30]	Esophageal tumors, gastric cancer, color cancer, thymoma, and retromediastinal tumors
Robotic paediatric surgery [31–34]	Pyelophasty for ureteropelvic junction obstruction, antireflux procedures for gastroesophageal reflux disease, and paediatric congenital heart diseases, such as ligation of patent ductus arteriosus

17.3 Surgical Robots

Laparoscopic surgeries result in better quality of life for patients, as shown in Figure 17.3. In case of laparoscopic surgery, surgical robots are broadly categorized into two types, which are listed in the following subsections.

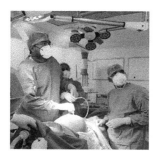 

FIGURE 17.3   Laparoscopic surgery.

17.3.1 Master–Slave Type

These kinds of robots have six degree of freedom (DOF) of motion. It has a four DOF arm outside the abdominal cavity and a two DOF wrist joint at the tip. At the time of surgical operation, using a master console, the surgeon can easily operate a remote slave with wrist joint. The robot enables an intuitive operation, since the slave arms in the abdomen reproduces the surgeon's six DOF hand motion at the console. Robots also enable telesurgery via network and microsurgery by adjusting the motion range between the master and the slave. The Da Vinci system is the most commonly used system, and in 2000 this system was approved by the Food and Drug Administration (FDA) as a first surgical system in United States for general laparoscopic surgery [28, 29, 31]. The master–slave robots have some drawbacks such as lack of haptics, large size, and high cost of surgery.

17.3.2 Handheld Robotic Forceps

The master–slave robot is not considered the best one for all surgical operations because it requires expansive space for the master console and has some other disadvantages discussed in the previous subsection. In contrast, handheld robotic forceps have a wrist joint at its tip and can be controlled from the interface installed on the forceps. Its translation operation is the same as the traditional one [20–25]. In this case, the system is small as compared to the master–slave due to the absence of the master console, and the setup time is also shorter. This system can be categorized into those controlled by actuators or mechanically driven.

17.4 Clinical Applications of Robotic Surgery

There are various clinical applications of robot surgery as shown in Figures 17.4 and 17.5 and listed in the following subsections.

17.4.1 Robotic Prostate Surgery

This surgery uses smaller incisions, which results in less damage to blood vessels and nerves and, consequently, in fewer complications and faster recovery time for patients compared to conventional surgery [25–28].

17.4.2 Robotic Kidney Surgery

This surgery may be needed in those cases where one kidney (robotic nephrectomy) or some portion of a kidney (robotic partial nephrectomy) needs to be removed. In conventional surgical procedures, a large incision up to 8 inches in length is required for the surgical procedure [29, 30]. In case of robotic kidney surgery, the incision size is small, making it a minimally invasive procedure. This surgery has fast recovery time for patients, who can resume their daily routine relatively shortly post surgery.

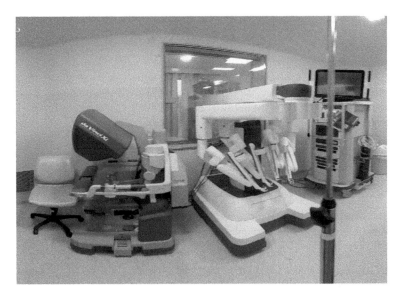

FIGURE 17.4   Da Vinci robot system.

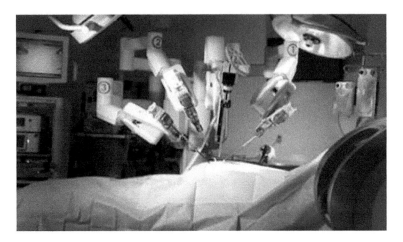

FIGURE 17.5   Robotic surgery.

17.4.3 Robotic Gynecological Surgery

This surgical system results in less blood loss during the operation and less postoperative pain for the patient. With this surgery, the size of the incision is small, leading to less blood loss, less pain, and shorter patient recovery time. In contrast, conventions surgery results in a protracted recovery time and physical as well as emotional complications for the patient [31, 32].

17.4.4 Robotic Gallbladder Surgery

It involves cholecystectomy (removal of the gallbladder). With the help of this systems, surgery can be done using a single incision. For the patient, this means smaller blood loss, smaller scars and shorter recovery time, with the subsequent quicker return to activities of daily life, compared to the same following conventional surgery [12, 18, 20].

17.4.5 Robotic Colorectal Surgery

This surgery system is beneficial in cases where the patient needs colectomy (for colorectal cancer) or in circumstances such as a benign tumor. Using this surgical system allows to operate effectively with smaller surgical incisions, which reduces the risk of nerve damage and complications associated with it. It also makes tissue cutting easier and supplies health care providers with other means of conducting colon surgery not available with conventional surgeries. Last but not least, postoperative recovery is also significantly shortened [34, 35, 37] .

17.5 Advantages of Robotics Surgery

This surgery offers several benefits to patients as compared to traditional or conventional or open surgery [31–36, 38]. Some of the important advantages are:

- During this surgery, surgeons can have a better understanding of the current situation.
- This surgery results in less tissue damage.
- This surgery has minimal blood loss, and therefore results in speedy recovery for patients.
- This system is more advantageous because patients can return to their daily routine in much less time.
- Chances of infections are also considerably smaller compared to the open surgery.
- Chances of neurovascular damage are lower and tissue cutting is easier with this system. Incision is small and recovery is faster.
- Minimal amount of blood loss in robotic surgery, which results in less need for a blood transfusion.
- Precise surgery; surgeons can access the area where meticulous operation is to be done with high-resolution pixel ratio.
- Complex surgery can be done in an efficient manner using robotics, which is not possible in a traditional procedure.
- Chances of wound infection are also lower.
- Surgeons have more operational flexibility as compared to conventional methods.
- This surgery is more precise because of instruments and precision of robot as compared to traditional methods.

17.6 Challenges of Robotic Surgery

The risk associated with modern methods are like traditional methods, but the risk factor is smaller. One limitation for any minimally invasive surgical approach is the surgeon's inability to be in direct contact with the tissue being operated upon. In case of robotic surgery, surgeon relies only on computer motion. Training in robotic surgery remains a blockade for various surgeons, and improvements in simulation and training modules would make a considerable impact [30–35, 37, 38]. The integration of machine learning (ML) with databases can provide step-by-step guidance to surgeons. Some of the other challenges are listed below:

- Patients some time have bleeding, which may lead to the operation becoming an open or traditional-method procedure.
- Additional training is required to train the surgeons.
- Cost of maintenance and implementation is high compared to traditional surgery.
- The lack of haptic feedback to the operator.

17.7 Future of Robotics surgery

Since being approved by the FDA in 2000, approximately 5 million robotic surgeries have been performed using the intuitive Da Vinci platform. Even in China, there is increasing demand for minimally invasive surgery, which has also pushed an R&D division of a China-based company to build a medical robot. In 2008, a collaborative project was also started to develop a novel robotic system for various types of surgery. Nowadays, robotic surgery has become a sophisticated tool for surgeons with added benefits such as accurate surgery, less amount blood loss and tissue damage, small incision, speedy recovery, and tremor-free surgery. This kind of surgery basically involves the surgeon sitting at a console and operating using two joysticks. Wrist movements of the surgeon are replicated within the body of a patient by instruments that allow surgery to be done [34, 35]. Overall, all the robotic instruments are under control of the primary surgeon.

17.8 Conclusion

This chapter provides an overview of laparoscopic surgery and its revolution where various open surgery was adapted to minimal access surgery (MAS) and its benefits to patients including less hospital stay, less postoperative pain, less blood loss, and speedier recovery with less chance of infection and improved outcomes. In this chapter practical use of robotics surgery, its current trends, various open challenges, and its future have been discussed. This chapter also describes different types of surgical robots with their drawbacks, literature review, and clinical applications of surgery.

REFERENCES

1. Overview of Robotic Surgery. Available at: https://www.narayanahealth.org/robotic-surgery/#:~:text=Robotic%20surgeries%20provide%20faster%20recovery,fewer%20complications%20due%20to%20surgery. [accessed on May 30 2020].
2. Overview of Robotic Surgery. Available at: https://www.ncbi.nlm.nih.gov/pmc/articles/PMC1681689/ [accessed on April 20 2020].
3. Overview of Robotic Surgery. Available at: https://bmcbiomedeng.biomedcentral.com/articles/10.1186/s42490-019-0012-1 [accessed on April 20 2020].
4. Overview of Robotic Surgery. Available at: https://pubmed.ncbi.nlm.nih.gov/12731212/ [accessed on April 20 2020].
5. Kim HB, Lee JH, Park Do J, Lee HJ, Kim HH, Yang HK. Robot-assisted distal gastrectomy for gastric cancer in a situs inversus totalis patient. *J Korean Surg Soc.* 2012;*82*:321–324.
6. Matsuhira N, et al. Development of a functional model for amaster-slave combined manipulator for laparoscopic surgery. *Adv Robot.* 2003;*17*(6):523–539.
7. Focacci F, et al. *Lightweight hand-held robot for laparoscopic surgery.* In: *Proceedings – IEEE International Conference on Robotics and Automation*; 2007. pp. 599–604.
8. Bensignor T, et al. Evaluation of the effect of a laparoscopic robotized needle holder on ergonomics and skills. *Surg Endosc.* 2015;*30*(2):446–454.
9. Zahraee AH, et al. Toward the development of a hand-held surgical robot for laparoscopy. *IEEE/ASME Trans Mechatron.* 2010;*15*(6):853–861.
10. Gomez G. *Sabiston Textbook of Surgery.* 17th ed. Philadelphia, PA: Elsevier Saunders; 2004. Emerging Technology in surgery: informatics, electronics, robotics.
11. Hazey JW, Melvin WS. Robot-assisted general surgery. *Semin Laparosc Surg.* 2004;*11*:107–112.
12. Himpens J, Leman G, Cadiere GB. Telesurgical laparoscopic cholecystectomy [letter]. *Surg Endosc.* 1998;*12*:1091.
13. Hanly EJ, Talamini MA. Robotic abdominal surgery. *Am J Surg.* 2004;*188*:19S–26S.
14. Hubens G, Ruppert M, Balliu L, Vaneerdeweg W. What have we learnt after two years working with the da Vinci robot system in digestive surgery? *Acta Chir Belg.* 2004;*104*:609–614.

15. Brunaud L, Bresler L, Ayav A, et al. Advantages of using robotic Da Vinci system unilateral adrenalectomy: early results. *Ann Chir*. 2003;*128*:530–535.
16. El-Hakim A, Tweari A. Robotic prostatectomy – a review. *MedGenMed*. 2004;*6*:20.
17. Spaliviero M, Gill IS. Robot-assisted urologic procedures. *Semin Laparosc Surg*. 2004;*11*:81–88.
18. Phillips CK, Taneja SS, Stifelman MD. Robot-assisted laparoscopic partial nephrectomy: the NYU technique. *J Endourol*. 2005;*19*:441–445.
19. Guillonneau B, Cappele O, Martinez JB, Navarra S, Vallancien G. Robotic assisted, laparoscopic pelvic lymph node dissection in humans. *J Urol*. 2001;*165*:1078–1081.
20. Advincula AP, Falcone T. Laparoscopic robotic gynecologic surgery. *Obstet Gynecol Clin North Am*. 2004;*31*:599–609.
21. Falcone T, Goldberg J, Garcia-Ruiz A, Margossian H, Stevens L. Full robotic assistance for laparoscopic tubal anastomosis: a case report. *J Laparoendosc Adv Surg Tech A*. 1999;*9*:107–113.
22. Margossian H, Falcone T. Robotically assisted laparoscopic hysterectomy and adnexal surgery. *J Laparoendosc Adv Surg Tech A*. 2001;*11*:161–165.
23. Marchal F, Rauch P, Vandromme J, et al. Telerobotic-assisted laparoscopic hysterectomy for benign and oncologic pathologies: initial clinical experience with 30 patients. *Surg Endosc*. 2005;*19*:826–831.
24. Nifong LW, Chitwood WR, Pappas PS, et al. Robotic mitral valve surgery: a United States multicenter trial. *J Thorac Cardiovasc Surg*. 2005;*129*:1395–1404.
25. Chitwood WR, Jr Current status of endoscopic and robotic mitral valve surgery. *Ann Thorac Surg*. 2005;*79*:S2248–S2253.
26. Argenziano M, Oz MC, Kohmoto T, et al. Totally endoscopic atrial septal defect repair with robotic assistance. *Circulation*. 2003;*108*(suppl1):II191–II194.
27. Wimmer-Greinecker G, Deschka H, Aybek T, Mierdl S, Moritz A, Dogan S. Current status of robotically assisted coronary revascularization. *Am J Surg*. 2004;*188*:76S–82S.
28. Morgan JA, Ginsburg ME, Sonett JR, Argenziano M. Thoracoscopic lobectomy using robotic technology. *Heart Surg Forum*. 2003;*6*:E167–E169.
29. Melfi FM, Menconi GF, Mariani AM, Angeletti CA. Early experience with robotic technology for thoracoscopic surgery. *Eur J Cardiothorac Surg*. 2002;*21*:864–868.
30. Ballantyne GH. Robotic surgery, telerobotic surgery, telepresence, and telementoring. Review of early clinical results. *Surg Endosc*. 2002;*16*:1389–1402
31. Bentas W, Wolfram M, Brautigam R, et al. Da Vinci robot assisted Anderson-Hynes dismembered pyeloplasty: technique and 1 year follow-up. *World J Urol*. 2003;*21*:133–138.
32. Lorincz A, Langenburg S, Klein MD. Robotics and the pediatric surgeon. *Curr Opin Pediatr*. 2003;*15*:262–266.
33. Suematsu Y, Del Nido PJ. Robotic pediatric cardiac surgery: present and future perspectives. *Am J Surg*. 2004;*188*(suppl):98S–103S.
34. Cannon JW, Howe RD, Dupont PE, Triedman JK, Marx GR, del Nido PJ. Application of robotics in congenital cardiac surgery. *Semin Thorac Cardiovasc Surg Pediatr Card Surg Annu*. 2003;*6*:72–83.
35. Sharma L (Ed.). (2021). *Towards Smart World*. New York: Chapman and Hall/CRC, https://doi.org/10.1201/9781003056751
36. Sharma L, Garg P (Eds.). (2020). *From Visual Surveillance to Internet of Things*. New York: Chapman and Hall/CRC, https://doi.org/10.1201/9780429297922
37. Sharma L, Garg PK. "Smart E-healthcare with Internet of Things: Current Trends Challenges, Solutions and Technologies", *From Visual Surveillance to Internet of Things*, Taylor & Francis Group, CRC Press, Vol. *1*, pp. 215.
38. Sharma L, Garg PK, Agarwal N. "A foresight on e-healthcar Trailblazers", *From Visual Surveillance to Internet of Things*, Taylor & Francis Group, CRC Press, Vol. *1*, pp. 235.
39. Sharma L, Garg PK. "IoT and its applications", *From Visual Surveillance to Internet of Things*, Taylor & Francis Group, CRC Press, Vol. *1*, pp. 29.
40. Sharma L, Singh S, Yadav DK. "Fisher's linear discriminant ratio based threshold for moving human detection in thermal video", *Infrared Physics and Technology*, Elsevier, March 2016.

18

Brain-Computer Interface: State-of-Art, Challenges, and the Future

V. Jokanović
ALBOS Ltd., Vinca; Institute of Nuclear Sciences, Belgrade, Serbia

B. Jokanović
Werner-von-Siemens-Straße 18, Meitingen, Germany

CONTENTS

18.1 Introduction

Conquering new possibilities through the BCI is one of the unfulfilled dreams of modern man. As the ancient Greeks imagined about flight, today we fantasize about machines solving the problem of human death, while we are horrified by the anticipation of a response that transcends the limits of human imagination. Recently, Elon Musk and Brian Johnson announced the imminent realization of such an interface, which made the question even more important of how close such a moment is and what the consequences of such a realization would be. How would that interface work and what are all the hidden dangers for us in that unnatural relationship? [1, 2].

DOI: 10.1201/9781003140351-18

Sophisticated bidirectional devices, used to record and stimulate the nervous system, have led to a radically new method of rehabilitating stroke and spinal cord patients by strengthening the link between different brain regions and the brain and spinal cord by redirecting information around the injury surface. They promote resuscitation of the paralyzed part of the body. This technology already has the ability to use the interface of the brain computer to enable the paralyzed movements, which are really much slower and less precise than the actual human movements. And yet, even though we are still far from the goal, it allows us to do some impressive things. Therefore, such first steps are encouraging, bearing in mind that neurons interact among them through a complex network of electrical signals and chemical reactions [1, 2].

Nevertheless, this electrochemical language can be explained through electronic circuits, for neurons do not make it easy to understand the message sent by such stimulation in the forest of all other sensations present during neural activity. The design of the interface itself puts a special problem of possible damage to neuronal tissues, since the brain tissue is soft and adaptable, while most electrically conductive materials, such as wires that connect to brain tissue, are very inflexible, which causes scarring and immune reactions that follow, resulting in a loss of interface efficiency over time [3, 4].

Despite all such challenges, the future should be looked at with more optimism, since the brain is astonishingly adaptable and able to learn to use the brain–computer interface in a similar way as it adopts new skills, such as driving a car. In addition, the brain can learn to explain new sets of sensory data, even if they are delivered noninvasively, via magnetic impulses. In addition, it is two-way adaptable, which means that just as electronics learn from the brain, in turn the brain learns from electronics, during the whole process of their interaction, which is necessary for building nerve bridges, i.e., mutually harmonious two-way brain–computer interfaces. Fascinating is the recent success in focusing of the treatment of disorders such as diabetes on application of the electroceuticals, experimental small implants, which medicate the illness without drugs by direct sending of instructions to the inner organs [5].

Elon Musk, who started Neuralink with the ultimate goal of creating a human–computer interface, expects that requested conditions will soon be created in which the human brain will increase its abilities, to the extent that they will far exceed current natural human abilities in the near future. Such a vision, although it seems distant and fairytale-like, has become so close and realistic that life without it will be unthinkable tomorrow, because as soon as the initial suspicion in the brain–computer interface is overcome, there will be an incredible jump of human abilities, well above of all its existing limitations. In a statement by Elon Musk, charged with emotions, he speaks about the unprecedented progress of neuro-engineering, because despite the emphasized healthy dose of skepticism, it is clear that such an endeavor is incomparably more complex than the colonization of Mars. Therefore, this represents one more reason not to hesitate to tackle all difficulties, persistently and courageously striving for the ultimate goal of connecting the human brain with artificial intelligence. Leaving aside the issue of concern, which imposes the intensive development of artificial intelligence, the essential idea of building a brain–computer interface is unavoidable in our near future, although it still smells like science fiction [6].

18.2 BCI Invasive and Noninvasive Devices

Brain computer devices allow control of the device by means of electroencephalographic (EEG) activity of the user's scalp or by monitoring the activity of one of the neurons present in the brain. The EEG has insufficient resolution and needs intensive learning, while imaging of only one brain neuron involves considerable clinical risk and has inadequate stability. Electrocardiographic (ECoG) activity registered from the brain area can allow clients to quickly and accurately menage a one-dimensional cursor, allowing them to identify signals related to different kinds of motor and speech inventions. In addition, open-loop examinations show that ECoG signals at a frequency of 180 Hz encode key evidences associated with two-dimensional joystick movement. This BCI method allows patients with hard impairments to have non-muscular correspondence that is potentially more stable and less traumatic than BCI that uses an electrode that penetrates the brain [5, 7].

The BCI uses noninvasive and invasive methods. Noninvasive BCI uses an EEG registered from the scalp. It is risk-free, low-cost, has low spatial resolution, requires extensive user training, while invasive BCI uses the signaling activity of neurons recorded in the brain. Invasive BCI shows higher spatial resolution and signal control, which have a greater degree of freedom, but also shows disadvantages regarding the aspect of maintaining a stable signal during long-term recording, because small places with high impedance lead to signal degeneration due to encapsulation. In addition, small dislocations of slightly incisive electrodes cause changes within the cortical layers containing an enormous number of neurons, which is characteristic of pyramidal neurons in the motor cortex layer. The BCI methodology based on cortical surface electrocardiographic activity (ECoG) is a good alternative to the previous EEG method. ECoG has a superior spatial resolution comparing with EEG (on the order of millimeters, instead of centimeters), a wider bandwidth (0–200 Hz, instead of 0–40 Hz), a larger amplitude (50–100 µV, instead of 10–20 µV) and a significantly less instability to artifacts than to electromyography (EMG), in which the electrodes do not penetrate the cortex, which affects better long-term stability and signal safety than single-neuron imaging [5, 8].

The BCI initiates a novel method of neurorehabilitation for patients with physical disabilities (paralyzed and with amputated limbs) and brain demages (stroke). Wireless recording, machine learning and real-time time resolution are the main goals of EEG-based BCI. Brain activity is observable by diverse neuroimaging methods. Improving the decoding of the kinematic movement of the upper limbs using an invasive electrode, enables precise regulation of prosthetic devices in 3D space. Unfortunately, invasive electrodes cause a high risk during surgery and progressive degeneration of the registered signals. Accordingly, noninvasive methods, like functional magnetic resonance imaging (fMRI), magnetoencephalography (MEG), near-infrared spectroscopy (NIRS), and EEG are more common methods in the medical practice [9, 10].

The motor imaginary paradigm describes image movement instead of actual executive movement, showing that imagination activates the area in the brain is responsible for each particular motion. The most commonly used models are sensorimotor rhythms (SMR) and imaginary body kinematics (IBK). Although high spatial resolution is fMRI, EEG is the most wiespread approach because of straightforward measurement of nerve activity, cheapness and easy portability for clinical application. The EEG measures the electrical activity of the brain produced by the flow of electric current during the excitation of the synapses of neural dendrites, especially in the cortex and deep brain architectures. These electrical signs are registered by locating electrodes on the scalp. These signs regulate facilities like wheelchairs and information assistance techniques. This method is also used to control auxiliary and rehabilitation equipments. EEG signs allow pathways from the brain to different outside devices, enabling the control of assistive devices of persons with disabilities and brain-directed rehabilitation equipments for clients with strokes and others neural disorders. It enables the analysis of the relationship between brain activity and bionic and cognitive processing [11, 12].

The SMR activity is the most promising method for poeple with paraplegia, spinal cord injury and amyotrophic lateral sclerosis (ALS). Volpav and everyone and McFarland used four-channel sensorimotor rhythms within the left and right central grooves to shift the cursor in 2D space to targets located in the four edges of the computer display. They then asked people with spinal cord injuries to move the right and left hand of the cursor to eight different objects on the side display device with the mouse. Finally, cursor control is required against targets placed in 3D space. In all these researches, patients are taught to adjust their SMR based on imagined their body parts, such as the arms and legs. SMR lacks the ability to directly extract kinematic parameters such as position, speed, and acceleration [13, 14].

Imagined body kinematics (IBK) is a motor imaginary method based on BCI invasive technology. This method can be noninvasive used only for low-frequency SMR signals (less than 2 Hz). This method assumes a training protocol and analysis in which patients is inquired to visualize the persistent movements of only one body part in multidimensional space. The obtained signals are decrypted in the time domain. Bradberry examined 2D cursor regulation with spontanous imagined motion of the right finger (deminishing so execise time to the value for invasive devices) and analyzed in time domain frequencies lower than 1 Hz [15].

18.2.1 Visual P300 and BCI Closed Loop

The P300 has a positive maximum in EPR (event-related potential) in the range of 5–19 mV, which is delayed between 220 and 500 ms after the event. EPR is determined as the average rise of the amplitude of the chronological orders of the most significant brain sigs on the midlines. If the interval between stimuli is less than 250–300 ms, then the signal may overlap with the next signal that is formed during that time. The visual P300 has a high degree of accuracy and can be calibrated in minutes, allowing users to use the device management system easily and quickly. The disadvantage is fatigue due to the need for a high degree of attention and the inability of visually impaired people to use this system due to the need for strong focus during use, although the application of this method significantly reduces training time. Strong attention is needed during the experiment, because the amplitude of the P300 depends on the number of occurrences of the target, the duration of the inter-attempts, the severity of the experiment, the patient's condition during attention, and common behavioral habits. This metaphor shows enhanced attention cognitive performance and quick memory replies. The most common application is related to the development of prosthetic keyboards in order to provide communication paths for patients with disabilities [16, 17].

Usually spelling facilities in BCI contain a matrix of letters, numbers, and symbols. The rows and columns of this matrix are in a row, and the focus of the patient's attention is focused on the designed characters, and then the speller edits them following his/her position within the given row and column. These facilities apply a statistical model based on the P300 in identifying the correct symbol throughout blinking. This method is useful for people with ALS and stroke, as well as for controlling humanoid robots and wheelchair navigation, controlling a computer cursor in 2D space by paralyzed people, and controlling an artificial hand in a virtual state in smart reality [18].

The BCI research shows significant advances in neurorehabilitation and ancillary equipment technology, showing an ability to manage exterior prosthetic devices for spinal cord damages and other kinds of the communication diseases such as amyotrophic lateral sclerosis (ALS) and multiple sclerosis (MS). One of the most important provocations of BCI is the necessary exercise time for the patient to achieve enviable skill in using such systems. Most patients take a long time to do this, which causes the subject to become tired. To further study this phenomenon, researchers investigate how subjects reflect behavioral variability between sessions to collect adequate data to calibrate patient behaviors at the beginning of each session. To avoid this problem to some extent, a learning transfer method was used to establish zero training as a generic BCI method that can be applied for most patients. Numerous methods of decoding, signal processing, and classification algorithms have recently been studied in detail, because the evidence derived from the EEG signal shows the signal-to-noise ratio insufficient to successfully control a device such as, for example, a multi-degree neuroprosthetic arm. This requires more robust, accurate, and faster network algorithms. To improve this performance, some investigators have recommended the application of pre-machine learning and deep learning methods, while others suggest adaptable classifiers and decoders to correct the non-stationary nature of the EEG signal [8, 9, 19].

The BCI closed loop is a system of adaptation and mutual learning where a human and a computer learn from each other while at the same time the process of adapting mathematical algorithms and neural circuits takes place. This is known as shared or hybrid control. A common BCI presupposes both low- and high-level control systems. The brain generates high-level commands, and conventional monitoring systems are responsible for unsatisfied command-monitoring functions. The desirable BCI system is a system with mutual communication, in which the executive regulation system is the main (in high-level control) while other parts of BCI system are used as its intelligent support (in low-level control). With cognitive supervision, the consumer acts as a supervisor of the exterior independent system rather than constantly communicating with the control instructions. EEG scalp is a cheap monitoring technology, which is why it has high potential for commercialization. Some investigations estimate behavioral modifications in reaction to sound signs in the drawing surrounding and identify correlations between brain waves and others sensory inputs, such as hepatic response [20].

The development of dry sensors does not require skin preparation or gel application, which facilitates the application of BCI, enabling, for example, better sleep quality and better efficiency of antidepressant application for the patient. One of the future directions of BCI application will be neuro-feedback as a process of self-regulation of brain waves to enhance different aspects of cognitive control, diminishing the negative adverse effects of drugs, such as headaches. This method could be used to help treat patients with addiction, obesity, autism, and asthma. Recent cognitive methods are focused to overcome neuro-rehabilitative cognitive defects, such as attention deficit hyperactivity disorder, anxiety, epilepsy, Alzheimer's disease, traumatic brain injury, and posttraumatic stress disorder [21, 22].

18.2.2 Machine Learning Algorithm

Many properties of BCI rely on the working algorithm of machine learning, which enables the given accuracy and speed of communication. The P300 and other BCI systems typically use the FLDA algorithm, when the numerous characteristics do not depend too much on the number of clients undergoing exercises. Numerous factors define the performance of the BCI system, like measured brain signs, signs-processing procedures used to extract characteristic sign performance, an algorithm that translates such commands, exit devices that implement such instructions, feedback offered to the user, and user-specific features. Future advancement needs systematic properly regulated investigations of evaluation and comparison of alternative signs and sign combinations, methods of extraction of alternative characteristics and translating algorithms, as well as alternative applications for correspondence and supervision in various clients' populations [23, 24].

The BCI 2000 system was developed to ensure a comprehensive platform that provides the assessment, comparison, and mixture of alternative brain signs, treatment procedures, implementations, and work protocols to reduce time, cost, and effort during examination of current designs, using a standardized offline processing data that allows increasing the efficiency of research and clinical application [25, 26].

The BCI system based on EEF for the development and study of EEG screening and establishment of the superior method of its translation into device driver is the main goal of modern research. An essential element of the system is the identification and application of EEG data, which is influenced by efficient communication between two adjustable regulators, the user who performs the EEG control, and the BCI system that converts that information into device regulation. The experimental system has its own topographic, spectral, and temporal characteristics. In practiced clients, the response to the instruction inside of 0.5 s is related to the rhythms of 18–26 Hz. The location and frequency that allow optimal control can vary on a daily basis or remain the same for several days. Another improvement is related to an algorithm that turns EEG regulation into device regulation using 3D filters correspondent to the spatial frequency of the consumer rhythm. This is achieved by autoregressive frequency analysis that enables better resolution than FTIR for short-time periods and better practical adaptations of the algorithms that convert EEG data into device regulation. Topographically different rhythms can be controlled at the same time if one of them increases and the other decreases. The EEG does not depend on muscle activity. It is also possible to control the time domain of EEF characteristics. Rhythm regulation may be connected to reduced activity of cortical potential [8, 27].

18.2.3 Brain–Computer Interface Speller

The BCI gives a new method of non-muscular reporting via a brain signal. BCI Speller P300 is one of the first BCI methods of applications, based on the monitoring of stationary visual evoked potential (SSVEP) and motor imaginary potential (MI). Various BCI approaches need particular characteristics of the encephalogram signal (EEG), which has accelerated the development of an appropriate graphical user interface (GUI). Motor neuron disease (MND) affects how the brain interacts with different organs in the body, affecting the neurological network and motor control of muscles. It is used in the treatment of amyotrophic lateral sclerosis (ALT), stroke, brain or spinal cord injuries, cerebral palsy, muscular dystrophy, and multiple sclerosis, as well as in the loss of the capability to regulate voluntary muscles, mainly consisted of skeletal muscles, responsible for functional and cognitive impairments. One of the communication systems intended for people who are not able to communicate or utilize their hands to work is using an eye-tracking spelling system dependent on the mobility of the eye controlled by the cursor on the virtual keyboard, while selecting the particular letters. Eye blink can also be applied as a conversation practice. Such and complementary systems are not convenient for people who have lost the ability to accurately regulate fine eye shifts or patients who suffer from uncontrolled head shifts [18, 28].

There are numerous ways for monitoring brain activity. One of the most acceptable among them is electroencephalogram (EEG). This is a noninvasive method that is extensively applied in recent BCI uses. It is more efficient than electrocorticography (ECoG), which needs a surgery through the skull for direct access to the brain structures. EEF device is an inexpensive, movable, and easy-to-install method. It gives a signal with a higher time resolution in comparison to alternative noninvasive procedures for registering brain activity. Magnetic resonances imaging (MRI) or positron emission tomography are also noninvasive BCI methods. They are suitable for examining the health of individuals and application. After checking and registration brain activity in BCI, individual signals are extracted and analyzed using a computer. The output potential of BCI can be used to restore, renew, improve, or enhance central nervous system functions. One of the most investigated utilizations is the BCI speller. It lets the user to contact with their surrounding by a graphical user interface (GUI) that displays letters, numbers, and special characters. Using a brain signal that records and analyzes the BCI, the client chooses the selected character and types it on the screen or other output display. The BCI speller also allows people to communicate directly by measuring and interpreting brain activity [5, 8, 29].

18.2.4 Event-Related Potentials (ERPs)

The ERPs are electrocortical signals that the EEG detects and measures during or after a sensory, motor, or psychological event. They generally have a limited delay in stimulus and various amplitudes in comparison with normal EEG activity. EPRs are rarer and more distinct than EEG registered signals. Various ERP signals can be triggered using various stimuli (events), where the ERP is defined by a certain time delay or position delay in relation to the moment when it is caused. The two most frequent ERPs are P300 and visually evoked steady state potential (SSVEP). The P300 wave is a kind of event-induced potential that appears in the human brain as a specific reflection that corresponds to a time delay of about 300 ms after the occurrence of certain events. The P300 signal is commonly amplified in the central parallel area of the brain, which can be registered by EEG. The event that initiates the R300 is known as the "unusual paradigm," which consists of three main assumptions. According to the first assumption, the subject is represented by a series of stimuli or events, belonging to one of the classes (wanted or unwanted event), while one of them occurs less frequently than the second class (a rare compared to a more frequent event). The patient must pay attention to one of the stimuli when an event appears (calculating the difference in flashing certain letter, corresponding to a rare event). The event induces a P300 signal in the brain. In doing so, it is possible to develop both a visual and auditory stimulus to initiate a P300 signal for various devices and uses, explaining it, for example, as visual potential (VEP) or auditory potential (AEP). One of the first P300 BCI speller developed by Farwell and Donchin in 1988, visually stimulates an "oddball paradigm" [16, 17, 30].

18.2.5 Movement Imagination

Sensorimotor rhythm (SMR) is processed by the motor cortex associated with any somatosensory areas. During motion, the SMR may decrease or increase. These opportunities are recognized as event-specific desynchronization (ESD) and event-specific synchronization (ESS). In the case of ESD, signal shifts become smaller than specific baselines, induced by desynchronization of activity in a particular brain area. On the other hand, ESS signal during motion is stronger in comparison to the baseline, which is characteristic of the signal at rest. The location of the signal varies depending on how it moves and on which part of the body a certain motion has occurred. It has been found that imagining a motion without its realization causes a similar weak EEG signal, compared to ERP and VEP. Nevertheless, it can be distinguished not only by legs but also by arms (left and right) [15, 31].

18.3 Quantum Brain Model

The brain is a complex physical system in continuous interaction with the outside surrounding. Generally, brain functions can be observed like this: (1) assume that the brain is in a state of jMi during an outside stimulus, for some time; (2) after removing the external stimulus, let the brain shift in a state of jNi, which in general should by the same type of coded (or recorded) information that is conducted by an exterior stimulus; and consequently, (3) held over (or call back) directly from the state of jNi, taking into account that (4) the brain did not inevitably go immediately from the state of jMi to jNi because many intermediate phases took place: jAi! jM1i! jM2i! ...! jNi, i.e., information (or messages) treated in the brain until it is remembered. There are several essential features that describe efficient brain functions: permanent steadness and non-localization, which is clearly indicated by the abundance of trials data. While permanent steadness is clear, non-localization, i.e., coherent neural activity in 3D distant cortical areas, makes a traditional approach to brain functions problematic [32].

At the same time, non-localization indicates the expediency of a quantum approach to such a problem. The connection between microscopic conditions and quantum treatment leads to the introduction of the so-called microscopic quantum states (MQS), which are often in the physical realm. Superconductivity, superfluidity, and magnetization are representative illustrations of MQS with distinct features: (1) particular structure and states and (2) a critical level of coherence that, when reached, causes (3) regulated states, which are very stable. As the example of magnetization of special structures of Weiss regions, tiny areas in ferromagnets in which electron spins are polarized in a particular course can be noticed. Since there are numerous tiny areas and polarizations, overall magnetization in the ferromagnetics is not noticeable. If, on the other hand, we use a powerful enough magnetic field B or if we reduce the temperature sufficiently (below the Curie point), i.e., under specific conditions, the ferromagnet shows magnetization, since now all electron spins in the entire macroscopic crystal are polarized following the same course, strictly mutually coupled, which leads to a considerably well-balanced, macroscopically coherent (or quantum) state, i.e., an arranged condition [33, 34].

In most physicists' interpretations, the transition from the unordered (numerous Weiss regions) to the ordered (magnetization) is named a "phase transition." The value of the essential parameters (magnetic field B or temperature T) at the transition point characterizes the phase transition and defines the critical point (Curie temperature). It is obvious that coherent states contain certain information (all electronic spins are polarized in the same way) that does not contain unordered states (randomly distributed electronic spins). In other words, disordered states are more symmetric (polarized electron spins in the same direction, with interruption of rotational symmetry), while ordered states have less symmetry (randomly distributed electron spins that are rotationally invariant, i.e. do not have a preferred direction). Hence, the coherent states consequently induce an automatic symmetry break leading to the phase transition. The features of the phase transition are: (1) universality – many various systems can be described by the same phase transition; (2) attractor – by changing certain system parameters, they can reach values close to the critical values, for the phase transition [35]. The critical point represents an attractor for everyone around you. Specifically, we do not feel the fine-tuning of the system instructions while reaching the coherent state. And finally (3) evolutionary equations – all the essential parameters of the phase transition (involving the previous two conditions) can be coded by a renormalization group of equations (RGE). They explain not only the deviation from the critical state but also other features of the phase transition. Macroscopically, quantum or coherent states possess very selective features: (1) very long steadness – they are highly changeless, there represent the long-term correspondence between basic constituents, like self-promoting stimulation loop, phonons, spin-waves, magnona, etc., which regulates the state different basic constituents, returning them into initial basic states induced by the disorder (this is symbolically denoted as the R + F property of MQS); (2) non-localization – it is clear that MQS can be found outside microscopic sites; (3) incidence – MQS has new features that are not existing at the basic elementary state. The recent features describe the states at the higher level in comparison with the levels at which the basic intercommunication between the basic elements occur [36, 37]

Namely, superconductivity is a new characteristic/peculiarity, which indicates a gathered examination of electrons under particular conditions while each electron follows fundamental rules of quantum electrodynamics. Let's remember the analogy between MQS and phase transitions with brain functions: (1) the uncoded brain is characterized by random signals and attenuated perception (this correlates with random polarization in the small Weiss regions of the ferromagnet), while (2) learning is characterized by the reception of external impulses, over several seconds, which amplifies and regulates disordered neural signs as ordered pieces of evidences, correspondent to a ferromagnet, when an exterior magnetic field B is applied or if the temperature is reduced below the Curie point. All this causes the interruption of multi-domain small structures with random polarization and leads to an ordered state, in which the electronic spins, through the entire ferromagnet, are fully harmonized at each point in the same direction [38].

In such conditions, it is a matter of a phase transition or a spontaneous break in symmetry. It is clear that such a process depends on the kind of exterior stimuli that interact with particular basic constituents and direct the system toward an appropriate MQS, or regulated state. Realistically speaking, in order to encrypt all qualitative diverse signals and produce an ordered unique impression of ourselves, an enormous number of qualitative various states is necessary, i.e., an almost unlimited number of qualitative diverse spontaneous symmetry breaks. Such symmetries are associated with a group of selection laws that serve as a physical filter against unwanted, insignificant stray signals. This is indeed a difficult task if we recall the fact that only visible spontaneous symmetry breaks, at the fundamental level, describe electroweak interactions. This is such a negligible number in relation to the incomparably large number of spontaneously broken symmetries [34–39].

Therefore, it is one of the problems that require string theory to be applied to them, while (3) encoded brain or memory results in a very balanced, ordered ring of a cluster of included neurons, which are not always confined and which correspond, when observed in analogy, to the ferromagnet of stability and macroscopic nature (including non-localization) of embedded magnetizing states. This type of naturally organized, coherent neural release, which is not always confined, can deliver an answer to the so-called problem binding. And finally (4) the recall process, during which a repetitive slight signal, similar to a learning signal, can provoke the currently coherent state, whereby its R + F feature is retranslated into the previous form. In that manner, the excitation of an orderly state is a process that warns us to revoke a given orderly state. This, by analogy with a ferromagnet, corresponds to the application of a weak magnetic field B_0, which is not completely parallel to the initial field B, which induces a forced oscillation of electron spins, at the moment, prior releasing them back into equilibrium, i.e., restoration of the coherent state, due to the R + F features of MQS. It is necessary to repeat (replicate) the signal, which is completely equal to the learning signal, completely revoking the information, due to the feature of the phase transition of the attractor, which has already been discussed [32, 40].

In the phase transition terminology, the memory recall mechanism corresponds to the action of the irrelevant operator. It must not elude our concern in the network of phase transitions, because R + F and the features of the attractor make it very easy to download information without the need to completely identify the replication signal and the learning signal. In other words, fine-tuning, which requires long periods of time, is irrelevant for information to be retrieved. Think of what would occur if it were necessary to observe all scenes of a fast approach of a starving lion, involving the contours of its jaws, before climbing a tree to escape it. That would not be really practical. The general picture of brain functions presented above has a promising character. But the question arises: Is there any experimental evidence for this? Such evidence is contained in the encephalogram (EEG). It is common to assume that EEG waves are formed by summing up the firing of local neurons, although this process is still somewhat more complicated [41].

According to some researchers, asynchronous ignition of randomly distributed neurons produces an overall net zero effect on bare electrodes. In contrast, researching electrical potentials during sensory stimulation and learning attempts, E. R. John proved that these potentials grow when large and scattered neural groups are fired in a completely different way from spontaneous random cortical activity. Time redistribution within neural groups is characterized by external stimuli of potential. Sayers et al. found similar independent evidence of strengthening time redistribution by examination EEG phase ordering. The frequency of the EEG spectrum during spontaneous cortical activity exhibits a random distribution of phase relationships, which move toward clear patterns of phase ordered exactly following sensory provocation. It is astonishing that the phase characteristics stimulated by the potential influence the spontaneous waveforms, due to which it is possible to replicate the shapes of the observed waves [42].

These time redistribution results were also confirmed by E. R. John, who showed that the classical expectations that the EEG data from the sum of neural firings are wrong, implying that the amplitude waves are the only difference between spontaneous and stimulated waveform [9, 10]. It seems that exterior stimulus adds energy to the brain and simultaneously directs it to organize itself in an ordered manner, as in the case of exterior field B in ferromagnetics. This confirms that the equivalence among brain functions and critical dynamic phenomena is completely justified. In such an approach, the effective mental world (W2) seems to actively interact with MQS so that the R + F feature of MQS and subsequently triggered by W2 MQS breakdown, providing a solution to an old problem dealing with the reasons for the strong connection of emotions with body action [43].

18.4 Specificity of the Architecture of Our Brain and the Brain Smart Activities

Our brains are composed of many neurons, single nerve cells linked into dendrites and axons. And we are only witnessing a perfect symphony, in which every time that mysterious world of neurons awakens and directs its attention through an indescribably complex world of choice, in which only those stimuli that carry the greatest energy in themselves have a chance to break out of the world of darkness, and come out into the world of clear insight, to feel something move us, as we return with a set or joy to our memories, and not dreaming that it all happened by activating our neurons, using small electrical signals transmitted from neurons to neurons, at a speed of about 250 m/h, whereby these same signals are generated due to differences in the electrical potential of the ions contained on the neural membranes. And although the pathways of these signals are trapped by the myelin sheath, some of them are lost along the way, giving scientists a tremendous opportunity to register, analyze, and then direct them to a device that allows a similar process to take place in the opposite direction. Figuratively, this means that when researchers understand what messages are sent from the optic nerve to the brain when someone sees red, they will be able, no doubt, to equip a camera that will send the same signals to the brain each time the camera "sees" red, warning a blind subject and helping them see without using their eyes [44].

It is important to note that one of the main challenges for BCI researchers is the design of the interface itself. The easiest and least invasive method would involve the use of a set of electrodes, such as electroencephalographs (EEGs), which would be connected to allow reading of brain signals. Since the skull blocks most of the electrical signal, and distorts those that still manage to break through, to get a better resolution signal, it is necessary to implant the electrodes directly into the gray matter, inside or on its surface, underneath the skull, in order to provide the most efficient guided reception of electrical signals and access to electrodes in a particular area of the brain, where they originate. This approach, while seemingly promising, requires invasive surgery to implant electrodes, which is very disadvantageous, because devices that stay in the brain for a long time most often cause brain damage, which increasingly blocks signals, independently of the position of the electrodes, measuring the differences in voltage between neurons. And when the signal is already received, it is relatively simple to amplify and select it, and after that, through the brain–computer interface, it will be interpreted using appropriate computer programs [5, 8, 45].

If the sensor sends a signal to the brain, via a computer–brain input interface, essentially the same thing happens, only in the opposite direction, so that the computer converts the video camera signal into the voltage to activate the neurons. The signals are then directed in a correct way to the implant fixed in the appropriate part of the brain, and if everything functions properly, the neurons ignite, after which a visual image of the object is obtained, which corresponds to the image from the camera.

Magnetic resonance imaging is also suitable for measuring brain activity, whose images showing high-resolution brain activity can be applied as part of a permanent or temporary BCI to better define certain brain functions or map sites. Accordingly, if researchers are trying to implant an electrode that will allow the patient's robotic arm to be controlled by his own thoughts, they would first have to record such a hand with magnetic resonance imaging and determine the extent to which it reminds the patient of his actual hand movements. Magnetic resonance imaging itself will present which part of the brain is active during such hand movements, providing a more exact picture of where the electrode should be placed [5, 46].

For years, the human brain has been viewed as a static organ. Recently, it is well known that from the earliest childhood, along with our growth and maturation, our brain changes and shapes in agreement with our new experiences, so that recent research has shown that such a process, although slow, takes place even in old age, testifying that thanks to its cortical plasticity, the brain adapts to new circumstances in incredible ways throughout life. In doing so, learning something new or participating in new activities creates new connections between neurons and reduces the occurrence of neurological problems associated with aging [47].

It is a fascinating discovery that if an adult experiences a certain brain injury, then automatically other parts of it take over the functions of the damaged part, which means that even an adult can learn to use the brain–computer interface, because the brain is always able to form new connections and adapt to new neuronal use. Thanks to that, an adult patient is able to treat the implanted part of the brain as his or her natural part, which opens the possibility of controlling smart machines with one's own thoughts. It is a divine gift for people with severe disabilities because it enables them to function normally. For quadriplegics, controlling computer cursors through mental commands would be a revolutionary improvement in their quality of life [5].

No matter how impressive such progress may seem, it is a much more difficult task to interpret brain signals to activate hand movements in a person who cannot move a hand, because in order to be able to do it on one's own, a person must be trained to use this type of aid. For such a thing, it is necessary to visualize the movement of the hands by the patient, so that finally, after numerous experiments, the software associated with the robotic hand can overcome this difficulty and interpret the signals associated with thoughts related to hand movements. Simply put, at the moment when the subject's thoughts are directed to a given hand movement, signals are sent to the robotic hand, after which it reacts to them in accordance with the issued instruction [5, 47].

18.5 The Basic Mechanism of Turning Thoughts into Computer or Robotic Action

Once the basic mechanism for converting thought into computer or robotic action is perfected, the potential application of this technology will be almost limitless. Instead of a robotic arm, users with disabilities will be able to have robotic straps connected to their limbs, which will allow patients to move them and to communicate directly with the surrounding, even without the robotic section of the device, as signals from the straps will be sent directly to appropriate control points motor nerves in the hands, bypassing the impaired part of the spinal cord and permitting real movements of one's personal hands. Sending relatively simple sensor signals is challenging enough, while we are still far from the moment when we will be able to send signals that could force someone to take some reluctant action [48].

The human brain consists of about 100 billion neurons and about 100 trillion synapses, with each neuron igniting about 100 times per second. If the brain model were presented as a simple neural network, it would be equivalent to a machine that performs 1,016 operations per second, which is equivalent to the capabilities of the best modern supercomputers. However, efficiency is much more important than speed, because each neuron has a complex structure that is connected to hundreds or even thousands of other neurons [49].

It is especially challenging to show the way the nervous system works with the help of computer simulations. In reptiles, the part of the brain responsible for unconscious behavior such as breathing, heartbeat, kidneys, etc. is described by Henry Markham's model, which simulates an IBM supercomputer. Assuming that each neuron has its own logic, and thus the corresponding mathematics, a 1.5 GHz G5 processor was needed for this type of simulation, while according to Markham, about 10,000 such processors are needed to build a working model of the reptilian brain, although the reptile's brain was unable to "think" or follow anything other than stimulation to ignite some synapses [50].

On the other hand, the neocortex, which is responsible for socializing, parenting, and reasoning, involves more than 100 trillion neurons and many more synapses, for a total of about 10^{12} latest-generation 2 GHz processors, which would be necessary for models with which it pretends to simulate brain behavior in a convincing enough way. Now there are realistic assumptions that, thanks to technological achievements, it is possible to significantly improve the cognitive outcomes of the human brain, from the point of view of its ability to learn and make decisions, using supercomputers. Since we have enough data on how the brain works, the computer approach is very promising, aware that with such a statement it enters the almost the territory of almost magical complexity that is the human brain. And although computers are already proving superior to the human brain when calculating or making decisions, the brain and the computer are two completely different architectures; they function perfectly, but each in its own domain, so that any comparison of them is meaningless [51].

In order, one day in the near future, for quantum computers to be able to fully simulate the work of the human brain, it is necessary to understand it well first. Although we are still far from that goal today, which is why the comparisons of the human brain to the computer are not proper, it brings us to some very interesting thoughts related to the memory capacity of the brain, which varies considerably (taking in account the average human brain, for data storage). Some estimates are very low, on the order of 1 terabyte, or 1,000 gigabytes, while others speak of a value of about 100 terabytes. According to Forest Wickman, since the human brain contains approximately 100 billion neurons (or, more precisely, about 86 billion), with each of the neurons being able to participate in about 1,000 potential synapses, if we multiply each of these 100 billion neurons by about 1,000 synapses, it will yield 100 billion, or about 100 terabytes of data [50, 51].

If so, then what makes the human brain so special? It should not be forgotten that the estimation of 100 terabytes itself has its drawbacks because it follows that each synapse saves only 1 byte of information. In reality, that number is probably higher, because synapses can be located in a larger number of intermediate states, and not only in the on or off state. A computer chip that mimics the human brain may, in the future, substitute our brain, even though it is made up of a vast network of about 100 billion neurons, which are constantly being edited, while our synapses form bridges, connecting two neurons, one presynaptic and the other postsynaptic, wherein presynaptic neurons release neurotransmitters that activate postsynaptic neuron receptors and postsynaptic cell membrane ion channels, while ion channels allow charged sodium, potassium, and calcium atoms to move in and out of the cell, playing an important role in regulating synaptic plasticity, or strengthening or weakening neural communications over time, assuming that, when neurons communicate with each other, their communication is not limited to simply turning the signal on and off [50, 51].

Most computer chips used to model brain activity do the following binary code, while the brain perhaps doesn't follow that direction, because synapses are generally interdependent and rely on each other to transmit some information. A few years ago, researchers wrote that the human brain produces about 6.4×10^{18} pulses per second, which is equivalent to the same number of transmitted instructions at the same time, from which it clearly follows that the human brain has an extremely large storage capacity. And although the comparison between computers and the human brain is a bit inappropriate, it is true that neurons combine so that each of them enhances many memories, so that the brain's memory capacity is close to 2.5 petabytes (1 petabyte ≈ 1,000 terabytes), which means that we could fit all the pictures that a television set would produce, working continuously for more than 300 years, to fully fill the human brain's memory capacity [50–52].

In that game of miraculous numbers, where is the real truth? Is it in order of: 1 terabyte, 100 terabytes or 2.5 thousand terabytes? Are the forgotten memories erased, or are some forgotten parts of the memory still present in the lost parts of our awareness? Does deeply embedded subconscious occupy more space than a passing illusion? Can the capacity of memory storage of the human brain be measured at all? [50–52]

Although the connection between the brain and the computer is a metaphor for cognitive psychology, there are still many significant differences between them, which are crucial for comprehension of the mechanisms of neural information processing and creating artificial intelligence. The brain works using analog code, unlike computers that use digital code [53].

Essentially, it is easy to imagine that neurons are binary, since they "ignite" the action potential if they reach a certain gap, or do not ignite it otherwise. This apparent resemblance to the digital code "0 and 1" does not disrupt a wide range of continuous and nonlinear processes that directly affect the processing of neural signals. Namely, one of the basic processes of information transfer is the speed at which neurons ignite, which is essentially continuously variable. Similarly, networks of neurons can ignite relatively synchronized or disordered, with their coherence affecting the strength of the signals received by the neurons [54].

In addition, computer information is accessed through a precisely defined memory address, whereas the brain uses addressed memory for a given content so that information can be accessed through "analog stimulus," because it is enough to think of a fox to automatically activate memories related to other images, ideas, and experiences associated with it, such as hunters who hunt it, or attractive members of the opposite sex, whose cunning and intelligence capture our attention. It basically reminds us that our brain has a type of "built-in internet," so just a few keywords are enough to evoke numerous memories [55].

Of course, computers work in a similar way, searching huge files of stored data, and although this difference between the brain and computers is seemingly insignificant, it has a great impact on neural computing. Therefore, in cognitive psychology, there is current debate about whether information is lost from memory because it disappears by itself or whether it happens due to the interference of other information, although, frankly, such a debate is partially based on the wrong premise that these two kinds of opportunities are mutually exclusive (following the way of thinking typical of computers), as a result of which this debate itself represents a false dichotomy [56].

18.6 Brain Modularity

The assumption of modularity, which is characteristic of computers, when it comes to the brain, is not sustainable, because "memory" areas (like the hippocampus) are significant not only for memory but also for imagination, spatial navigation, and other various functions. In doing so, the rate of neural information processing is subject to various limitations, such as the time required for passing of electrochemical signals through axons and dendrites, axonal myelin, and the propagation time of neurotransmitters through the synaptic cleft. In addition, it depends on the difference in synaptic efficiency, coherence of nerve stimuli, ongoing accessibility of neurotransmitters, and former records of neuronal inflammation. Although there are particular differences in "information processing speed," it is significant to remind that this is a very complicated issue, since all signals are probably indexed as heterogeneous combinations of all the above speed limits. A special question is whether the central clock is present in the brain, about which opinions are divided, because it is unclear what part of the brain actually resembles a clock in its operation/function. Although it is often accepted that the cerebellum calculates information that includes precise time, such as the time required for some subtle movements, some current observation implicates that time in the brain shows more resemblance to pond waves than to a common digital clock [57].

Besides numerous similarities between RAM and short-range or "working" memory, to many earlier cognitive psychologists, recent research has revealed astonishingly significant differences. Although both RAM and short-range memory request adequate power supply, short-range memory appears to contain only "signposts" of long-term memory, while RAM contains isomorphic data similar to data stored on a hard disk. Contrary to RAM, the limit capacity of short-range memory is not fixed. It varies depending on the "speed of information processing": experience and knowledge. Although for years the brain was conceived as hardware on which the "smart program" or "smart software" executes, such approaches completely ignored the essential fact that the mind originates directly from the brain, and that because of this, variations in the mind are always induced by variations in the brain [57, 58].

Another pernicious metaphor is related to the computer properties of the brain. It suggests that the brain works on the principle of receiving and transporting electrical signals, or more specifically action potentials, that travel together in individual logic circuits. Unfortunately, this is only partially correct, because the signals that are transported along the axons are naturally electrochemical, which implies that they pass much slower than the electrical signals in a computer. In addition, they can be adjusted in many ways. And while computers process information stored in memory by appropriate processors and then write the obtained results back into memory, neurons, besides processing information, are able to change that information on their way through their synapses, causing memories to change somewhat, becoming stronger over time but also more inaccurate [57–59].

Experience deeply and instantly influences the essence of neural information processing, in a very different manner than in conventional microprocessors, because the brain is a system inclined to self-regulation, as evidenced by "trauma-induced plasticity" often occurring after injury, which induces a multitude of diverse interesting changes. Among them, some lead to the release of untapped brain potentials, known as acquired savantism, but also those that can lead to deep cognitive dysfunction, typical of traumatic brain injuries [56–59].

It is an extremely important fact that the brain has a body at its disposal. Jeremy Wolf, through his imaginative experiments, showed that even after hundreds of repetitions of simple geometric forms displayed on a computer screen, people continued to look for the displayed images on screen, relying on physical sight rather than their memory [60]. This is seemingly unusual, but to successfully simulate brain function, appropriate biological models would have to encompass 225 million billion interactions between various cell types, neurotransmitters, neuromodulators, axonal branches, and dendrites' spines. To all this confusion should be added the influence of dendrites' geometry and about 1 trillion glial cells that can, although need not, be significant for neural data processing. The brain serves an incredibly ambitious set of needs, which include improving the tools for recording and manipulating the cerebral cortex when a person is healthy, but also when he or she is ill. In addition, as recent research shows, the human brain is able to store more data in its memory than is contained on the entire Internet [60, 61].

Fascinatingly, unlike classic computers that use information codes 0 and 1, brain cells use twenty-six different ways to encode information. Based on these facts, it is estimated that the brain can store 1 petabyte, or a quadrillion bytes of information. This is an exciting fact, because the current estimation of brain memory capacity increases by a factor of 10, compared to earlier conservative estimates. The brain can store using energy of only about 20 watts, which can barely light a small light bulb, while a computer with the same memory and processing rate requires 1 gigawatt of power, or the power of an entire nuclear power plant [60, 61].

It has been shown that the more the brain circuit is trained – that is, the more networks of neurons are activated – the greater the chances that a neuron will ignite in that circuit when a signal is sent. It follows that the process of reinforcing the neural network probably enhances intercommunication of the synapse, multiplying the quantity of neurotransmitters released, so that if neurons essentially chat with each other through synapses, then brain cells communicate through more synapses louder than when there are fewer synapses. If the size of the synapses can vary by a factor of 60, and the number of the synapses differs by about 8 percent, then it follows that there are a number of different values of synaptic quantities, which are in a given range [60–62].

Depending on the resolution of the brain, the success with which the brain will register the difference between the signals depends. And because neurons are in 26 different size ranges, it essentially reveals a whole scale of different "voice" volumes that neurons use to chat with each other, which helps determine how information is transmitted between any two neurons. Computers store information as bytes, which can have two possible values between 0 or 1. But in neurons, this binary message (lit or not) can be produced by 26 different types of "voices" of neurons. Converting the number 26 into bytes means that 2 needs to be raised to n to get 26, this is 4.7. From this value it follows that this capacity is about 10 times greater than previously believed [63].

Recent investigations highlight how information is stored in the brain. In fact, most neurons do not ignite in response to incoming signals, although body translates these signals into physical forms very precisely, clarifies partly why the brain is more productive than computers: Majority of its cells do nothing most of the time. Despite, even if the numerous brain cells are passive 80 percent of the time, that doesn't explicate why a computer needs 50 million times more energy to solve the same problems as the human brain. The second question is related to the biochemistry efficiency in comparison with electrons efficiency, because computers use electrons to perform their computations. It is obvious that computers show much greater energy losses during their calculations through various conductors, while biochemical pathways in the brain show perfect efficiency [63, 64].

According to Ray Kurzweil [65] and Paul Zakati Myers [66], the time needed to create an artificial mind ranges from quite optimistic expectations of only 10 years, to more moderate than 20 or 100 years, to completely pessimistic, that this will never happen. Although the processing power of computers has been growing exponentially for years, it is unlikely that such growth will continue in the years to come. Accordingly, there are already plans to develop a supercomputer in the next three years, which will perform at least as many operations as the human brain. With all this in mind, the fact that data is not the same as human thoughts and 10^{16} operations per second does not mean that the computer will reach the human brain, when it comes to its more subtle functions, such as cognition or creativity. So the question remains, will we be able to use this processing power to accurately model a system that will fully mimic the brain and create artificial intelligence based on that model.

Ray Kurzweil says that it will definitely happen, but Myers claims that it will not happen in the near future. Ray Kurzweil's position is interesting, from the aspect how it is possible to build an equivalent structure to the human mind, which would function efficiently in both directions, faithfully simulating or copying human brain processing techniques. Bearing in mind, that essential information related to the functioning of the human brain is contained in the genome, whereby the genome has a total storage capacity of approximately 50 megabytes of memory, with 25 megabytes reserved for brain function. The data that describes our entire behavior includes about a million lines of genetic code, based on which Kurzweil concludes that the brain, in all its complexity, can be described with about a million lines of code. Opposite him, Myers believes that this way of thinking is too simplistic, pointing to the complexity of protein folding, protein-protein and cell-cell interactions, and all other molecular biology systems that are probably necessary for the development of the human brain. Since scientists still do not understand the behavior of all these systems, it is obvious that modeling that could simulate the functioning of the brain requires much more effort to develop adequate computer programs, with sufficient speeds of information processing and, in addition, the way the brain works indisputably reminiscent of the work of a computer [65, 66].

In the recent past, this way of thinking about the brain has been useful, because the brain actually stores and processes information similar to a computer, so that it is possible to draw even some rough parallels between parts of the brain and computer components. However, in an important sense, the brain seems to function quite differently from a computer, since the processing power of a computer is highly centralized in one or at most two processors, so that while the processor does everything related to computing in data form, it stores the hard drive. This means that the data is constantly rotated back and forth from the hard drive to the processor (using RAM as an intermediary, to avoid the hard drive doing so, as such data transfer would be too slow, inefficient and create a bottleneck that would limit maximum speed on which the computer can run) [65–67].

The brain, on the other hand, works quite differently. Because specific brain functions are placed in regions of the cerebral cortex, each area appears to have its own ability to calculate and store the information it needs, at least temporarily. Theoretically, this is much more efficient, because it does not require any movement on the data. How the brain really does this is a mystery, of course [68].

Researchers describe a new type of electronic component that can mimic the dual functions of brain neurons, using a "phase change" of material to enable the processor to perform all four basic arithmetic operations (addition, subtraction, division and multiplication), as well as to store data that precisely define the state of crystallization of the material. This is a phenomenal idea, although it is a serious step towards a computing system similar to the human brain, which is why researchers intend to connect about a hundred such chips soon and try to create neural networks for some simple tasks, such as image recognition. This all makes it clear that future computers will be significantly different from current ones [68].

As recent studies show that the human mind is not a classical computer, it is becoming increasingly apparent that it cannot be fully reduced to any type of computer, due to the non-algorithmic nature of mental processes, because most mental processes are governed by the rules of quantum physics, including and control processes involved in most mental operations, which is why the control process itself is a kind of meta-thinking, whose logic resembles some of the quantum meta-languages, which describes the highest level of mental processing, which includes reasoning, decision making, reminder, etc. [69].

If we adopt this theoretical framework, it follows that if we have at our disposal a new type of brain-computer interface, which enables quantum computer-quantum computer communication, we could use the human mind to control it, through quantum meta-language. The whole system composed of a human subject and an artificial quantum computer, under the control of the quantum meta-language of the subject itself, is a new type of human-computer hybrid, which enables better integration between the human mind and artificial devices. Conceptual problems underlying the design of a command language interpreter require the solution of a number of serious problems, such as knowledge of the essential characteristics of brain signals associated with certain types of intentions or mental states, selecting the best detection techniques in the presence of noise and various artifacts and finding the best way to online implement detection sequences and action performance [65–70].

18.7 Soft Computing Algorithms

So far, all these problems have been solved by relying on experiments with people who direct their thoughts to perform a certain action, and then on the application of soft computing algorithms, which allows artificial neural networks to learn, relying on previous examples. This strategy is based on the hypothesis that mental states are fully characterized by a specific functional pattern of brain neuron activation, meaning that the same mental state is not necessarily associated with the activation of the same brain neurons, because although the same mental state is associated with different patterns neural activations, all of these patterns, are characterized by a kind of invariant "signature," similar to the electromagnetic signals emitted by the brain, which are detected by electroencephalogram (EEG), so that all these details, can be determined through analysis of the appropriate EEG [71].

However, as this analysis is extremely complex, the best strategy is to establish the largest possible number of connections between the EEG signal and the neural network, to ensure that by training, through the learning process based on known examples, the artificial neural network produces output only when the EEG – corresponds to the given intention. Thus, if learning were to be successful, the neural network would automatically implicitly describe the procedure used to analyze the EEG signal, thus deciphering the essence of the "signature" of intent. Without going into the technical details of implementing this strategy, it is obvious that it is increasingly being used to design the actual brain–computer interface [71, 72].

However, this step also contains serious conceptual difficulties, because the performance of the network after the completion of the initial phase of testing depends crucially on the examples used during learning, and learning itself should cover a large number of different possible situations. Unfortunately, the amount of data available is always severely limited for practical reasons, which are related to the way human experiments are performed, and because we do not know, and may never know, how the EEG is structured, since mental states are not defined only expected input-output associations. Moreover, the number of different inputs and outputs associated with a given intent is virtually unlimited, as the number of different possible contexts is unlimited, while the word "context" includes not only the state of the environment, but also the emergence of others, currently present, mental states. Obviously, such a strategy of designing a brain–computer interface is unsuitable for recording the appearance of intentional states in the heads of subjects, which means that the main goal of introducing a brain–computer interface will never be achieved in this way, which is why possible alternatives are necessary [73, 74].

In the history of psychology, the concept of meta-thinking is not very popular, which is why Flavel introduced the analogous term "meta-cognition" in the 1970s, while some authors have recently begun to introduce computer models to simulate the work of the prefrontal cortex. Meta-thinking processes aim to maintain some kind of balance or, even better, coherence. Quantum theory, as a tool with powerful coherence mechanisms, leads to the superposition of various probability distributions correspondent to diverse mental states [75, 76]. It implies that various descriptions of the same mental system can be nonequivalent, because a given mental can go through different types of transformation from a given phase to another phase, which means that this theory can only describe the emergence of meta-thinking. The quantum logic of the mind describes quantum computation in cubits, entities that are in the superposition of two quantum states, conventionally denoted as "0" and "1". Each cubit can be viewed as carrying some kind of implicit double potential, which can lead to the effect of the projection of the operator, which leads to the collapse of the cubit state. Thus the normal functioning of the human mind, in many cases, can be seen as the equivalent of convenient quantum computer manipulation, which leads to the collapse of the cubic state. Thus the normal functioning of the human mind, in many cases, can be seen as the equivalent of convenient quantum computer manipulation [74–77].

18.8 Molecular Machines

Unlike microscopic systems, molecular systems show significant dynamic motions, according to Brown's law. Macroscopically observed, many conventional machines work with often negligible friction, but in molecular systems this friction is incomparable larger. Feynaman's strategy of machine, founded to use Brown's movement is not opposed to it. Similar to macroscopic machines, molecular machines would possess moving parts. But while we are surrounded by microscopic machines in everyday life, an analogous strategy for molecular machines is impossible, because the dynamics of the process on a large and small scale are completely different [78, 79].

Cell membranes are an outstanding example of such molecular machines, in which the lyophilic barrier through numerous transport mechanisms enables the movement of specific bio-proteins and minerals from one part of the cell to another [78, 79].

The approach to the construction of a nanomachine implies the preliminary construction of its elements, which are then assembled into the required machine according to the given request. As the basic parts of every machine are the motor and switches, in molecular machines they would be molecular switches and molecular motors. The switch defines the state of the system (on/off), and the motor affects the trajectory of the system. The switch performs a translational movement, releasing the flow of energy with its movement, while the motor uses that energy to function reproductively [78, 79].

18.8.1 Synthetic Machines

The main constructors of molecular machines would be chemists. The simplest such machines could contain only one molecule. However, the most common constructions would be very precisely defined architectures of elements/molecules that are interconnected according to the principle- "mechanically interlocked molecular architectures." Examples of such structures are rotaxenes and catenates. Molecular motors are consisted from molecules prone to rotation when they are given energy from an external source. Many molecular machines have been designed using the light to react with other molecules. A molecular propeller is a molecule that can push fluid during rotation, thanks to its unique form and design analogous to macroscopic propellers. There are several knives on the molecular scale attached at some angle of inclination around the periphery of the nanodimensional spear. A molecular switch is a molecule reversibly moving between two or more stable states in response to varieties in pH, temperature, electric current, microenvironment, the attendance of ligands [80].

A molecular switch is a molecule capable of connecting molecules or ions between two close locations. Molecular tweezers are molecules able to hold between two shoulders, using non-covalent bonds, hydrogen bonds, or metallic coordinate, hydrophobic forces, Van der Waals forces, π-π interactions, or electrostatic effects. An example of a molecular twister is the construction of DNA, like the DNA of a machine. A molecular sensor is a molecule capable to interact with an analyte to cause an observable variation. They combine a molecular recognition with some form of reporting. A molecular logical threshold is a molecule able to make logical operations with correspondent logical inputs, transforming it in a logical "output" [81–83].

18.8.2 Biological Molecular Machines

The most complex molecular machines, located inside the cells, comprise protein motors, like myosin liable for muscle contact, kinesin, responsible for transport into cell and outside of the cell nucleus along microtubules, and dynein, liable for the axonemal rhythm of cilia and flagella: These proteins and their nanoscale behavior are much more sophisticated than any artificially made molecular machine. The basic mechanism of ciliary motion was described by J. L. Ross in his study [82, 83]. The high level of abstraction within his concluding remarks assumes that cellular nanomachines are composed of more than 600 proteins in a molecular complex, with many parts of such machines being considered independent nanomachines. The construction of more sophisticated molecular machines is today at the center of numerous theoretical researches. Numerous molecules have theoretically already been investigated, although methods for synthesizing such molecules are still in the experimental research phase. Namely, for the practical design of such sophisticated molecular machines, it is necessary to master the constructions of specific machines, which could successfully perform the role of molecular assemblers [84–88].

18.8.3 Nanorobots

It is expected that the introduction of nanorobots in medicine will totally change the world of medicine. Their function would be to repair damaged organs and treat infections when they are introduced into the patient's body. The size these nanorobots should be between 0.5 and 3 μm in size, because the maximum size of blood capillaries has this value. Among the materials that are candidates for the construction of such a robot, carbon materials (diamond/fularene-nanotube composites) are in the first place, due to their strength and other physical characteristics (thermal conductivity-diamonds and electrical conductivity-nanotubes). Nanorobots would be fabricated in a desktop nanofactory specializing in such purposes. Nanorobots would be fabricated in a desktop nanofactory specializing in such purposes [84].

During operation, nanodevices could be controlled by the magnetic resonance method, especially if its components were made of carbon 13C (because 13C has zero nuclear magnetic moment). Such astonishing medical nanodevices can be inserted into a specific organ or tissue. The diagnosis of the disease itself would be much more precise, because such diagnostic nanodevices could correct the target region. The doctor's task would be to scan the desired section of the body and see the nanodevice in the vicinity of the target (tumor) and thus assess with certainty whether the medical therapy for the treatment of the diseased tissue (cancer-affected tissue) was successful. Nanorobot technology is the technology of creating machines or nanorobots on a scale close to the nanometer. More specifically, nanorobots represent a widespread engineering discipline in the design and fabrication of nanorobots, devices whose dimensions are of the order of 0.1–10 μm in size designed of nanoscale using molecular constituents. There is still no artificial biological robot, so for now it is just a theoretical idea. The names; nanorobots, nanoids, nanites, or nanomites are also applied to delineate these hypothetical nanodevices [90–92].

Other definitions are sometimes used for nanorobots that permit precise interplays between nanoscale devices or manipulation with them, using nanoscale resolution. Due to a such approach, larger devices such as AFM can be taken for nanorobotic instruments, because they enable nanomanipulation. Therefore, currently, nanodevices are one of the most important topics. Certain such crude molecular machines have already been constructed. One of them is a sensor that has a "switch" cross section of 1.5 nm, which can count particular molecules in a chemical specimen. The first nanomachine ever built in medical technology was a machine for identifying and killing cancer. Another potential application of nanomachines would be related to working with toxic chemicals and measuring their concentrations in the atmosphere and water. At Rice University, monomolecular circuits were demonstrated that were developed by a chemical process involving fullerenes as wheels. They are put into operation by controlling the ambient temperature and are positioned within a "scanning" tunnel microscope [93].

Potential applications of nanorobots in medicine include diagnostics, targeted delivery of cancer drugs, biomedical instruments, surgery, pharmacokinetics of diabetes monitoring, and health care. In future plans, medical nanotechnologies will use robots injected into patients to perform treatment at the cellular level. Such robots do not replicate because replication leads to device complexity and reduces availability. Medical nanorobots would be made in hypothetical nanofactories, in which nanoscale machines could be integrated into a putative "desktop" scaled machine to create a macroscopic product. A more detailed theoretical discussion of nanorobots, including specific design, communication, navigation, manipulation, locomotion, and computerization, in the medical context, was first presented by Robert Freitas [94, 95].

In the near or far future, the construction of micro and nanoscale robots capable to assemble other nanomachines and move into the body to deliver drugs or robots that would perform microsurgery functions is expected [94, 95].

18.8.4 Cell Repair Machines

The task of doctors in nanomedicine is to use drugs and nanosurgery to enable tissues to "encourage" them to reconstruct themselves. Due to molecular machines, such reconstructions would be successful. Cellular reconstructions will be based on some already proven principles, which are fully proven in living systems. Admission to cells is already attainable now, because citologists today can inject needles into cells without destroying them. Hence, the first task of molecular machines would be to enter the cell. Based on already known facts in the field of biochemistry and microbiology, related to the problems of biochemical interactions within cells, it is to be expected that such molecular systems-molecular machines will be able to recognize each other by touch, creating the conditions for proper diagnosis and "thought" plan building damaged sequences of each molecule, could completely reassemble damaged molecules. Eventually, the cells themselves could replicate such a molecular system, so that reproductive self-assembly of any system is possible, which in this way interacts with the cell. Following the basic principles of nature, related to cell repair at the molecular level, soon, nanoscale-based devices will be designed to be able to move into the cell, sensing imperfections inside of healthy cells and modifying their structure on the desired way [96].

The potential of cell repair machines constructed in this way will obviously be very impressive. Compared to the dimension of a virus or bacterium, their close packed parts make them to be very sophisticated. In the first generation, such machines will be highly personalized. When they open and close the cell membrane and pass across tissues and penetrate cells and viruses, nanomachines will be able to repair some personal molecular defect, like DNA impairment or enzyme lack. After this first step in the development of such machines, advanced cellular machines will be programmed to possess, in addition to all the numerous possibilities already described, the potential to substantially improve the human autoimmune system. To run such complex machines will require the prior development of very powerful nanocomputers, whose function will be to examine and register the site of damage, access it and rebuild in a regular way the damaged molecular structure, thus translating them from a state of disease to a state of health. After repairing the individual cells, they will, in interconnected communication and division of labor, move on to organ repair, organ by organ, restoring all body. Cells hurt to the level of complete inactivity will be make from the nutrients available to them (e.g., from starch). Therefore, the development of such advanced nanomachines for cell repair will liberate medicine from support on the self-repair, which is immanent to the process of self-healing of the organism [97, 98]

18.8.5 Neuro-electronic Interfaces

The construction of neuro-electronic interfaces is a fantastic intention in the design of nanomachines that will allow computers to connect and link to the person's nervous system. Such a construction needs the construction of molecular elements that enable control and registration of a nerve impulse by an outside computer. Accordingly, computers will be capable to annotate record and answer to the body signal, arising in response to a given sensation. Real constructions of these types of structures would be of great importance, because they could solve many severe neurosomatic diseases that result in complete breakdown of the nervous system (ALS and multiple sclerosis). Also, in this way, numerous injuries could be repaired, in which the nervous system was damaged, which results in its dysfunction and paraplegia [99, 100]

If computers could control a patient's nervous system, then the effects of illness and serious injury could be overcome. There are two strategies for approaching such problems, under symbolic names: the "refueling" strategy and the "refueling" strategy. In a "refueling" strategy, energy in the system is replenished constantly or periodically from an external sound, chemical, magnetic or electrical source. In the "discharge" strategy, all the power from the system is extracted from the internal energy storage, until all the energy is depleted. The only limitation of this innovation is the fact that in the process of transmitting information from the computer to the patient's nervous system, electrical interference, electric fields caused by electromagnetic pulses-EPM, and stray electric fields are quite possible from other surrounding electrical devices/sources. Also, insulation problems are very serious in order to avoid leakage currents, and to provide high electrical conductivity of electrical pulses in "in vivo" media, which lead to pronounced risks of sudden losses of pulse power and popping. Finally, thicker wires would be needed to conduct a substantial power level without overheating [101].

The current state-of-the-art is still insufficiently developed to provide such signal transmission structures. In addition, wire netting of such structures is extremely difficult and its precise positioning in the patient's nervous system, so that the computer monitor is able to record and answer to the patient's nerve signal. Finally, an additional requirement is that the interface structures must be in harmony with the person's immune system to be able to remain inside the patient's body long enough without rejection. In addition, such structures must be very sensitive to changes in the intensity and direction of ionic currents emitted by the patient's nervous system. Based on all the above, it follows it follows that although the potential of these structures is incredible, it is not yet possible to predict from afar when such structures will actually be technically and realized [102, 103].

18.8.6 Quantum Robot

The concept of the quantum robot, first proposed by Paul Benioff, was conceived in the form of a mobile system containing a quantum computer, and all the necessary auxiliary systems, whereby the quantum robot interacts with the surrounding of the quantum system. Although original quantum robot is not aware of its environment and is unable to make decisions, it is an extremely intriguing question whether quantum robots will ever be aware of their environment in the near future so that they can perform experiments on their own without human participation. If that happened, it would mean that robots could even become aware and have "free will". Such a quantum robot, obviously, should be equipped with some kind of "internal observer", which is able to control itself by internal, quantum calculations. Such an observer would operate using a quantum meta-language adapted to control quantum language, expressed in the form of cubit manipulations. It is natural to assume that the core of quantum meta-language consists of internal measurements of operators, such as those used in quantum mechanics. Unfortunately, such a language is not suitable for controlling a quantum robot, which means that traditional "quantum logic" would not correctly describe the internal measurements necessary for a quantum robot [97].

As a possible way out of this problem, Aharonov introduced the term so-called "weak measurements", which are based on the measurement of various physical quantities, which describe the quantum system in interaction with the dissipative environment. Accordingly, the physical processes occurring in the brain, at least in terms of quantum aspects expressed through dissipative quantum field theory, in the absence of brain dynamics would be characterized not only by six unique ground states but also by many different ground states adapted to many different ones memory states, which inevitably occur in the brain, which solves the problem of decoherence in quantum robots [105–108].

This would open the way for the implementation of a quantum computer in which the human subject, through quantum meta-language, controls the quantum computer, using a brain–computer interface, much more powerful than before, thus transforming human intentions into actions in a more efficient way. This type of quantum computer paves the way for a deeper connection between humans and computers, showing that the quantum approach allows the mind to act directly on matter, which opens up the possibility of projecting a new type of brain computer interface, causing a revolutionary change in our real way of thinking, on the base of the tacit assumptions that our thoughts have direct effects on the world around us [105–108].

18.9 Future Visions of the Brain Computer Interface

So far, most research on the connection between the brain and computers has focused on restoring lost bodily functions in a surviving neurological disease or injury. Accordingly, cochlear implants have been designed to enable people with hearing loss to hear, and retinal implants, which enable those who are blind to see, while neuroprostheses decrypt the movement desires given by brain instructions and are used to control computer cursors or robotic arms of people with paralysis [109].

There is no longer any doubt that we are on the verge of achieving the goal of connecting the brain with artificial intelligence, which will lead to incredible progress, which is not even imagined in the hottest imagination. That what will happen is immeasurably bigger than anything imaginable, because if Neuralink achieves its ultimate intentions, it will bring into question what man really is. Of course, a lot of effort is needed, but the fact that some things cannot be done with today's technology does not mean that it is impossible, although the human brain is a much harder nut to crack than anything known so far, because we still do not know how the brain works. However, it is certain that such progress of the brain–computer interface will soon occur, which will make this issue pointless [110].

The power of modern computers is growing faster, with a better understanding of our brains, turning science fiction into reality. It is now possible to transmit a signal directly from a computer to the human brain, which allows him to see, hear or feel, a specific change caused by a change in the input sensor signal. What is the reality that arises from our desire to give divine or demonic power to machines to manipulate our thoughts, which is extremely important for people with severe disabilities? That is why the development of the brain–computer interface has become one of the most important technological advances in the entire history of the human race [5, 56].

18.10 Recent Application of BCI Technology

The current BCI can be categorized into two groups from the aspect of features of the signals use as input. Some are dependent on user manage of endogenous electrophysiological activity, like amplitude in a certain frequency range in the EEG registered via the sensorimotor cortex, while others depend on activity induced by particular stimuli or better ability to manage with given model because a sufficiently well-trained user has excellent control over the environment. Such BCI requires very intensive training. Exogenous BCI does not reqwest intensive training but it frequently requires a more or less ordered surraunding (stereotypical visual input). The BCI appliances have possible application in the verbal communication, during daily activities, environmental protection, movement and training. The BCI uses can be adapted for each individual customer or group of customers, in a standardized and objective way. Through conduct analysis that addresses the requirements, wishes, and basic motives of the consumer and his caregivers, BCI progress includes not only techno;ogical and electrophysiological rules, but also well-established learning rules. The current BCI records electrophysiological signals applaying noninvasive or invasive methods. Noninvasive BCI uses scalp imaging of EEG impulses or enveloped potentials, while invasive BCI recorded personal activities in the cortex or an EEG that shows a subdural imaging. They show an information transfer rate of 5–25 bpm and can be applied to synhronize the motion of the cursor or for selection certain letters or icons [15, 111, 112].

Like similar information systems, inputs, outputs, and translation algorithms that convert previous to subsequent BCI signals are assumed to depend on the intercommunication of two adjustive controllers, the consumer brain generating the input, through its electrophysiological action determined by the BCI and the system transforming those activities. in output in the form of specific commands that act on a word that comes from outside. Succesful BCI functioning requests that the consumer develop and mentain his creativity, not only for muscle control but also control of EEG. BCI inputs involve slow cortical potentials, P300 evoked potentials, μ and β rhythms from the sensorimotor cortex, and the activity of a specific unit originating from the motor cortex. The recording methodology comes down to maximizing the signal-to-noise ratio. Noise comprises from ECG, and other activities induced from ecternal sources and brain activities other than specific rhythms or evoked potentials contained in the BCI input. Different of time and space filters decrease noise, amplifaying the signal-to-noise ratio. [5, 8, 111, 112].

18.11 Conclusion

In this chapter, all aspects of BCI technology improvement are analyzed. Over the years, technologists and scientists have tried to make this system faster, more accurate, and more useful. It has been shown that such a BCI system based on the P-300 requires only light training in order to achieve satisfactory precision in the execution of commands, which the brain communicates with its characteristic EEG signals to one of such devices. Users could use SSVEP with coded frequencies for this purpose with little training. This method has been shown to enhance the quality of life of users suffering from stroke or some kind of neurological deseases. The chapter particularly emphasizes the progress in repairing various neurological injuries, as well as the physiology of neurological rehabilitation, neurophysiology of the brain machine, training for the neural networks, the cutting edge in the interfaces of neural and brain machines, neurorehabilitation strategies, and future expectations of progress in this area. Deep neural network architecture and deep learning algorithms enable a new quality of information processing and tool classification in the context of better EEG signal definition. Movement control with noninvasive BCI enables the improvement of its speed and accuracy by expanding the adaptive algorithm. This control was also analyzed from the aspect of potential inclusion of suplementary EEG registering sites, frequency range, and EEG characteristics of the time domain in cursor motion.

REFERENCES

1. Jokanović V, (2020), A man of the fifth dimension, Prometej, Novi Sad, (accepted).
2. Krucoff MO, Rahimpour S, Slutzky MW, Edgerton VR, Turner DA, (2016), Enhancing nervous system recovery through neurobiologics, neural interface training, and neurorehabilitation. *Front Neurosci. 10*: 584. doi:10.3389/fnins.2016.00584
3. Peterka DS, Takahashi H, Yuste R. (2011), Imaging voltage in neurons. *Neuron. 69* (1): 9–21.
4. Mazzatenta A, Giugliano M, Campidelli S, et al. (2007), Interfacing neurons with carbon nanotubes: electrical signal transfer and synaptic stimulation in cultured brain circuits. *J Neurosci. 27*, (26): 6931–6936.
5. Shih JJ, Krusienski DJ, Wolpaw JR, (2012), Brain-computer interfaces in medicine. *Mayo Clin Proc. 87*, (3): 268–279.
6. Pisarchik AN, Maksimenko VA, Hramov AE, (2019), From novel technology to novel applications: Comment on "An Integrated Brain-Machine Interface Platform With Thousands of Channels" by Elon Musk and Neuralink. *J Med Internet Res. 21*, (10): e16356, doi:10.2196/16356
7. Nicolas-Alonso LF, Gomez-Gil J, (2012), Brain computer interfaces, a review, *Sensors (Basel). 12*, (2): 1211–1279.
8. Mak JN, Wolpaw JR. (2009) Clinical applications of brain-computer interfaces: Current state and future prospects. *IEEE Rev Biomed Eng. 2*: 187–199.
9. Tariq M, Trivailo PM, Simić M, (2018), EEG-Based BCI control schemes for lower-limb assistive-robots. *Front Hum Neurosci. 12*: 312, doi:10.3389/fnhum.2018.00312
10. Lazarou I, Nikolopoulos S, Petrantonakis PC, Kompatsiaris I, Tsolaki M, (2018), EEG-based brain-computer interfaces for communication and rehabilitation of people with motor impairment: A novel approach of the 21st century. *Front Hum Neurosci. 12*: 14. doi:10.3389/fnhum.2018.00014
11. Yuan H, Liu T, Szarkowski R, Rios C, Ashe J, He B, (2009), Negative covariation between task-related responses in alpha/beta-band activity and BOLD in human sensorimotor cortex: an EEG and fMRI study of motor imagery and movements. *Neuroimage, 49* (3): 2596–2606.
12. Baxter BS, Edelman BJ, Sohrabpour A, He B, (2017), Anodal transcranial direct current stimulation increases bilateral directed brain connectivity during motor-imagery based brain-computer interface control. *Front Neurosci.* https://doi.org/10.3389/fnins.2017.00691
13. Wolpaw JR, McFarland DJ, Vaughan TM, Schalk G, (2003), The Wadsworth Center Brain–Computer Interface (BCI) research and development program, *IEEE Trans Neural Syst Rehab Eng, 11*, (2): 204–207.
14. Wolpaw JR, McFarland DJ, (2004), Control of a two-dimensional movement signal by a noninvasive brain–computer interface in humans. *PNAS. 101*, (51): 17849–17854.
15. Yuan H, He B. (2014), Brain-computer interfaces using sensorimotor rhythms: current state and future perspectives. *IEEE Trans Biomed Eng. 61*, (5): 1425–1435.
16. Woodman GF, (2010), A brief introduction to the use of event-related potentials in studies of perception and attention. *Atten Percept Psychophys. 72*, (8): 2031–2046.
17. Sur S, Sinha VK, (2009), Event-related potential: An overview. *Ind Psychiatry J. 18* (1): 70–73.
18. Rezeika A, Benda M, Stawicki P, Gembler F, Saboor A, Volosyak I, (2018), Brain-computer interface spellers: A review. *Brain Sci. 8*, (4): 57. doi:10.3390/brainsci8040057
19. Nicolas-Alonso LF, Gomez-Gil J, (2012), Brain computer interfaces, a review. *Sensors 12*, (2): 1211–1279.
20. Luu TP, He Y, Brown S, Sho N, Contreras-Vidal JL, (2015), A closed-loop brain computer interface to a virtual reality avatar: Gait adaptation to visual kinematic perturbations. *Int Conf Virtual Rehabil. 2015*: 30–37.
21. McFarland DJ, Daly J, Boulay C, Parvaz M, (2017), Therapeutic applications of BCI technologies. *Brain Comput Interfaces (Abingdon). 47*, (1–2): 37–52.

22. Al-Taleb MKH, Purcell M, Fraser M, Petric-Gray N, Vuckovic A, (2019), Home used, patient self-managed, brain-computer interface for the management of central neuropathic pain post spinal cord injury: usability study. *J Neuroeng Rehabil. 16*, (1): 128. doi:10.1186/s12984-019-0588-7

23. Rashid M, Sulaiman N, Abdul Majeed A, et al. (2020), Current status, challenges, and possible solutions of EEG-based brain-computer interface: A comprehensive review. *Front Neurorobot. 14*: 25. doi:10.3389/fnbot.2020.00025.

24. Hoffmann U, Vesin J-M, Touradj E, Diserens K, (2008), An efficient P300-based brain–computer interface for disabled subjects. *J Neurosci Methods 167*: 115–125.

25. Schalk G, McFarland D, Hinterberger T, Birbaumer NR, Wolpaw J, (2004), BCI2000: a general-purpose Brain-Computer Interface (BCI) system, *IEEE Trans Biomed Eng. 51*, (6): 1034–1043.

26. Milsap G, Collard M, Coogan C, Crone NE, (2019), BCI2000Web and WebFM: Browser-Based Tools for Brain Computer Interfaces and Functional Brain Mapping. *Front Neurosci.* https://doi.org/10.3389/fnins.2018.01030

27. Jonathan R, Wolpaw J, Birbaumerc N, McFarland DJ, Pfurtschellere G, Vaughan TM, (2002), Brain–computer interfaces for communication and control. *Clin Neurophysiol. 113*: 767–791.

28. Padfield N, Zabalza J, Zhao H, Masero V, Ren J, (2019), EEG-based brain-computer interfaces using motor-imagery: Techniques and challenges. *Sensors (Basel). 19*, (6): 1423. doi:10.3390/s19061423

29. Schalk G, Leuthardt EC, (2011), Brain-computer interfaces using electrocorticographic signals. *IEEE Rev Biomed Eng. 4*: 140–154.

30. Al-Ezzi A, Kamel N, Ibrahima F, Gunaseli E, (2020), Review of EEG, ERP, and Brain Connectivity Estimators as Predictive Biomarkers of Social Anxiety Disorder. *Front Psychol. 11*: 730. doi:10.3389/fpsyg.2020.00730

31. Ramos-Murguialday A, Birbaumer N, (2015), Brain oscillatory signatures of motor tasks. *J Neurophysiol. 113*, (10): 3663–3682.

32. Nanopouls DV, (1995), Theory of Brain Function, Quantum Mechanics and Superstrings, Act-08/95 Cern-Th/95-128 Ctp-Tamu-22/95 hep-ph/9505374.

33. Bakas I, Kiritsis E, (1992), Beyond the Large N Limit: Non-linear W(infinity) as symmetry of the SL(2,R)/U(1) coset model. *Int J Mod Phys. A7*, (Suppl. 1A): 55.

34. Zamolodchikov AB, (1986), Ireversibility of flux of the renormalization group in 2D field theory. *JETP Lett. 43*, (1986), 730; *Sov. J Nucl Phys.* 46 (1987), 1090.

35. Clements EM, Das R, Li L, et al. (2017), Critical behavior and macroscopic phase diagram of the mono-axial chiral helimagnet $Cr_{1/3}NbS_2$. *Sci Rep. 7*, (1): 6545. doi:10.1038/s41598-017-06728-5

36. Ellis J, Mohanty S, Nanopoulos DV, (1989), Quantum gravity and collapse of the wave function. *Phys Lett B. 221*, (2): 113–119.

37. Feynman RP, Vernon FL Jr. (2000), The theory of a general quantum system interacting with linear dissipative. *Ann Phys (NY). 281*, (2): 547–607.

38. Goodstein D, Goodstein J, (2000), Richard Feynman and the history of superconductivity. *Phys Perspect. 2*: 30–47.

39. Hawking S, (1976), Breakdown of predictability in gravitational collapse. *Phys Rev. D14*: 2460–2473.

40. Louis J, Mohaupt T, Theisen S, (2007), String theory: An overview. *Lect Notes Phys. 721*: 289–323.

41. Kelso JA, (2012), Multistability and metastability: Understanding dynamic coordination in the brain. *Philos Trans R Soc Lond B Biol Sci. 367*, (1591): 906–918.

42. Ito D, Tamate H, Nagayama M, Uchida T, Kudoh SN, Gohara K, (2010), Minimum neuron density for synchronized bursts in a rat cortical culture on multi-electrode arrays. *Neuroscience. 171*, (1): 50–61.

43. Korn H, Faure P, (2003), Is there chaos in the brain? II. Experimental evidence and related models. *Comptes Rendus Biologies. 326*, (9): 787–840.

44. Jokanović V, (2013), How our cells lives and died, Institute of Nuclear Sciences "Vinča", Belgrade.

45. Lotte F, Bougrain L, Clerc M, (2015), Electroencephalography (EEG)-based Brain Computer Interfaces. Wiley Encyclopedia of Electrical and Electronics Engineering, Wiley, pp. 44, ff10.1002/047134608X.W8278ff. ffhal-01167515f

46. Borikar SS, Kochre SR, Zade YD, (2014), Brain Computer Interface, National Conference on Engineering Trends in Medical Science – NCETMS – 2014.

47. Byrne R, (1996), Relating brain size to intelligence in primates, in P. Mellars and K. Gibson Eds., *Modelling the Early Human Mind*, McDonald Institute Monographs, Cambridge, pp. 49–56.

48. Al-Quraishi MS, Elamvazuthi I, Daud SA, Parasuraman S, Borboni A, (2018), EEG-based control for upper and lower limb exoskeletons and prostheses: A systematic review. *Sensors (Basel)*. *18*, (10): 3342. doi:10.3390/s18103342

49. Ackerman S, (1992), *Discovering the Brain*, The Institute of Medicine, National Academy of Sciences, Washington, DC.

50. Buzsáki G, (2006), *Rhythms of the Brain*, Oxford University Press.

51. Tierney AL, Nelson CA, (2009), Brain development and the role of experience in the early years. *Zero Three*. *30*, (2): 9–13.

52. Skarda CA, Freeman WJ, (1987), How brains make chaos in order to make sense of the world. *Behav Brain Sci*. *10*: 161–195.

53. Kriegeskorte N, Douglas PK, (2018), Cognitive computational neuroscience. *Nat Neurosci*. *21*, (9): 1148–1160.

54. Gros C, (2012), Complex Adaptive Dynamical Systems, a Primer, Institute for Theoretical Physics Goethe University Frankfurt.

55. Schweizer S, Satpute AB, Atzil S, et al. (2019), The impact of affective information on working memory: A pair of meta-analytic reviews of behavioral and neuroimaging evidence. *Psychol Bull*. *145*, (6): 566–609.

56. Signorelli CM, (2018), Can computers become conscious and overcome humans? *Front Robot AI 5*: 121. doi:10.3389/frobt.2018.00121

57. Zaytseva Y, Fajnerová I, Dvořáček B, et al. (2018), Theoretical modeling of cognitive dysfunction in schizophrenia by means of errors and corresponding brain networks. *Front Psychol*. *9*: 1027. doi:10.3389/fpsyg.2018.01027

58. Markram H, Gerstner W, Sjöström PJ, (2011), A history of spike-timing-dependent plasticity. *Front Synaptic Neurosci*. *3*: 4. doi:10.3389/fnsyn.2011.00004

59. David JC, (2003), *Building Better Health: A Handbook for Behavioral Change,* PAHO, Washington, DC.

60. Wolfe JM, (1998), Visual memory: What do you know about what you saw? *Curr Biol*. *8*, (9): 303–304.

61. Lovinger DM, (2008), Communication networks in the brain: Neurons, receptors, neurotransmitters, and alcohol. *Alcohol Res Health*. *31*, (3): 196–214.

62. Sejnowski TJ, (2018), *The Deep Learning Revolution: Artificial Intelligence Meets Human Intelligence*, MIT Press, Cambridge, MA.

63. Kuhl PK, (2010), Brain mechanisms in early language acquisition. *Neuron*. *67*, (5): 713–727.

64. Mayford M, Siegelbaum SA, Kandel ER, (2012), Synapses and memory storage. *Cold Spring Harb Perspect Biol*. *4*, (6): a005751. doi:10.1101/cshperspect.a005751

65. Kennedy P, (2014), Brain-machine interfaces as a challenge to the "moment of singularity". *Front Syst Neurosci*. *8*: 213. doi:10.3389/fnsys.2014.00213

66. Haushalter JL, (2018), Neuronal testimonial: Brain-computer interfaces and the law. *V and L Rev*. *71*, (4): 1365–1400.

67. Crick FHC, (1979), Thinking about the brain. *Sci Am*. *241*, (3): 219–233.

68. Koziol LF, Budding D, Andreasen N, et al. (2014), Consensus paper: The cerebellum's role in movement and cognition. *Cerebellum*. *13*, (1): 151–177.

69. De Sousa A, (2013), Towards an integrative theory of consciousness: part 2 (an anthology of various other models). *Mens Sana Monogr*. *11*, (1): 151–209.

70. Duncan R, (2006), *Types for Quantum Computing*, Merton College, Oxford, PhD Thesis.

71. Schmidt J, Marques MRG, Botti S, Marques MAL, (2019), Recent advances and applications of machine learning in solidstate materials science. *NPJ Comput Mater*. *5*, 83–118

72. Schirrmeister RT, Springenberg JT, Fiederer LDJ, et al. (2017), Deep learning with convolutional neural networks for EEG decoding and visualization. *Hum Brain Mapp*. *38*, (11): 5391–5420.

73. Roy Y, Banville H, Albuquerque I, Gramfort A, Falk TH, Faubert J, (2019), Deep learning-based electroencephalography analysis: A systematic review, *J Neural Eng*. *16*, (5), 051001–051037.

74. Montavon G, Samek W, Müller K-R, (2018), Methods for interpreting and understanding deep neural networks. *Digit Signal Process*. *73*: 1–15.

75. Lewis MD, (2005), Bridging emotion theory and neurobiology through dynamic systems modeling. *Behav Brain Sci*. *28*: 169–245.

76. Eichbaum QG, (2014), Thinking about thinking and emotion: The metacognitive approach to the medical humanities that integrates the humanities with the basic and clinical sciences. *Perm J*. *18*, (4): 64–75.

77. Plotnitsky A, (2007), Prediction, repetition, and erasure in quantum physics: Experiment, theory, episte-mology. *J Mod Opt. 54*, (16–17): 37th Winter Colloquium on the Physics of Quantum Electronics.

78. Erbas-Cakmak S, Leigh DA, McTernan CT, Nussbaumer AL, (2015), Artificial Molecular Machines. *Chem Rev. 115*, (18): 10081–10206.

79. Hänggi P, Marcheson F, (2009), Artificial Brownian motors: Controlling transport on the nanoscale. *Rev Mod Phys. 81*: 387–442.

80. Baroncini M, Casimiro L, de Vet C, Groppi J, Silvi S, Credi A, (2018), Making and operating molecular machines: A multidisciplinary challenge. *Chem Open. 7*: 169–179.

81. Ramezani H, Dietz H, (2020), Building machines with DNA molecules. *Nat Rev Genet. 21*, (1): 5–26.

82. Roberts AJ, Kon T, Knight PJ, Sutoh K, Burgess SA, (2013), Functions and mechanics of dynein motor proteins. *Nat Rev Mol Cell Biol. 14*, (11): 713–726.

83. Ross JL, Shuman H, Holzbaur ELF, Goldman YE, (2008), Kinesin and dynein-dynactin at intersecting microtubules: Motor density affects dynein function. *Biophys J. 94*, (8): 3115–3125.

84. Klarner FG, Kahlert B, (2003), Molecular tweezers and clips as synthetic receptors. Molecular recogni-tion and dynamics in receptor-substrate complexes. *Acc Chem Res. 36*: 919–932.

85. Sumerin V, Schulz F, Atsumi M, Wang C, Nieger M, Leskelä M, Repo T, Pyykkö P, Rieger B, (2008), Molecular tweezers for hydrogen: Synthesis, characterization, and reactivity. *Am Chem Soc. 130*, (43): 14117–14119.86.

86. Jokanović V, (2012), *Nanomedicine, the Greatest Challenge of 21st Century*, Monograph, Data Status, Belgrade, Serbia.

87. Saadeh Y, Vyas D, (2014), Nanorobotic applications in medicine: Current proposals and designs. *Am J Robot Surg. 1*, (1): 4–11.

88. Zia K, Siddiqui T, Ali S, Farooq I, Zafar MS, Khurshid Z, (2019), Nuclear magnetic resonance spectros-copy for medical and dental applications: A comprehensive review. *Eur J Dent. 13*, (1): 124–128.

89. Saxena S, Pramod BJ, Dayananda BC, Nagaraju K, (2015), Design, architecture and application of nanorobotics in oncology. *Indian J Cancer. 52*: 236–241.

90. Patra JK, Das G, Fraceto LF, et al. (2018), Nano based drug delivery systems: recent developments and future prospects. *J Nanobiotechnol. 16*, (1): 71–104.

91. Li J, Esteban-Fernández de Ávila B, Gao W, Zhang L, Wang J, (2017), Micro/nanorobots for biomedi-cine: Delivery, surgery, sensing, and detoxification. *Sci Robot. 2*(4): eaam6431. doi:10.1126/scirobotics. aam6431

92. Martel S, Mohammadi M, Felfoul O, Lu Z, Pouponneau P, (2009), Flagellated magnetotactic bacteria as controlled MRI-trackable propulsion and steering systems for medical nanorobots operating in the human microvasculature. *Int J Rob Res. 28*, (4): 571–582.

93. Martins NRB, Angelica A, Chakravarthy K, Svidinenko Y, Boehm FJ, Opris I, Lebedev MA, Swan M, Garan SA, Rosenfeld JV, Tad H, Freitas RA Jr., (2019), Human brain/cloud interface. *Front Neurosci*, 112–124.

94. Freitas RA Jr., (2010), Comprehensive nanorobotic control of human morbidity and aging, in G.M. Fahy et al. Eds., *The Future of Aging*, 685 C, Springer Science Business Media B.V. doi: 10.1007/978-90-481-3999-6_23

95. Alberts B, Johnson A, Lewis J, et al. (2002), *Molecular Biology of the Cell*, 4th edition, Garland Science; Protein Function, New York. https://www.ncbi.nlm.nih.gov/books/NBK26911/

96. Tang SKY, Marshall WF, (2017), Self-repairing cells: How single cells heal membrane ruptures and restore lost structures. *Science. 356*, (6342): 1022–1025.

97. Silva GA, (2018), A New Frontier: The Convergence of Nanotechnology, Brain Machine Interfaces, and Artificial Intelligence. *Front Neurosci. 12*: 843. doi:10.3389/fnins.2018.00843

98. Pancrazio JJ, (2008), Neural interfaces at the nanoscale. *Nanomedicine (Lond). 3*, (6): 823–830.

99. Liu G, Zhao P, Qin Y, Zhao M, Yang Z, Chen H, (2020), Electromagnetic Immunity Performance of Intelligent Electronic Equipment in Smart Substation's Electromagnetic Environment. *Energies. 13*, (5): 1130–1149.

100. Lenarz T, (2018), Cochlear implant – state of the art. GMS Curr Top Otorhinolaryngol Head Neck Surg. 16: Doc 04. doi:10.3205/cto000143

101. Prochazka A, (2017), Neurophysiology and neural engineering: a review. *J Neurophysiol. 118*, (2): 1292–1309.

102. Benioff P, (1999), Quantum robots and quantum computers, in A.J.G. Hey, Ed. *Feynman and Computation*, Perseus Books, 155–176.
103. Bechtel W, (2008), *Mental Mechanisms: Philosophical Perspectives on Cognitive Neuroscience*, Erlbaum, New York.
104. Chiara DML, Giuntini R, Leporini R, Toraldo di Francia G, (2008), Quantum Computational Logics and Possible Applications. *Int J Theoret Phys. 47*: 44–60.
105. Lieberoth A, Pedersen MK, Marin AC, Planke T, Sherson JF, (2014), Getting Humans to do Quantum Optimization – User Acquisition, Engagement and Early Results from the Citizen Cyberscience Game Quantum Moves. *Human Comput. 1*, (2): 221–246.
106. Cavalcanti A, (2003), Assembly Automation with Evolutionary Nanorobots and Sensor-Based Control Applied to Nanomedicine. *IEEE Trans Nanotech. 2*, (2): 82–87.
107. Betthauser JL, Thakor NV, (2019), Neural Prostheses, in J. Webster Ed. *Wiley Encyclopedia of Electrical and Electronics Engineering*, John Wiley & Sons. doi:10.1002/047134608X.W1424.pub2
108. Yarovyy A, (2010), *Applied Realization of Neural Network and Neurolike Parallel-Hierarchical System Based on GPGPU, 10th International Conference on Development and Application Systems*, Suceava, Romania.
109. Wolpaw JR, (Guest Editor), Birbaumer N, Heetderks WJ, McFarland DJ, Peckham PH, Schalk G, Donchin E, Quatrano LA, Robinson CJ, Vaughan TM, (2000), Brain–Computer Interface Technology: A Review of the First International Meeting, *IEEE Trans Rehabil Eng. 8*, (2): 164–173.
110. Jokanović V, (2020), Smart healthcare in smart cities, in L. Sharma Ed. *Towards Smart World: Homes to Cities using Internet of Things*, 1st edition, Chapter 4, Taylor & Francis Group.

19

Artificial Intelligence: Challenges and Future Applications

Pradeep Kumar Garg
Indian Institute of Technology Roorkee, India

Lavanya Sharma
Amity Institute of Technology, Amity University, Noida, India

CONTENTS

19.1 Introduction

The AI has become a main driving force for industrial transformation around the world. The AI technology is driving and shaping the future of tomorrow, and is impacting our lifestyle. With the developments of AI, it is believed that many tasks can be efficiently and intelligently managed by AI. It includes the concept and development of computing systems capable of performing the given tasks, e.g., decision-making, speech recognition, visual perception, and translation of languages using the algorithms Grace and Mantha (2019). These algorithms may be used for data processing, computation, automated reasoning, and prediction as well as reducing the wastages to streamline supplies and maintaining the optimum

DOI: 10.1201/9781003140351-19

inventory. The AI-based predictive analytics may be used in industry to improve utilization of various plants by anticipating their demands and taking suitable steps to match the production with the anticipated demands (Markets & Markets, 2018). The AI-based inventory management ultimately may help in establishing a new pricing strategy for the manufactured products.

The artificial neural network (ANN) allows modeling of nonlinear approaches, and it has become an essential tool for solving many complex problems. Developments in ANN have further enhanced the use of AI in aerospace, travel, health care, manufacturing, and automotive (Grand View Research, Inc, 2020b). Further, the research in computer vision with AI has given a new dimension in digital image processing for applications in security and surveillance, transportation, etc. The ANNs with machine learning (ML) systems are evolving much accurate computer vision and digital image analysis techniques. For example, images and videos taken in low light or poor resolution can be transformed into high-definition quality images by employing these techniques. The ML has a great scope in companies to improve users' experience (Shankar, 2020). Such emerging ML methods are expected to bring changes the way AIs are trained and deployed for various applications.

The AI applications include Google's search algorithms, IBM's Watson, self-driving cars, and robots Brighterion Inc. (2019). Current AI is also known as narrow AI (or weak AI), which is primarily designed to perform a narrow task (e.g., only facial recognition task or only driving a car task). However, the long-term goal of development is to create general AI (or strong AI). Narrow AI may be superior to humans in performing its specific task (e.g., playing chess or solving complex equations. For example, Google is using it to diagnose heart disease more quickly and accurately, and American Express is deploying AI-based robots to serve its customers online.), whereas strong AI potentially can be better than human beings at nearly every cognitive task Adixon (2019), Mike Thomas (2020), Mordor Intelligence (2019).

Ransbotham et al. (2017), based on a survey of more than 3,000 executives, managers, and analysts in 112 countries in 2017, observed that 75% of executives believe that AI will help their companies move into new business areas, while an even larger majority (84%) says that AI will better serve their companies to obtain or sustain a competitive advantage (Figure 19.1). Despite this, only 20% of companies are using AI or selling products that incorporate AI, and just 5% are making extensive use of the AI.

AI cannot replace all the given tasks. It also has certain limitations. It remains just a tool that strengthens and boosts performance and efficiency of humans and improves productivity of various tasks. Its potential also reaches the possibility of creating new job profiles that don't currently exist.

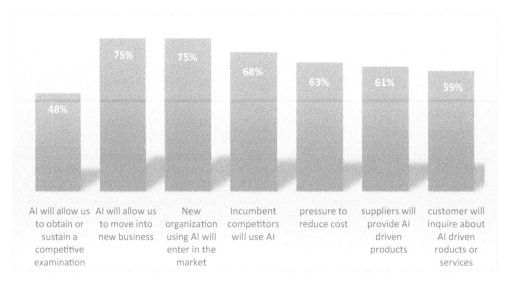

FIGURE 19.1   Reasons for adopting AI.

19.2 Challenges of AI

The use of AI is not limited to innovation and research labs, but it is also being used to radically transform the businesses and products. However, businesses are required to understand main challenges before they can use this technology to their fullest potential. In business, the input data is the key element of AI solutions, which can present several challenges. Most companies have data-related issues, as the AI system used will be only as good as the data quality. The AI holds enormous potential, apart from having certain challenges. The main problem with AI systems is that they perform as good or bad as the quality and quantity of data they are trained on. The cost, quality, dependability, and process are all important parameters for the development of AI-based applications.

In 2018, Deloitte carried out a survey of 1,900 IT business executives from various companies that were early users of AI, representing seven countries; Australia (100 respondents), Canada (300), China (100), Germany (100), France (100), the United Kingdom (100), and the United States (1,100) (Budman et al., 2019). Table 19.1 shows their views country-wise on Maturity, Urgency and Challenges.

AI *maturity* is found to be generally low, and <25% within each country, and is qualified as Seasoned adopters with US at 24% (see first row in the matrix). Strategic *maturity*, having a comprehensive, company-wide AI strategy, is low with China and UK having higher values, i.e., 46% an 41%, respectively. In *urgency*, the executives found that AI is "very" or "critically" important for the success of

TABLE 19.1

Maturity, Urgency, and Challenges to Adopt AI in Businesses

Factors	Parameters	Overall	Australia	Canada	China	France	Germany	UK	USA
Maturity	Percentage that are 'seasoned" AI adopters	21	17	19	11	16	22	15	24
	Have a comprehensive, company-wide AI strategy	35	34	27	46	28	26	41	37
Urgency	Believe AI is very or critically important to company's success now	63	56	58	54	49	46	61	69
	Achieve strong competitive advantage with AI	37	22	31	55	27	47	44	37
	Believe AI will transform their business within three years	56	51	51	77	63	60	55	55
Challenges	Major or extreme concern about AI risks	43	49	44	16	48	29	35	46
	Cyber-security vulnerabilities of AI are a top-three concern	49	46	42	54	49	51	44	50
	Moderate-to-extreme AI skill gaps	68	72	72	52	57	62	73	68

Note: All figures are in percentage.

their company. As critically important, AI will surge over the next two years, and some countries are expected to make a higher growth than the others. A majority of respondents within each country, however, believes that AI might transform their business within next three years, with China exhibiting 77% (see fifth row of the matrix). They further expressed different concerns about the AI risks. For example, 49% from Australia and 48% from France showed a major or extreme concern, as compared to 16% from China (see sixth row in the matrix). Some countries, like the UK, Germany and China appeared to be "fully prepared" for the AI-related risks than the other countries. Another common challenge involves cyber-security vulnerabilities of AIs. The executives rated this as a top-three concern; maximum of 54% in China (see seventh row in matrix). Looking at specific challenges, there is AI skills gap ranging from 51 to 73% between the countries (Budman et al., 2019).

The biggest problems with AI implementation are not primarily technical, but rather how to achieve value from the technology (Davenport, 2020). There are several factors that are preventing AI implementation and having a substantial return from new organizations, as discussed below:

19.2.1 Reengineering

Business processes and tasks involving AI are required to be redesigned, considering the processes and tasks that machines will do best, and which are best suited to humans. The industry must overcome the computing power challenges necessary to process the big data to build an AI system, and utilizing techniques, like ML and deep learning (DL).

19.2.2 Data Quality and Quantity

The AI system requires large training datasets and will learn from available information in a similar way as humans do. It is generally observed that the firms are still facing difficulties to acquire big data which is input to the AI big data analytics. The types of data required by the algorithm may normally not be available; some may be missing, some must be collected or procured, while some data may still be quite tedious to obtain, e.g., clinical data. In such cases, synthetic data may be created to train the AI-based model or use open data or Google dataset to train the model for predictions.

The quality of results from any AI system depends on the quantity and quality of data that is fed into it. However, it needs much more data than what humans may require for interpretation as well as to identify the accurate patterns (Polachowska, 2019). The more quality data is available to AI system, the better outcomes it is likely to provide. In order to train the ML algorithms, large and quality data sets, with minimal biases, are required. Most of this data may not be readily available for using straightaway, because it is either available in unstructured form, or inadequate quantity, or stored in a different format. As a result, many companies must invest more in creating the infrastructure to effectively collect and store the massive data, and to recruit trained manpower to process the data and make the data usable and productive.

The capabilities of AI and its reliability for a particular application would depend directly on the accuracy of supervised and labeled input data used for training and learning the algorithm (Harkut et al., 2019). Labeled data, though in scarcity, is organized to make it understandable for machines to learn. Approaches are being developed to make AI models learn themselves, despite the scarcity of quality-labeled data using transfer learning, active learning, deep learning, and unsupervised learning.

19.2.3 Integration of Data

Data can be collected in a variety of forms, such as text, audio, images, and videos for various applications. Several platforms used to collect these data adds to the challenges of AI. In order to successfully implement AI, all these data are integrated and transformed into useful results. Integrating AI into the existing systems is a more complicated process than adding a plugin to the existing browser. The interface and elements to address the needs of industry must be set up, such as data infrastructure needs, data storage, labeling, and feeding the data into the system. In addition, model is required for training and testing the reliability of the developed AI system, and thus creating a feedback loop to continuously improve the developed model based on actions and data sampling, mainly to reduce the quantity of data and run the model in a faster way, while still producing the accurate results. When AI implementation is done step by step, the risk of failure is minimized, and therefore more people can be educated to use the model.

19.2.4 Data Privacy and Security

The main challenge for utilization of AI in businesses is the data privacy. The decision-making capability of AI are based on large amount of data; often sensitive and may be personal (Harkut et al., 2019). These systems can learn from the data and improve themselves continuously. Due to continuous learning, these ML systems can be exposed to data breach and theft. Mostly, companies and government generally global AI-based network, and thus it can make the intruders difficult to intrude them.

AI can greatly benefit the society with its proper implementation. However, there is a great risk that commercial use of AI may have a negative impact on human rights. Applications of these technologies demand generation, collection, processing, and sharing of large data, both about individual and groups. Some of the information, like spam or suggesting an item for online shopping may be acceptable, whereas other information can have more serious consequences, and may even pose threats to the right to privacy and the right to freedom of expression and information. Its use can also impact the number of other rights, including the right to an effective remedy, and freedom from discrimination) Anderson et al. (2018), Mani (2019).

The ML, DL, Natural language processing (NLP), facial recognition, and emotion detection technologies requires the input data to provide results which can suggest actions (Markets and Markets, 2018). These technologies extract required data from large amount of data, and help making critical decisions. It is a challenge for firms to safeguard the data and information, as their customers have an apprehension about their personal data, (such as their location, phone number and marketing trends) being collected and used by AI. Many nations are complying the regulations to safeguard their citizens' privacy and avoid any misuse of data. To keep the public trust intact and ethical use of data by AI, a robust AI-based framework is required to be developed. For example, the European Union (EU) has already implemented the General Data Protection Regulations (GDPR) that ensure full protection of personal data, as the citizens have increasing awareness regarding large number of machine-made decisions.

19.2.5 Algorithms and Data

AI is all about data and algorithms. Algorithm design is the main technical feature of AI systems, particularly those which are based on ML. It is not possible to separate out the data from algorithms, and ML algorithms learn all from the data. The biggest drawback to effectively implement the algorithm is either inadequate, or poor quality, or unlabeled data. Many ML algorithms are not transparent or easy understandable to humans, particularly DL algorithms, which may have large number of variables. The lack of transparency in the algorithm may create a low confidence by users, executives, regulators, consumers and other stakeholders.

19.2.6 Software Malfunctioning

The decision making in AI is mostly automatic, as the machines and algorithms control AI. Through automation, however at times it is impossible to detect the root cause of errors or malfunctions, if present in the results. In addition, it is the human nature to relax learning and understanding of working in an automated complex system, since they don't have control over the system (Harkut et al., 2019). In case a software or hardware crashes, it is difficult for a human to pinpoint what has gone wrong (Lath, 2018). A recent example is the self-driving car that took the life of a pedestrian. On the other hand, the tasks performed by humans can easily be traced back.

19.2.7 Algorithm Bias

It is mostly derived from biased datasets or samples. Biased results may be obtained generally with poorly designed algorithms. These may be poorly designed, i.e., the purpose of creating the algorithms is not clear, or they may have loose fit to the data, etc. AI purely makes the decisions based on the available data-sets, as it doesn't provide its own opinions. As it learns from the opinions of others, the biasness might occur in the decisions (Polachowska, 2019). Biasness can occur due to several factors, e.g., the way data is collected, sampling method used, etc. For example, if the data is collected from a survey published in a magazine, it might represent only a limited social group reading that magazine, and the dataset may not be a representative of the entire population.

The main problem with AI systems is that their level of accuracy depends on the datasets they have used for training. Bad/biased data is often linked with biasness. A biased algorithm can take crucial deci-sions which could lead to wrong results (Harkut et al., 2019). Therefore, there is an urgent need to work on AI which is trained with unbiased data. In future, such biases can be eliminated as many AI systems are to be trained on the big data. Unethical and unfair results may be obtained if data used for training is associated with racial, gender, communal, or ethnic biases which might impact the decision-making process. Such biasness will be more prominent, if AI systems has been trained using the bad data.

19.2.8 Scarcity of Field Specialists

Both the technical knowledge and business understanding are important to successfully develop AI-based solutions. Unfortunately, in reality, it is either of the two. The implementation of AI technology requires a deeper understanding of current AI technologies and their limitations (Polachowska, 2019). For busi-nesses, there is an acute shortage of advanced skills. There are several concerns about AI, ranging from the need of hiring experts to managing robots. The lack of AI know-how may, on the one hand, motivate working toward impossible goals, while on the other hand it may hinder AI adoption in many fields.

Many executives of the companies also lack the technical knowledge for AI adoption, while many data scientists developing the models may not be very interested to know real-life applications of these models. Most of companies beyond the FAMGA group (Facebook, Apple, Microsoft, Google, Amazon) are finding difficulty to attract the best talent in AI. Small and medium firms may have budget crunches for the adoption of AI-based solutions. However, outsourcing an AI team may prove to be a better option.

The shortage of data science skills within humans is a big challenge to get maximum output from AI. The AI–human interface has a big shortage of trained manpower with skills of data analytics and data science to get optimal output from AI (Polachowska, 2019). As the advancements in AI are rising, businesses would require more skilled manpower who can meet the requirements to work with this tech-nology. Industries therefore are required to provide training to their manpower professionals to derive the benefits of this technology. According to statistics, about 55% experts feel that the biggest challenge is the changing role of humans when everything will be automated.

19.2.9 Lack of Investment

AI is less expensive technology in the long run, but many companies can't invest large funds in the beginning for computing power and infrastructure needed to run the AI models effectively (Lath, 2018). In addition to the investment, it is important how these investments are being managed effectively. Most companies don't invest much yet, and the investment levels are large by many large firms only. One major factor for lack of return from AI is simply the failure to invest large amount on it in the beginning. Few companies are doing return on investment analysis both before and after the deployment of AI (Harkut et al., 2019). Although the adoptability of AI is increasing, but still the firms who have incorporated AI are still in nascent stage, and thus are not able to correctly estimate the cost-benefit.

19.2.10 Building Trust

AI is all related to science and algorithms, which requires technology. People who are completely una-ware of these algorithms and technology working behind AI find it difficult and complicated to understand its functioning (Polachowska, 2019). It is a basic tendency of humans that they often neglect something which they don't understand and try to stay away. It is also true with the AI technology; being related to huge data, data science and algorithms, and thus users do not easily understand the concepts used (Harkut et al., 2019). AI therefore can face trust issues by people, despite its ability to complete the tasks speedily.

AI is like a black box for people, and they are uncomfortable if they don't understand the process through which AI has taken the decision. Therefore, AI must create trust among people. The best solution for this may be that the people themselves see and realize that the AI technology really produces better results. In addition, AI is providing a lot of opportunities to derive more accurate and faster predictions. For example, the banking sector, financial service and insurance companies, including financial analysis, risk assessment, and investment/portfolio management, where AI has played an important role due to its high demand for risk management.

19.2.11 Implementation Strategies

One challenge of using AI is the lack of strategy for its clear implementation when applied to transform the processes of the industry/organization. A strategic approach is required to implement the AI technology, which may include, identify areas of improvement, and a continuous improvement feedback system (Tiempo Development, 2019). Organizations/industries will need to have a good understanding of AI technology, which would enable them to identify areas that can be improved further using AI (Polachowska, 2019).

19.2.12 Legal Issues

One of the concerns of AI includes the legal issues being raised that the implementing organizations must consider (Tiempo Development, 2019). If the AI-based system is collecting sensitive/private data, it is considered as the violation of privacy laws, even if the data is not utilized to reveal the privacy of public (Harkut et al., 2019). The organizations should be careful of any possible impact that might adversely affect them, in case the data is stolen/hijacked and used by a third party.

The legal system fails with the AI technology, as currently there are no rules that clearly define what must be done if AI causes havoc, may be due to fault of AI, or something is damaged. Another issue is the GDPR where data is taken as a commodity which is required to be handled with care, and this may offer a challenge of data collection and handling big data (Polachowska, 2019). The sensitive data may pose legal issues to an organization/company in case of leakage and threaten the position of company.

19.2.13 Higher Expectations

The AI may not be able to fulfill all the expectations of the people around. People, in general, are not fully aware how AI works, and hence they have higher expectations, which may not be even possible to perform (Harkut et al., 2019). Humans generally have high expectations for new technology and tools. However, like any other technology, AI will also have certain limitations. In fact, AI has seen much publicity in the market, but the fact is that AI is still in its initial phase.

19.2.14 AI Can Be Dangerous

AI can be potentially harmful if machines started thinking better than the humans, as it may put them in positions of control over humans. Humans have the knowledge to diffuse what they have created, and they know how to do it, but AI-based devices do not (Shankar, 2020). It is believed that artificial superintelligence is much smarter and better than the best human brains, and the autonomy of AI and robots may become a potential threat to humanity. A super intelligent AI might become a risk, as the AI-based autonomous weapons may be programmed to devastate or kill. These systems may be used to kill innocents by the terrorists. AI is although programmed to do useful activities, but it might end up doing destruction (Thomas, 2020). For example, if you want to go to airport in an intelligent car as fast as possible, it might get you there with lots of jerks, as the car has literally done what you have asked it to do. Therefore, a super-intelligent AI might be perfect in achieving its goals, but if those goals aren't aligned with our objectives, it might cause a serious problem. The main goal of AI safety research is to protect the risk of humanity.

19.3 AI as a Job Creator

There are controversies about the future of AI, such as, AIs will impact the job market if human-level AI is developed. It may lead to an intelligence explosion, which could be accepted or rejected (Shankar, 2020). One hypothesis is that AI will take us to an era where humans are no longer required to work. Repetitive jobs, such as customer services are already being controlled by AI, and therefore the loss of jobs is one of the biggest concerns during such automation. AI may have positive impact on every industry and work that demand innovation, creativity and faster decision-making.

According to Mr. Sundar Pichai, the CEO of Google,

> everything in Google is going to be AI centered which could mean that AI will create numerous job opportunities that require creativity, critical thinking skills and much more. Most people and organizations stand to benefit from collaborating with AI to augment tasks performed by humans. Jobs are going to be in plenty as AI can never replace humans.

Although, a machine may think on its own, but it still requires some instructions from humans. For example, the smartphone is smart, but is still controlled by humans. AI will rather become a colleague/friend than a replacement. Machines will always provide you a solution, but if you are not familiar about the working of algorithm, you might consider it to be the correct answer.

During the Fourth Industrial Revolution, the physical, digital, and biological worlds will merge (Kiser & Mantha, 2019). The AI jobs are expected to be in high demand as AI will transform the global economy. According to the World Economic Forum (WEF), AI is expected to create more jobs than it loses, and by 2022, about 75 million jobs may be displaced due to automation. AI ultimately may lead to a creation of 58 million new jobs globally (Johnson, 2020). The WEF estimates that about 54% of all employees around the world will require necessary retraining by 2022, though this gap is much wider in some regions of the world. The WEF highlights that around 37% of workers in Europe don't have even the basic digital skills. According to a new IBM Institute for Business Value study (Brown, 2019), in next three years about 120 million workers in the world's twelve largest economies (including 11.5 million in the US) are required to be reskilled to use AI and automation.

AI is a complex system, and companies require a workforce who are well-versed with the technologies for developing, managing and implementing AI systems. The big challenge remains that the workforce in AI and, most importantly, the youth of today are to be trained for the jobs of tomorrow. There are limited engineers with AI skills worldwide, but millions may be needed. Five top countries that train AI experts and are also leading employers: USA, China, UK, Canada, and Germany, accounting for 72% experts (Markets & Markets, 2018). The shortage of AI skills is the single biggest hurdle to the technology for its adoption and implementation across businesses.

According to Indeed (2019), top ten AI skills in demand from companies for jobs are given in Table 19.2. The demand of ML engineering job was the highest (75%) in 2019. These engineers can develop devices and software that use predictive technology, such as Apple's Siri or weather-forecasting apps. The DL engineers were next in the demand (i.e., 61%), as they can develop programs that mimic the brain functions, among other given tasks. These engineers are contributing in three important fields: autonomous driving, facial recognition and robotics. The global facial-recognition market alone is expected to grow from US$3.2 billion in 2019 to US$7 billion by 2024 (Johnson, 2020). The demand of other jobs, such as Computer vision engineers, Data scientists, Algorithm developers, range between 58 and 47%. The highest paid jobs in descending order were identified as Machine learning engineer, Data scientists, Computer vision engineer, Data warehouse architect and Algorithm developers.

A report from Gartner (2019) shows that the business applications for AI have grown from 10% in 2015 to 37% in 2019, increasing the job demand that surpasses the current supply of trained people. There is 344% growth of jobs for ML engineers, 128% Robotics engineers, 116% Robotics engineers, and 78% Data scientists from 2015 to 2019 (Indeed, 2019). The reason for this increase is that the firms are more willing to implement the AI technology that has matured significantly. The report says that 52% of telecom-based organizations adopt chatbots, while 38% of health care providers use

TABLE 19.2

Top Ten Jobs Involving AI or ML Skills

Rank	Job	% of AI or ML required
1	Machine learning engineers	75
2	Deep learning engineers	60.9
3	Senior data scientists	58.1
4	Computer vision engineers	55.2
5	Data scientists	52.1
6	Algorithm developers	46.9
7	Junior data scientists	45.7
8	Developer consultants	44.5
9	Director of data science	41.5
10	Lead data scientists	32.7

TABLE 19.3

Predicted Shifts in Skill Sets, US versus Western Europe

Skills	United States, all sectors		Western Europe, all sectors	
	Hours worked in 2016 (billions)	Change in hours worked by 2030 (%)	Hours worked in 2016 (billions)	Change in hours worked by 2030 (%)
Physical and manual skills	90	−11	113	−16
Basic cognitive skills	53	−14	62	−17
Higher cognitive skills	62	+9	78	+7
Social and emotional skills	52	+26	67	+22
Technological skills	31	+60	42	+52
Total	287		363	

computer-assisted diagnostics. Other operational use cases for AI could include fraud protection and consumer fragmentation.

The McKinsey Global Institute Report (2017) suggested that intelligent agents and robots may replace about 30% of the world's current human labor by 2030. Table 19.3 presents the comparison of workforce needs between US and western Europe on five basic skills: (1) physical and manual skills, (2) basic cognitive skills, (3) higher cognitive skills, (4) social and emotional skills, and (5) technological skills. According to this report, automation can displace between 400 million and 800 million jobs by 2030, requiring about 375 million people to switch jobs, depending on various adoption scenarios. However, number of jobs related to programming, robotics, engineering, etc., will increase as these skills are required to improve and maintain AI and automation. With more than double demand for trained engineers in last few years, the opportunities for professionals who want to work on the cutting edge of AI research and development will be tremendous.

19.4 Next-Generation AI

AI is expected to change the activities in the world in the history of mankind (Thomas, 2020). It is going to impact the future of virtually every human being and industry. The AI algorithms have been used for almost 60 years (Figure 19.2); in the early stages, algorithms were simply "I repeat" capabilities (Oblé et al., 2018). These have progressed over time, and today "I learn to learn" algorithms can be used to create smarter machines that learn by themselves. The next generation "I contribute, I exchange" algorithms can make distributed AIs a reality, although these already exist in the most advanced research labs.

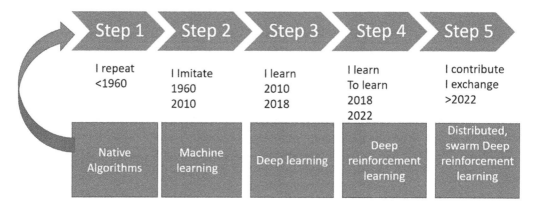

FIGURE 19.2   Algorithm maturity level.

In past, data collection and their analysis have increased significantly mainly due to good connectivity, the IoT-based devices and sensors, and higher speed of data processing. Some sectors are just beginning their AI journey, while others have now well tested. However, both the sectors have a long journey to undertake, regardless of the impact AI on our lives (Thomas, 2020). The major industries have already adopted modern AI, or "narrow AI," which use trained-data models, and employ ML or DL. The major players in the AI market include Atomwise, Inc.; Lifegraph; Sense.ly, Inc.; Zebra Medical Vision, Inc.; Baidu, Inc.; H2O AI; IBM Watson Health; NVIDIA; Enlitic, Inc.; Google, Inc.; Intel Corporation; and Microsoft Corporation (Johnson, 2020). The next generation AI and ML must be intelligent, self-learning, and adaptive.

Adoption of big data and AI in businesses remain a big challenge, as the absence of adequate quality data affects the results from AI. The AI applications successfully developed may not be implemented successfully due to lack of data. The next phase of innovation is therefore required to combine the crowd-sourced data in the cloud with AI capabilities to create new business opportunities. AI is the main driver of emerging technologies, such as big data, robotics and IoT, and it will continue to act as a technological innovator for the future. Some future areas include as given below (Thomas, 2020):

Transportation: Sensors on vehicles and roads can send continuous data about traffic that could be used to predict the problems and subsequently optimize the flow of traffic. The autonomous cars will soon be a reality although it could take a decade or more to perfect them.

Manufacturing: The AI-powered robots can be used to perform tasks such as assembly and stacking.

Health care: In health care, diseases can be diagnosed accurately, and drug discovery is streamlined. Virtual assistants can monitor the patients like a nurse. The health care segment has been benefitted with AI, such as robot-assisted surgery, dosage reduction, clinical trial, hospital workflow management, preliminary diagnosis, and automated image diagnosis (Grand View Research, Inc, 2020b). AI can be used to reduce the burden on clinicians and provide an efficient tool to paramedical staff for undertaking their tasks in a better way. For instance, AI voice-enabled symptom checkers can better access a situation and assist patients to the emergency department to receive immediate treatment. The AI is expected to contribute to around 20% of health care demands.

Education: Complete textbook can be digitized with the help of AI. Virtual tutors can assist students. Facial analysis can judge the emotions of students whether they are getting the lecture or not, and accordingly AI tutor can tailor the matter as per individual needs.

Media: Journalism is benefitting from the AI. For example, Bloomberg is using Cyborg technology to analyze complex financial reports.

Customer service: AI is helping provide customer services: for example, Google is working on an AI-based assistant that can make human-like calls to fix appointments, say, to your dentist. Additionally, the system can also understand context and nuance.

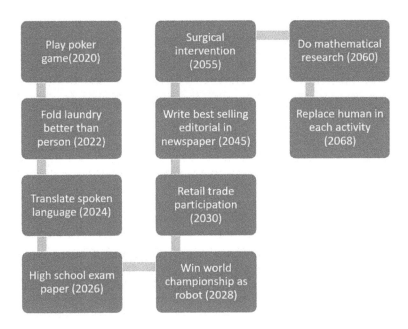

FIGURE 19.3 Anticipated developments of AI.

Cognitive network management represents one of the most important emerging network infrastructure opportunities. The AI, in combination with Software-Defined Networking (SDN) and advanced analytics, is expected to take autonomous, and intelligent network operation. The large economic and social benefits may be derived as networks can achieve a new level of self-awareness, self-configuration, self-optimization, self-healing, and self-protection. This will greatly benefit the existing networks, IoT systems, and 5G networks. More than 50% organizations will leverage AI technology for networking, and the market is expected to reach US$5.8 billion by 2023, the health care, manufacturing, and retail being the largest revenue earning industries (Mind Commerce, 2018).

For network security, several new programs are available. In most cases, once the vulnerability is found, a module is developed rapidly to counter it. Hackers can easily do the reverse engineering of the module. Programs are therefore required to be developed that automatically detect new malware attacks against specific targets, individual computer or networks. Flexibility should be there to change input parameters of the programs to work even in new environments.

Latest developments in the AI techniques result in further devising the range of applications for the AI. The anticipated developments in AI are shown in Figure 19.3.

It is expected that by 2026, machines will be capable of writing essays; by 2027 self-driving trucks will be operational; by 2031 the AI will overtake the retail sector; by 2049 the AI could be the next Stephen King and by 2053 the next Charlie Teo. It is expected that all human jobs will be automated by 2065, and the AI will revolutionize the modification of our genomes (Thomas, 2020). Scientists would be able to edit human DNA with beneficial genes, in the same way as an editor corrects a manuscript, replacing the weaker sections with the strong one (Talty & Julien, 2019). By 2065, humans would be on the verge of freeing themselves from the biology that created them.

19.5 Market Potential of AI

The AI can be considered as a revolutionary technology, and its integration for large number of applications is one of the major factors driving the market growth. Worldwide, human-centric industries, such as financial services and retails are expected to be the biggest investors in implementation of AI, followed by manufacturing, energy and utility, transport industries, etc., (Figure 19.4). In business, efficiency,

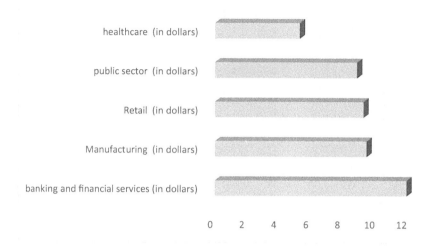

FIGURE 19.4   Projected AI spending by various industries in 2021 (ATOS, 2020).

improved customer services and consumer satisfaction, cyber-security, analytics, the AI is growing fast. It will have a greater impact as 40% of new industrial applications are expected to spend US$45 billion by including smart machine technologies by 2021 (ATOS, 2020).

Growing investments in research and development by leading industries is going to play a critical role for enhanced applications of AI technologies, which includes tagging, clustering, categorization, hypothesis generation, alerting, filtering, navigation, and visualization. Increased use of cloud-based platforms and hardware equipment for the safe & secure storage of huge volume of data has given clue for the development of analytics platform (Armstrong, 2016). The Zinnov consulting firm estimated that worldwide firms have spent over US$470 billion on digital media in 2017, while digital spending is likely to increase at a CAGR of more than 20% to reach US$1.1–1.2 trillion by 2022 (Krishna, 2017). Worldwide market revenue expected up to 2025 from AI is given in Figure 19.5.

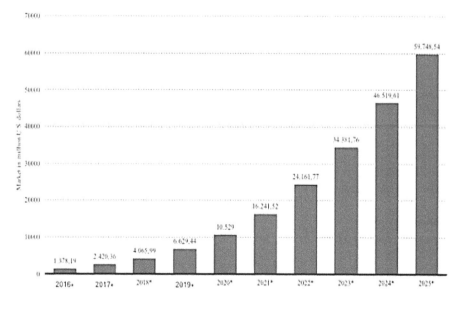

FIGURE 19.5   The AI market revenue worldwide, 2016–2025.

The major factors for the development of AI market are availability of big data, cloud-based applications and services, and intelligent virtual assistants. New launches by large firms, such as NVDIA (US), Intel (US), Xilinx (US), Samsung (South Korea), Micron (US), IBM (US), Google (US), Microsoft Corporation (US) and Amazon Web Services (US) will help in the growth of AI market in coming years (Markets & Markets, 2018). Many countries are taking the opportunity of AI revolution to promote domestic economic and technological development. The AI technology can be used to increase business productivity by up to 40%. North America was holding the major share in the global market in 2019, however it is anticipated that the Asia Pacific will supersede North America and will emerge as the leading regional market by 2025. It may be all due to various reasons, such as tremendous improvements in the storage capacity, high computing power, and parallel processing.

Industrial robots create a large volume of data, which can be used for training these robots. Countries, such as Japan, China and South Korea, are considered the largest market for industrial robots. Figure 19.6 shows that the rating of market growth potential in robotics; robo-advisors, followed by self-driving vehicles, analytics, logistics, and industrial robots. According to Grand View Research, Inc. (2020a), the AI-based robots will automatically identify the area to be cleaned, and will distinguish between the dirt and other objects. The growth of cleaning robot's market is due to automation in technology and the higher manual labor costs for cleaning large areas.

The ML and DL will attract significant investments in AI. Globally AI market size is expected to touch US$390.9 billion by 2025 and US$733.7 billion by 2027 to expand at a CAGR of 46.2% from 2019 to 2027, according to Grand View Research, Inc. Report (2020a). The goals of many companies extend well beyond their specific application of AI, which can become a possible source of investment and generate support for the AI technology in future. The number of AI start-ups worldwide since year 2000 has grown 14 times.

The AI assets are expected to gain value (funds) as time passes, so new financial metrics will assess the "Return on AI," which could include the funds generated from the AI algorithm. Loucks et al. (2018) found that 82% industries have successfully gained a good return from their AI investments. For all industries, the median return on investment from cognitive technologies was about 17% (Figure 19.7). Some industries are more experienced than others to convert investment into financial benefits. The figures on return on investments show that industries are getting value from cognitive technologies. They are also the driving force for companies, such as Google, Microsoft, and Facebook, to develop the market and improve the services. For example, Netflix by using AI found that customers get improved search results for a movie in few seconds, which earlier used to take more than 90 seconds and thus, losing the search interest of customers. It has earned Netflix US$1 billion a year more as compared to using without AI.

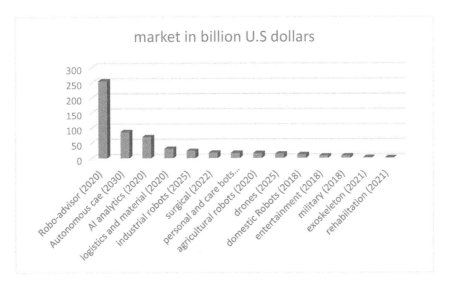

FIGURE 19.6 AI-robot market estimates between 2018 and 2035.

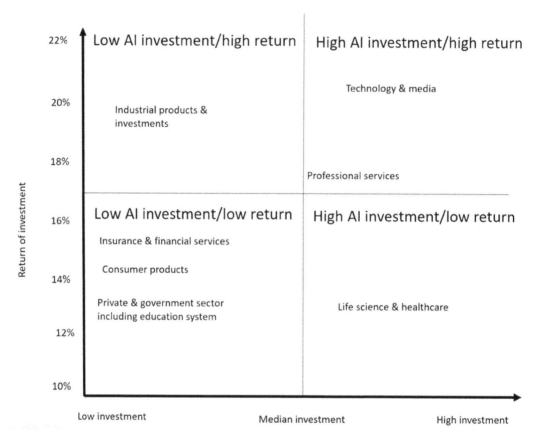

FIGURE 19.7 The AI investments and return on investments: relative landscape of industries. (*Note: the dotted lines represent the median return on investments and median AI investments, cross-industry.*)

Advances in image recognition and voice recognition are other important areas which are driving the market growth. Improved image recognition technology is critical in intelligent drones, self-driving cars, and robotics (Grand View Research, Inc., 2020a). The combination of image/vision related AI applications is expected to represent 30% of the total AI market by 2025, according to Tractica recent study (Iyar, 2018). In terms of revenue earned, the AI usage ranking was higher for the detection of vehicles, machines, and objects; static image recognition; and patient data processing. The future applications considered with most AI market potential in terms of revenue expected are given in Figure 19.8. With expected cumulative revenue of over US$ 8 billion, "static image recognition, classification and tagging" is forecast to lead, followed by "algorithmic trading strategy performance improvement" (US$7.5 billion) and "efficient, scalable processing of patient data" (US$7.4 billion (Armstrong, 2016).

With image processing and analysis becoming an important application of AI in various industries, it is expected to witness tremendous growth. The Asia-pacific would exhibit the highest growth. The AI image recognition market providing service and solutions is highly competitive in national and international market. The main industries operating in these markets are continuously developing their products on the basis of customers' feedback, and this is providing them sustainable competitive advantage.

With companies spending collectively nearly US$20 billion on AI products and services annually, big players, like Google, Apple, Microsoft and Amazon are investing huge amount to develop those products and services, and universities are including AI as an important subject in their curricula (e.g., MIT alone is providing US$1 billion on a new college committed solely to computing with focus on AI), the demand of AI is going to increase manifolds (Thomas, 2020). The more objective the job is, such as separating things into bins, washing dishes, picking fruits and answering customer service calls, etc.,

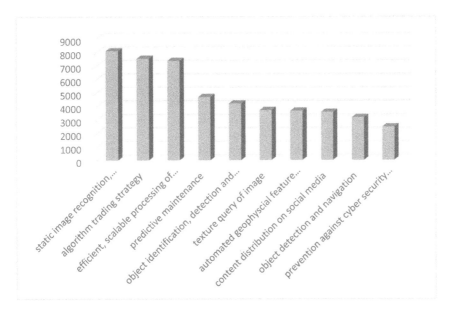

FIGURE 19.8   Forecast cumulative global AI revenue from 2016 to 2025 in various applications.

the more demanding of AI technology would be, as there are many tasks that are repetitive and routine kind. In next 5–15 years, such repetitive and routine nature jobs will be replaced by the application of AI. For example, Amazon, the warehouses of online shopping has deployed more than 100,000 robots working, but picking and packing jobs are still performed by humans, which is expected to change with time (Sahún & Riocerezo, 2018).

19.6 Summary

The demand of AI is growing in the society and businesses. The AI is very important and useful in today's world due to its several capabilities. An extension to ML, the DL can solve complex problems and work on huge dataset to create several new business scenarios (Shankar, 2020). Exponential growth of data generation by IoT devices, cloud-based storage capabilities, high computational power, and advancements in ML, all have contributed to increasing demand of AI.

The AI has drawn the attention of industries and organizations, but the technology is still at an early stage of implementation. Along with 5G and the IoT, AI is expected to grow in many fields. It will transform the world in a big way, improve the efficiency of many sectors, and create higher-value services that can lead to overall economic development. In future, the AI technology will contribute more toward economic growth (Budman et al., 2019).

Today's world is completely technology-driven, such as driverless cars, automization, algorithms responding to customers' enquiries, automated business intelligence on one tap (Budman et al., 2019). Robotics is one of the most promising areas of AI which will make a revolution in the future. The news of job losses due to induction of robots is not correct (Sahún & Riocerezo, 2018). Research is continuing in Robotics to create such robots who can behave exactly like humans, interact like humans, and think like humans. Self-driving cars are yet to give 100% confidence to its users.

The AI technology is evolving daily, and is expected to shape the future of a variety of industries, such as transportation and logistics, digital health, finance and insurance. The use of DL in AI is yet to be fully explored (Shankar, 2020). It is suggested to apply the technology to businesses, automize the processes, focus on improving processes, and customer satisfaction. The future of AI is thus extremely promising which needs trained manpower having creative thinking and reasoning. The main aim of AI is to provide support humans by increasing the productivity and making the job faster and more accurate. In future, the AI is going to be better friend and partner for you than any other human (Koenig, 2020).

REFERENCES

Adixon, Robert (2019), Artificial Intelligence Opportunities & Challenges in Businesses, July 24, https://towardsdatascience.com/artificial-intelligence-opportunities-challenges-in-businesses-ede2e96ae935

Anderson, Janna, Rainie, Lee and Luchsinger, Alex (2018), Artificial Intelligence and the Future of Humans, December 10, https://www.pewresearch.org/internet/2018/12/10/artificial-intelligence-and-the-future-of-humans/

Armstrong, Martin (2016), The Future of AI, https://www.statista.com/chart/6810/the-future-of-ai/)

ATOS (2020), Artificial intelligence, For your business, right now, https://atos.net/en/artificial-intelligence

Brighterion Inc. (2019), Artificial intelligence and machine learning: The next generation, Brighterion Mastercard, https://brighterion.com/wp-content/uploads/2019/05/Artificial-Intelligence-And-Machine-Learning-The-Next-Generation.pdf

Brown, Tony (2019), The AI Skills Shortage, IT Chronicles, https://itchronicles.com/artificial-intelligence/the-ai-skills-shortage/

Budman, Matthew, Hurley, Blythe, Bhat, Rupesh and Gangopadhyay, Nairita (2019), Future in the balance? How countries are pursuing an AI advantage, Deloitte Insights, https://www2.deloitte.com/us/en/insights/focus/cognitive-technologies/ai-investment-by-country.html

Chan, Tim (2017), How AI boosts industry profits and innovation, https://www.accenture.com/sg-en/company-news-release-businesses-singapore-benefit-ai

Davenport, Tom (2020), Return On Artificial Intelligence: The Challenge and the Opportunity, March 27, https://www.forbes.com/sites/tomdavenport/2020/03/27/return-on-artificial-intelligence-the-challenge-and-the-opportunity/#bb9ec036f7c2

Gartner (2019), https://www.gartner.com/en/newsroom/press-releases/2019-01-21-gartner-survey-shows-37-percent-of-organizations-have

Grand View Research, Inc. (2020a), Artificial Intelligence Market Size Worth $390.9 Billion by 2025: February 5, https://www.prnewswire.com/news-releases/artificial-intelligence-market-size-worth-390-9-billion-by-2025-grand-view-research-inc-300999236.html

Grand View Research, Inc. (2020b), Market Analysis Report: July 20, https://www.grandviewresearch.com/industry-analysis/artificial-intelligence-ai-market?utm_source=prnewswire&utm_medium=referral&utm_campaign=ict_5-feb-20&utm_term=artificial-intelligence-ai-market&utm_content=rd1

Harkut, Dinesh G., Kasat, Kashmira and Harkut, Vaishnavi D. (2019), Introductory Chapter: Artificial Intelligence - Challenges and Applications, https://www.intechopen.com/predownload/66147

https://www.indeed.com/lead/best-jobs-2019

Indeed (2019), Here Are the Top 10 AI Jobs, Salaries and Cities, Indeed Editorial Team June 28, https://www.indeed.com/lead/top-10-ai-jobs-salaries-cities

Iyar, Ananth (2018), Artificial Intelligence, April 22, https://witanworld.com/article/2018/04/22/witan-sapience-artificial-intelligence/

Johnson, Reece (2020), Jobs of the Future: Starting a Career in Artificial Intelligence, May 14, 2020, https://www.bestcolleges.com/blog/future-proof-industries-artificial-intelligence/

Kiser, Grace and Mantha, Yoan (2019), Global AI Talent Report 2019, https://jfgagne.ai/talent-2019/

Koenig, Stephen (2020), What does the future of artificial intelligence mean for humans? July 28, https://techxplore.com/news/2020-07-future-artificial-intelligence-humans.html

Krishna, Vishal (2017), 7 Indian industries that will be impacted by AI: The dawn of an era of tectonic change, https://yourstory.com/2017/10/7-indian-industries-that-will-be-impacted-by-ai?utm_pageloadtype=scroll

Lath, Anuja (2018), 6 challenges of artificial intelligence, *BBN Times*, May 22, https://www.bbntimes.com/companies/6-challenges-of-artificial-intelligence

Loucks, Jeff, Davenport, Tom and Schatsky, David (2018), *State of AI in the Enterprise*, 2nd Edition, Deloitte Insights, Deloitte Development LLC.

Mani, Chithrai (2019), The Next-Generation Applications of Artificial Intelligence and Machine Learning, *Forbes*, November 6, https://www.forbes.com/sites/forbestechcouncil/2019/11/06/the-next-generation-applications-of-artificial-intelligence-and-machine-learning/#34814ff547bc

Markets & Markets (2018) Artificial intelligence market, https://www.marketsandmarkets.com/Market-Reports/artificial-intelligence-market-74851580.html

McKinsey Global Institute Report (2017) https://www.mckinsey.com/featured-insights/future-of-work/jobs-lost-jobs-gained-what-the-future-of-work-will-mean-for-jobs-skills-and-wages#

Mike Thomas, Mike (2020), The Future of Artificial Intelligence – 7 ways AI can change the world for better or worse, April 20, https://builtin.com/artificial-intelligence/artificial-intelligence-future

Mind Commerce (2018), Next Generation Networking AI market 2018–2023, https://mindcommerce.com/reports/next-generation-networking-ai/

Mordor Intelligence (2019), AI image recognition market – growth, trends, and forecast (2020–2025).

Oblé, Frédéric, Monchalin, Eric and Lefebvre, Guillaume (2018), Reinvent your future with the promise of Artificial Intelligence, ATOS, https://atos.net/en/blog/reinvent-future-promise-artificial-intelligence

Polachowska, Kaja (2019), 12 challenges of AI adoption, June 6, https://neoteric.eu/blog/12-challenges-of-ai-adoption/

Ransbotham, Sam, Kiron, David, Gerbert, Philipp and Martin, Reeves (2017), Reshaping Business With Artificial Intelligence Closing the Gap Between Ambition and Action, MIT Sloan Management Review and The Boston Consulting Group, September 2017.

Sahún, Enrique and Riocerezo, Gorka (2018), nae, March 17, https://nae.global/en/market-expectations-for-artificial-intelligence/

Shankar, Ramya (2020), Future of Artificial Intelligence, April 9, https://hackr.io/blog/future-of-artificial-intelligence

Talty, Stephan and Julien, Jules (2019), What Will Our Society Look Like When Artificial Intelligence Is Everywhere? https://www.smithsonianmag.com/innovation/artificial-intelligence-future-scenarios-180968403/

Thomas, Mike (2020), The Future of Artificial Intelligence – 7 ways AI can change the world for better … or worse, April 20, https://builtin.com/artificial-intelligence/artificial-intelligence-future

Tiempo Development (2019), Artificial intelligence's biggest challenges, July 22, https://www.tiempodev.com/blog/artificial-intelligence-challenges/

Index